ESSENTIALS OF

MODERN OPEN-HOLE LOG INTERPRETATION

ESSENTIALS OF

MODERN OPEN-HOLE LOG INTERPRETATION

JOHN T. DEWAN

PennWell Publishing Company
1421 South Sheridan Road/P.O. Box 1260
Tulsa, Oklahoma 74101

Library of Congress Cataloging in Publication Data

Dewan, John T.
Modern open-hole log interpretation.

Includes index.
1. Geophysical well logging. I. Title.
TN871.35.D45 1983 622'.3382 83-4228
ISBN 0-87814-233-9

Printed in the United States of America

3 4 5 87 86 85

DEDICATION

This book is dedicated to my former colleagues at Schlumberger, who did much to advance the science of well logging, and to my present associates and my family, who patiently endured its preparation.

CONTENTS

INTRODUCTION

The aim of this book is to present modern log interpretation as simply and concisely as possible. The book is written for the geologist, petrophysicist, reservoir engineer, or production engineer who is familiar with rock properties but has little experience with logs. It will help him specify good logging programs with up-to-date tools and hand-interpret zones of interest with the latest techniques. The book will also familiarize him with computer-processed logs generated by the service companies at the wellsite.

Accordingly, obsolete logging tools are mentioned only in perspective. Very brief descriptions of the instruments in common use indicate how they apply to different logging conditions. Salient features of new tools, including Spectral Gamma Ray, Litho-Density, Dual Porosity Neutron, and Long-Spacing Sonic, emphasize how these tools fit into the everyday logging picture. The interpreter need not be overly concerned how a tool operates. What is important is the instrument's response to the various formation and borehole variables.

In a similar vein, interpretation equations and charts are kept to the minimum needed for routine evaluation of logs. Fundamental principles are stressed, rather than mechanical application of formulae. This is particularly true in the chapter on shaly formation interpretation where an effort has been made to draw together the latest concepts in this ever-changing field.

This book addresses the normal well situation where a standard set of logs is run in a liquid-filled open hole to locate hydrocarbons in place and where promising zones are then tested to evaluate their producibility. Abnormal situations such as empty hole, water well, and geothermal and mineral logging are not included.

To provide a little perspective, well logging is in its third major development stage. The first 20 years, from 1925–1945, saw the introduction and gradual worldwide acceptance of the so-called ES (Electrical Survey) logs. These logs were run with simple downhole tools and, while quite repeatable, were often difficult to interpret.

The second phase, from 1945–1970, was a major tool development era, made possible by the advent of electronics suitable for downhole use. Focused electrical devices were introduced, having good bed resolution and various depths of penetration. A variety of acoustic and nuclear tools were developed to provide porosity and lithology information. There was a progression through second- and even third-generation tools of increasing capability and accuracy. Simultaneously, much laboratory and theoretical work was done to place log interpretation on a sound, though largely empirical, basis.

The third and current phase, which began about 1970, may be called the log processing era. With the advent of computers, it has become possible to analyze in much greater detail the wealth of data sent uphole by the logging tools. Log processing centers, providing sophisticated interpretation of digitized logs transmitted by telephone and satellite, have been set up by service companies in strategic locations. Logging trucks have been fitted with computers that permit computation of quick-look logs at the wellsite. At the same time logging tools have been combined to the point that a full set of logs can be obtained on a single run.

The present state of the art is that logs are adequate to determine hydrocarbons in situ in medium- to high-porosity formations but are pushed to their limits in low-porosity, shaly, mixed-lithology situations. More precise determination of the matrix makeup, including amounts and types of clay present, is needed. Promising developments are underway.

Advances are being made in predicting the producibility of hydrocarbons found in place, but the critical factor, a continuous permeability log, is still lacking. Meanwhile, point-by-point permeability and pressure values can be obtained by repeat formation testing, a technique that is finding increased use.

Developments in the testing stage promise to provide more precise lithology information, better movable oil determination, and additional mechanical properties of formations. Unquestionably, answers obtainable from logs will continue to become more accurate and broader in scope.

Chapter 1

THE LOGGING ENVIRONMENT

Relatively little is learned about the producing potential of a well as it is being drilled. This is a surprise to the uninitiated, who have visions of early gushers. But the drilling mud actually pushes hydrocarbons, if encountered, out of the way and prevents their return to the surface. Examination of returned cuttings indicates the general lithology being penetrated and may reveal traces of hydrocarbons, but it allows no estimates of the amount of oil or gas in place.

Well logs furnish the data necessary for quantitative evaluation of hydrocarbons in situ. Modern curves provide a wealth of information on both the rock and fluid properties of the formations penetrated. From the point of view of decision-making, logging is the most important part of the drilling and completion process. Obtaining accurate and complete log data is imperative. Logging costs account for only about 5% of completed well costs, so it is false economy to cut corners in this phase.

THE BOREHOLE

When the logging engineer arrives at the wellsite with his highly instrumented logging unit, he finds ready to be surveyed a borehole that has the following characteristics:

- an average depth of about 6,000 ft but which may be anywhere between 1,000 and 20,000 ft
- an average diameter of about 9 in. but which can be between 5 in. and 15 in.
- a deviation from vertical that is usually only a few degrees on land but typically 20–40° offshore
- a bottom-hole temperature that averages about 150°F but may be between 100°F and 350°F
- a mud salinity averaging about 10,000 parts per million (ppm) but which can vary between 3,000 and 200,000 ppm; occasionally the mud may be oil based
- a mud weight averaging about 11 lb/gal but which can vary from 9 to 16 lb/gal
- a bottom-hole pressure averaging perhaps 3,000 psi but which can be as low as 500 and as high as 15,000 psi

- a sheath of mud cake on all permeable formations that averages about 0.5 in. in thickness but may be as little as 0.1 in. and as much as 1 in.
- an invaded zone extending a few inches to a few feet from the borehole in which much of the original pore fluid has been displaced by drilling fluids

Even more severe conditions are occasionally encountered. In any case it is a challenging environment from which to derive accurate information about the state of the formations as they were prior to any drilling disturbance.

LOGGING PROCEDURE

Accustomed to the challenge, the logging crew proceeds to align the truck with the well, spool the logging cable through the lower and upper sheave wheels, and connect the logging tools. The engineer performs the surface checks and calibrations. After this, the logging array is dropped to bottom as quickly as practicable. Once on bottom the downhole calibrations are carried out, recording scales are set up, and the crew "comes up logging" (Fig. 1–1). Survey speed is maintained constant, between 1,800 and 5,400 ft/hr, depending on the logging tools.

The logging string is typically 3⅝ in. in diameter and 20–50 ft long. It usually consists of several different tools in tandem. The most important is the tool that measures the electrical resistance of the formation because increased resistance occurs when water is replaced by hydrocarbons. Accompanying the resistivity tool is at least one tool that measures porosity and one that distinguishes permeable from nonpermeable zones. The basic logs may be obtained on a single run in the hole or may require two runs with different logging tools. Operating power for the tools is sent down one pair of insulated conductors inside the armored logging cable, and logging data are transmitted to the surface on the remaining five conductors.

In recent years the major service companies have been replacing older surface instrumentation with completely computer-controlled systems that are much more versatile and easier for the engineer to operate (Fig. 1–2).* Logging data are digitized and fed into the computer where they are

---

*Denoted Cyber Service Unit (CSU) by Schlumberger, Computer Logging System (CLS) by Dresser Atlas, Digital Logging System (DLS) by Welex, and Direct Digital Logging (DDL) by Gearhart.

processed and output to paper, film, and magnetic tape recorders. The engineer controls the system almost entirely with commands from the keyboard. At the same time, he monitors the output on a screen which, during

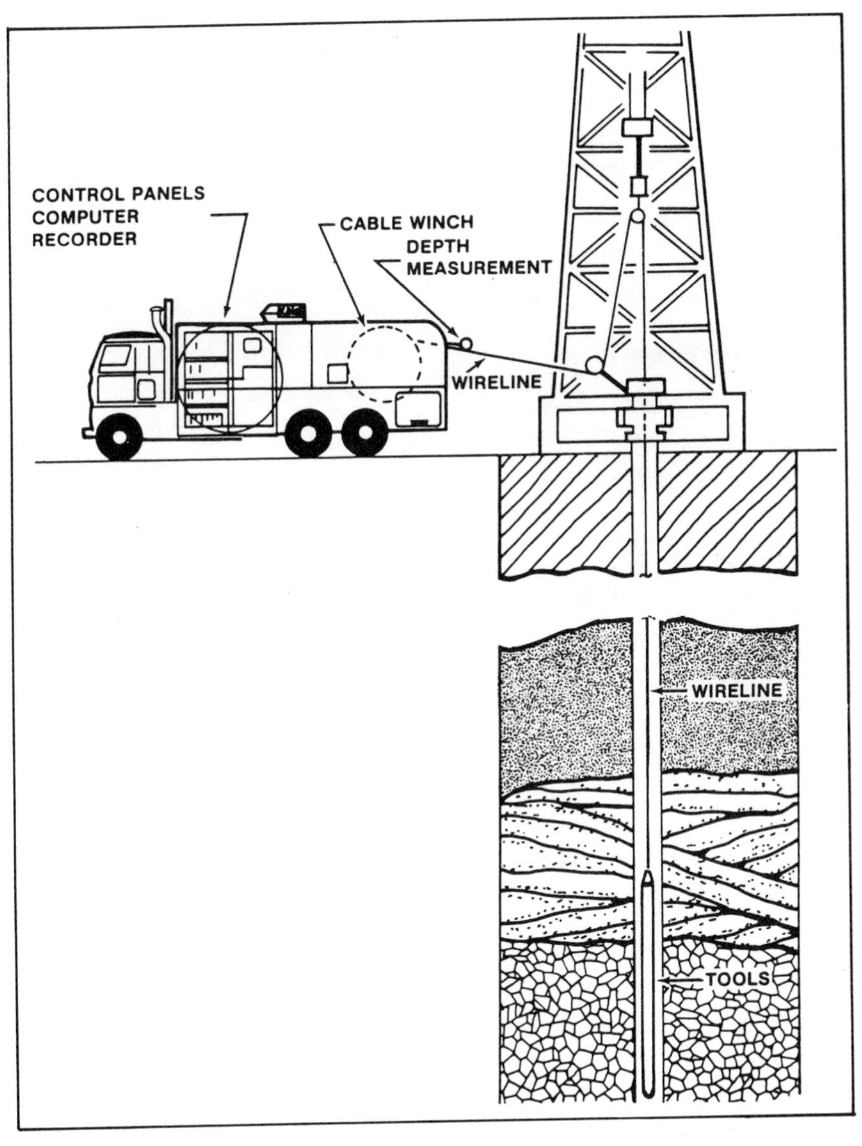

Fig. 1–1 Wellsite setup for logging (courtesy Gearhart)

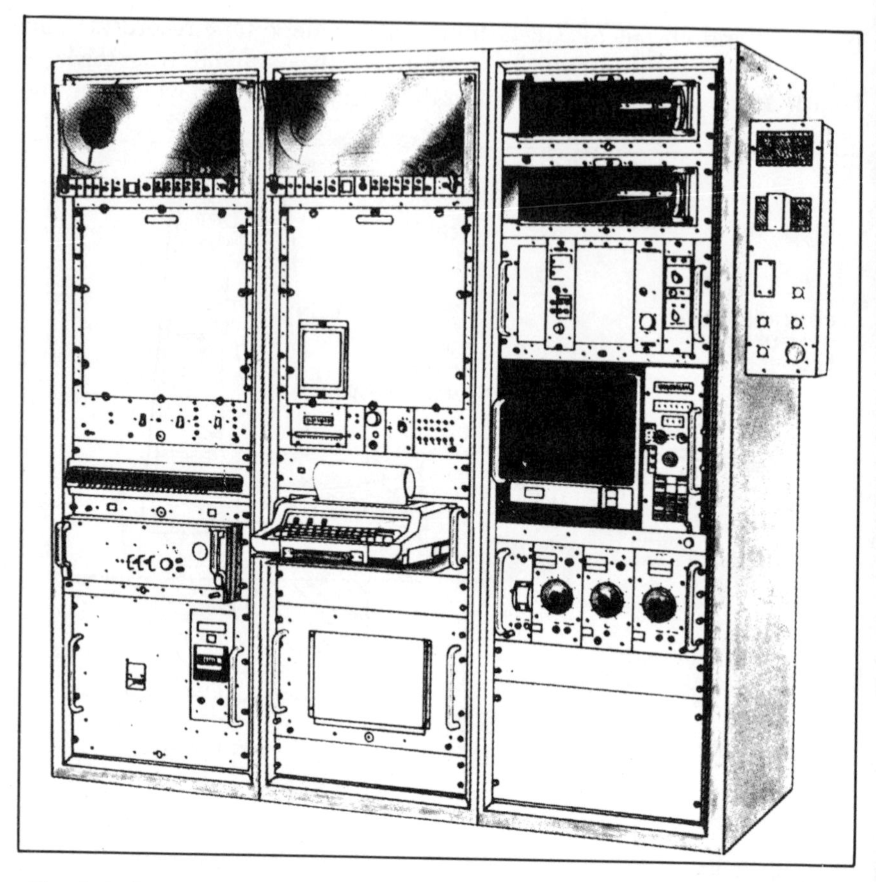

Fig. 1–2 Computerized surface instrumentation (courtesy Schlumberger)

logging, continuously displays the last 100 ft logged. The options in calibrating, depth shifting, averaging, computing, and scaling the logs are virtually unlimited with the computer. Further, after the logging is completed, the taped information can be played back and the various logs edited, combined, and run through interpretation programs to provide fully interpreted logs at the wellsite.

THE UNDISTURBED RESERVOIR

An idealized view of porous hydrocarbon-bearing reservoir rock is shown in Fig. 1–3. The rock matrix consists of grains of sand, limestone, dolomite, or mixtures of these. Between the grains is pore space filled with water, oil, and perhaps gas. The water exists as a film around the rock grains and as hour-glass rings at grain contacts; it also occupies the very fine crevices. The water forms a continuous path, although very tortuous, through the rock structure. Oil occupies the larger pore spaces. If gas is present, it will occupy the largest pores, leaving oil in the intermediate spaces.

The rock properties important in log analysis are porosity, water saturation, and permeability. The former two determine the quantity of gas or oil in place, and the latter determines the rate at which that hydrocarbon can be produced.

Porosity

Porosity, denoted ϕ, is the fraction of total volume that is pore space. In unconsolidated formations porosity depends on grain size distribution, not on absolute grain size. It will be high, in the range of 0.35–0.4, if all grains are close to the same size. It will be lower, down to about 0.25, if there is a wide range of grain sizes such that small grains fill the pore spaces between larger ones. At even lower porosities the matrix particles are generally cemented together with siliceous or calcareous material, resulting in consolidated formations. These may have porosities down to virtually zero.

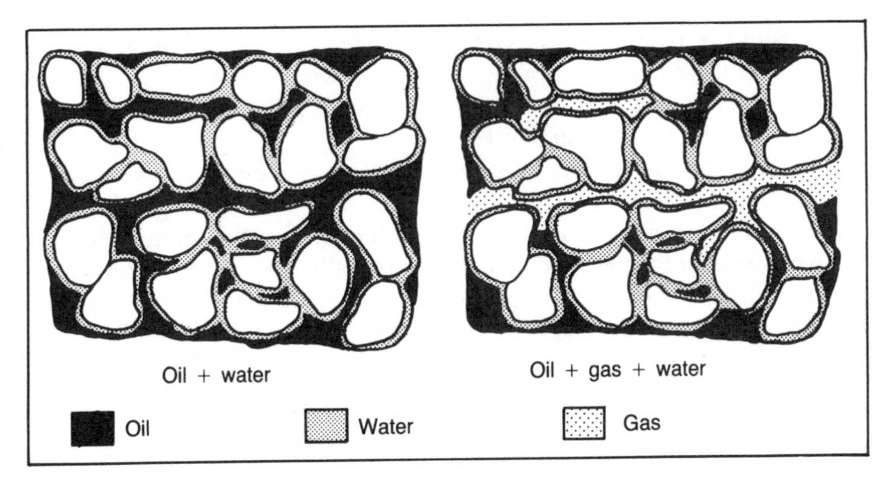

Fig. 1–3 Hydrocarbon-bearing rock

Water Saturation

The fraction of pore space containing water is termed *water saturation*, denoted S_w. The remaining fraction containing oil or gas is termed *hydrocarbon saturation*, S_h, which of course, equals $(1 - S_w)$. The general assumption is that the reservoir was initially filled with water and that over geologic time oil or gas that formed elsewhere migrated into the porous formation, displacing water from the larger pore spaces. However, the migrating hydrocarbons never displace all of the interstitial water. There is an *irreducible water saturation*, S_{wi}, representing the water retained by surface tension on grain surfaces, at grain contacts, and in the smallest interstices. Its value varies from about 0.05 in very coarse formations with low surface area to 0.4 or more in very fine-grained formations with high surface area. The irreducible water will not flow when the formation is put on production.

The fraction of total formation volume that is hydrocarbons is then ϕS_h or $\phi (1 - S_w)$. A major objective of logging is to determine this quantity. It can vary from zero to a maximum of $\phi (1 - S_{wi})$.

Permeability

Permeability, denoted k, is the flowability of the formation. It is a measure of the rate at which fluid will flow through a given area of porous rock under a specified pressure gradient. It is expressed in millidarcies (md); 1,000 md is a high value and 1.0 md is a low value for producing formations.

In contrast to porosity, permeability depends strongly on absolute grain size of the rock. Large-grained sediments with large pores have high permeabilities, whereas fine-grained rocks with small pores and more tortuous flow paths have low permeabilities.

Table 1–1 lists porosities and permeabilities of some well-known producing formations. Porosity varies only by a factor of 3, whereas permeability varies by a factor of about 4,000. We can infer that the Woodbine formation with extremely high permeability is exceptionally coarse sand, whereas the Strawn formation of the same porosity but low permeability is a very fine-grained sandstone.

Hydrocarbon-Bearing Rocks

Hydrocarbon-bearing rocks are primarily sands (SiO_2), limestones ($CaCO_3$), and dolomites ($CaCO_3 \cdot MgCO_3$). Most sands are transported by

TABLE 1-1 POROSITIES AND PERMEABILITIES OF SELECTED OIL SANDS

Sand	Porosity (%)	Permeability (md)
Clinch, Lee Co. VA	9.6	0.9
Wilcox, Okla. Co. OK	12.0	100
Cut Bank, Glacier, Co. MT	15.4	111
Bartlesville, Anderson Co. KS	17.5	25
Olympic, Hughes Co. OK	20.5	35
Woodbine, Tyler Co. TX	22	3,390
Strawn, Cooke Co. TX	22	81
Nugget, Fremont Co. WY	25	147
O'Hern, Duval Co. TX	28	130
Eutaw, Choctaw Co. AL	30	100

and laid down from moving water. The greater the water velocity (the energy of the environment), the coarser the sand will be. Because of this mechanism, sands tend to have fairly uniform intergranular-type porosity.

Limestones, on the other hand, are not transported as grains but are laid down by deposition from seawater. Some is precipitation from solution; some is the accumulated remains of marine shell organisms. Original pore space is often altered by subsequent redissolution of some of the solid matter. Therefore, porosity tends to be less uniform than in sands, with vugs and fissures, termed secondary porosity, interspersed with the primary porosity.

Dolomites are formed when magnesium-rich water circulates through limestones, replacing some of the calcium by magnesium. This process generally results in a reduction of the matrix volume. Therefore, dolomitization is an important mechanism in providing pore space for hydrocarbon accumulation.

Formations containing only sands or carbonates are called *clean formations*. They are relatively easy to interpret with modern logs. When such formations contain clay, they are called *dirty* or *shaly formations*. Such reservoir rocks can be quite difficult to interpret.

Clay and Shale

Clays are common components of sedimentary rock. They are aluminosilicates of the general composition $Al_2O_3 \cdot SiO_2 \cdot (OH)_x$. Depending on the environment in which they are formed, they may be of several basic types: montmorillonite, illite, chlorite, or kaolinite.

Clays have very small particle sizes—1 to 3 orders of magnitude less than those of sand grains. Surface-to-volume ratios are very high, 100–10,000 times those of sands. Thus, clays can effectively bind large quantities of water that will not flow but that do contribute to log response.

Shales are primarily mixtures of clay and silt (fine silica) laid down from very slow-moving waters. While they may have good porosity, permeability is essentially zero. Pure shales are therefore of little interest in hydrocarbon production, although they are source rocks for petroleum. On the other hand, sands or carbonates containing modest amounts of clay or shale may be important hydrocarbon producers.

Accounting for clay and shale when analyzing hydrocarbon-bearing formations substantially complicates log interpretation. Consequently, in chapters 2 through 6 we establish the principles of log interpretation for clean formations and in chapter 7 take up the analysis of shaly formations.

DISTURBANCE CAUSED BY DRILLING

The drilling process is illustrated in Fig. 1–4. A drill bit at the end of a long drillstring is rotated from the surface at speeds of 50–150 rpm. Simultaneously, weights of 10,000–40,000 lb are applied to the bit, and the combined action chews up the rock. The cuttings produced are swept from below the bit to the surface by drilling mud, which is pumped down the center of the drillstring, passed out holes in the bit, and returned up the pipe-formation annulus.

During the drilling process formations may erode or cave to diameters larger than bit size, drilling fluid may invade permeable zones, and mud cakes may build up on the same zones. Invasion in particular causes logging problems.

The Process of Invasion

The process of invasion is also illustrated in Fig. 1–4. During drilling the mud pressure in the annulus, P_m, must be kept greater than the hydrostatic pressure of fluid in the formation pores, P_r, to prevent a well blowout. The differential pressure, $P_m - P_r$, which is typically a few hundred psi, forces drilling fluid into the formation. As this happens solid particles in the drilling mud plate out on the formation wall and form a mud cake. Liquid that filters through this mud cake—the *mud filtrate*—passes into the formation and pushes back some of the reservoir fluid there. An invaded zone is formed adjacent to the borehole.

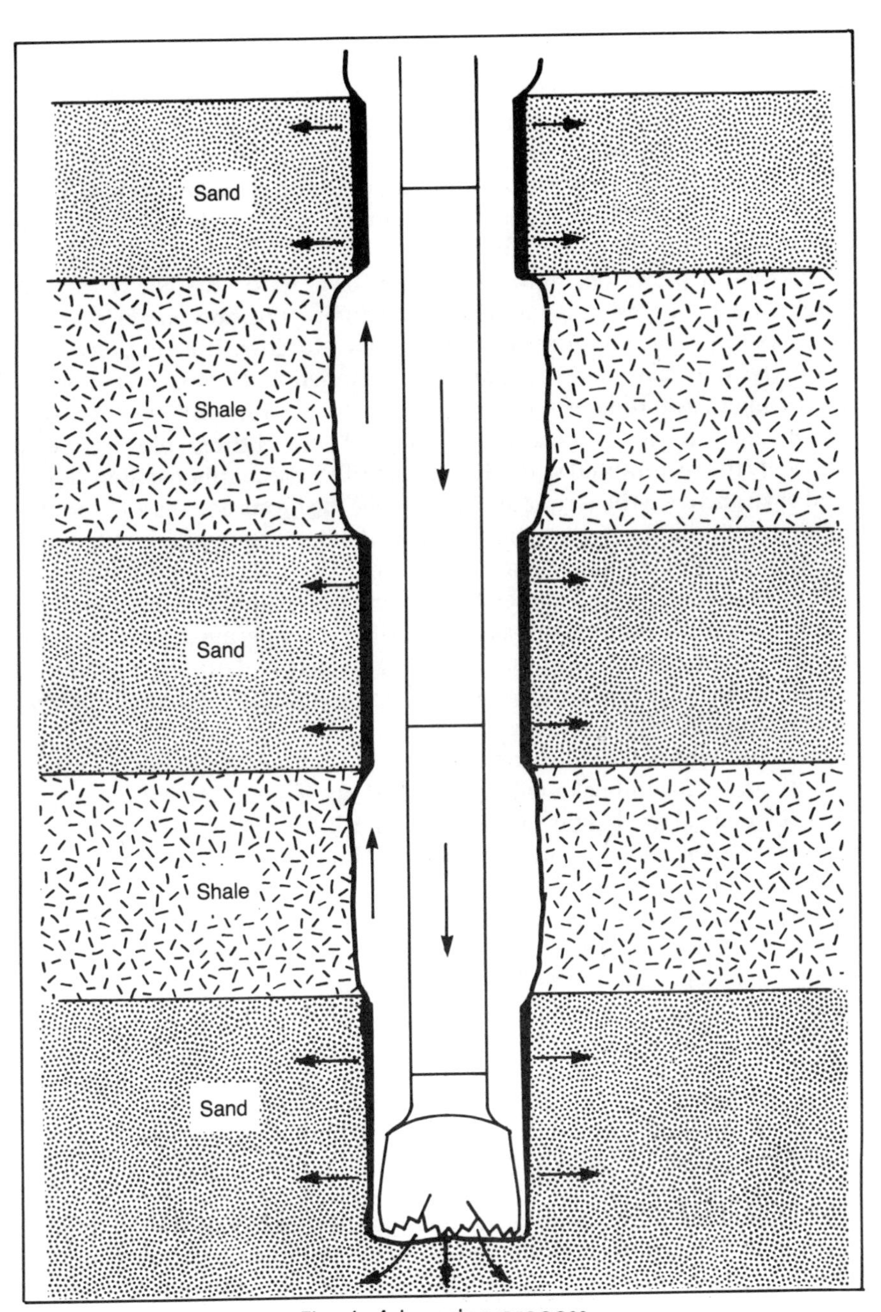

Fig. 1–4 Invasion process

Invasion involves a mud spurt, dynamic filtration, and static filtration.[1] As the bit penetrates a permeable bed, there is an initial spurt of drilling fluid into the freshly exposed rock. This lasts for a matter of seconds until small solid particles in the downcoming mud stream (or formed by grinding action of the bit) bridge the pore entrances in the rock. Bridging is most rapid when the particle size distribution in the mud is well matched to the pore entrance distribution in the rock.

As the bit passes on, mud cake begins to build on the newly formed borehole wall. Invasion is rapid in the beginning but slows quickly as the mud cake thickens and increases its resistance to flow. If conditions were static, the mud cake would continue to build indefinitely with filtration rate decreasing in accordance with $1/\sqrt{t}$, where t is the time following spurt.

During drilling, however, flow of mud and cuttings plus abrasion caused by the turning and whipping of the drillstring continuously erode the mud cake and even the formation itself. Once the formation ceases to erode, a dynamic equilibrium condition is reached where the mud cake thickness and the rate of filtration become constant.

When the drillstring is pulled to change the bit, the abrasive action is no longer present and mud cake resumes building at the permeable zones under static filtration conditions. When the string is run in and drilling is resumed, the soft outer mud cake just formed will erode away and dynamic equilibrium once again will be reached.

Finally, when the drillstring is pulled for logging, static filtration will resume and soft mud cake will again build up. The additional buildup is often evidenced by logging tools measuring hole diameter less than bit diameter in permeable zones near bottom. Mud cake is typically ⅛–¾ in. in thickness at the time of logging.

Fig. 1–5 shows schematically the rate of invasion, mud cake thickness, and depth of invasion at a given permeable bed as a function of time since the bed was penetrated. The depth of invasion increases rapidly during the spurt and formation erosion periods. Later it slows because of dynamic equilibrium and because the rate of increase in invasion depth, for a constant filtration rate, is inversely proportional to the invasion depth already reached.

Depth Of Invasion At Time Of Logging

The depth to which mud filtrate has penetrated a porous formation at the time of logging depends on several factors, principal of which are the filtration characteristics of the drilling mud and the differential pressure between mud and reservoir. Static filtration rate of a mud is given as a water loss figure on the log heading. This is the amount of filtrate (in cc) passing through a filter paper in 30 min at 100 psi differential pressure and 76°F in a standarized API test. A typical figure is 12 cc; 30 cc is considered poor wall building mud, and 4 cc is very good. Unfortunately, experiments have shown that there is little correlation between static filtration characteristics at surface temperature and dynamic filtration at borehole temperatures. Consequently, it is not possible to predict invasion depth from available mud and drilling information. The analyst must infer this from logs.

One can predict, nevertheless, how depth of invasion for a given mud relates to porosity. Once the mud cake has started to build, its permeability becomes low relative to that of the average formation so that almost all of the pressure differential $(P_m - P_r)$ is across the mud cake and little is applied to the formation. The mud cake therefore controls filtration rate. Consequently, in a given time the same volume of fluid will invade different formations, regardless of their porosities or permeabilities (unless permeability is below about 1.0 md). This means depth of invasion will be minimum at high porosity where plenty of pore space is available for invading fluid and maximum at low porosity where little room is available. It is

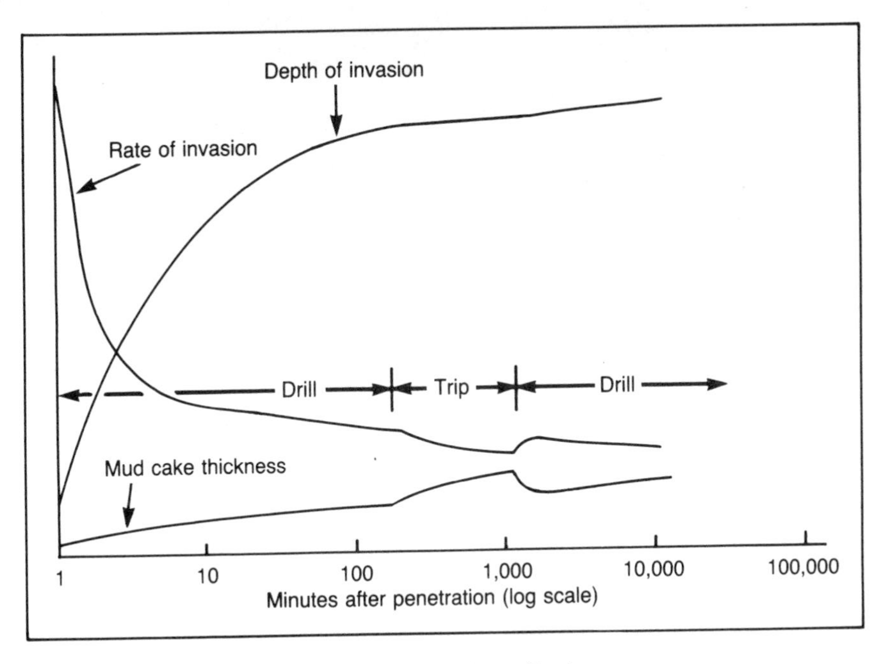

Fig. 1–5 Invasion effects

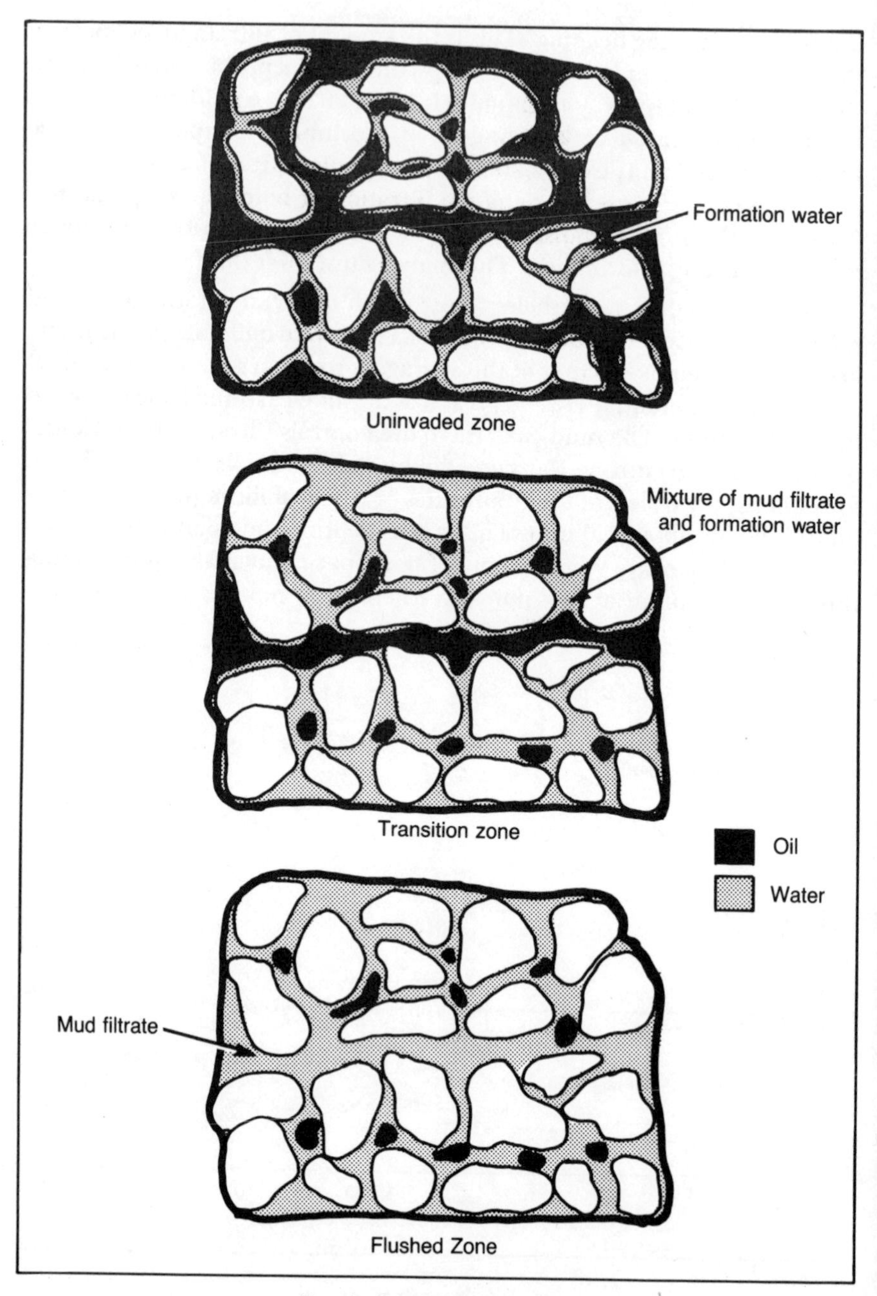

Fig. 1–6 Invaded rock

approximately proportional to $1/\sqrt{\phi}$ where ϕ is the porosity. Other things being constant, invasion depth will double as porosity reduces from 36% to 9%, for example. However, many other factors come into play. About all that can be said is that invasion depths can vary from a few inches to a few feet, with typical values perhaps 1–2 ft.

The Picture Of Invaded Rock

Fig. 1–6 illustrates invasion as it is pictured. Proceeding outward from the wall of the hole, there is first a flushed zone, then a transition zone, and finally the unperturbed formation. In the flushed zone it is generally assumed that all of the formation water has been replaced by mud filtrate (which may not be quite true). If the formation is hydrocarbon bearing, then some but not all of the hydrocarbons will be pushed back by invading filtrate. The residual hydrocarbon saturation remaining will normally be in the range of 10–40%. The saturation will depend on the initial hydrocarbon content and on the contrast between the mobility of the filtrate and that of the hydrocarbon. Water displaces medium-gravity oil fairly well but displaces high-viscosity heavy oil and low-viscosity light gas quite poorly. The water fingers through these media.

In the transition zone some of the virgin water and some of the hydrocarbons, if present, have been replaced by mud filtrate but to a lesser extent than in the flushed zone. The transition zone initially is close to the borehole but gradually progresses away from it. It may take a few days after a formation is drilled for the invasion pattern to reach a more-or-less equilibrium condition.

In extremely porous and permeable sands the invading fluid can gravity-segregate vertically as well as progress laterally. Low-salinity filtrate invading a high-water-salinity sand will tend to rise to the top of the bed; water invading an oil sand will tend to drop to the bottom of the oil. Successive logs may show this progression.

Shales, by virtue of almost zero permeability, do not invade or build up mud cake. More often, fresh water in the drilling mud will cause the clay in the shales to swell, resulting in sloughing and caving of those formations. Suitable mud conditioning can minimize this problem.

SUMMARY

BOREHOLES

- Depths 1,000–20,000 ft; average 6000 ft
- Diameters 5–15 in.; average 9 in.

- Deviations 0–5° onshore; 20–40° offshore
- Bottom-hole temperatures 100–350°F; average 150°F
- Mud weight 9–16 lb/gal; average 11 lb/gal
- Mud salinity 3,000–200,000 ppm; average 10,000 ppm
- Bottom-hole pressure 500–15,000 psi; average 3,000 psi
- Mud cakes 0.1–1 in.; average 0.5 in.
- Invaded depth 0.1–3 ft

ROCK PROPERTIES

- Porosity, ϕ: fraction of total volume that is pore space
- Water saturation, S_w: fraction of pore space occupied by water
- Irreducible water saturation, S_{wi}: fraction of pore space occupied by immovable water
- Permeability: flowability measured in millidarcies (md)

HYDROCARBON-BEARING ROCKS

- Sands, SiO_2: uniform porosity
- Limestones, $CaCO_3$: nonuniform porosity
- Dolomites, $CaCO_3 \cdot MgCO_3$: nonuniform porosity
- Clean if free of shale or clay; dirty if not

INVASION

- Creates mud cake on permeable formations
- Displaces formation water with mud filtrate in flushed zone
- Leaves 20–40% residual hydrocarbon saturation in flushed zone
- Extends a few inches to a few feet; generally greater at low porosities

REFERENCES

[1]C.K. Ferguson and J.A. Klotz, "Filtration from Mud During Drilling," *Petroleum Transactions* AIME, Vol. 201 (1954), pp. 29–42.

Chapter 2

EVALUATION OF HYDROCARBONS

The manner in which the presence of hydrocarbons in pore space is sensed is via the electrical resistance of the formation. For normal logging situations, the rock matrix is considered a perfect insulator; it conducts no electricity. All conduction is then via the fluid in the pores. At depths below 2,000 ft, the water found in formation pores is generally fairly saline, which makes it quite conductive. Water-bearing formations therefore tend to have high electrical conductivity or, the equivalent, low electrical resistivity since resistivity is the reciprocal of conductivity.

What happens when some of the saline pore water is replaced by hydrocarbons? Oil and gas do not conduct electricity. There is then less pore fluid available for conduction. The electrical current that does flow is forced to take a more tortuous path, weaving around the hydrocarbon that occupies the larger pore spaces. Both effects increase the electrical resistance of the formation.

Therefore, the basis of logging is to measure the actual electrical resistivity of the formation and to compare it with the resistivity that would exist if all of the pore space contained water. If the measured resistivity is significantly higher than the calculated resistivity, the presence of hydrocarbons is inferred. All of this is bound up in one simple equation: the Archie relation. If this equation is well understood and easily manipulated, interpretation of clean formations with modern logs can often be performed without a single chart.

FUNDAMENTAL INTERPRETATION RELATIONS

To establish the relations for hydrocarbon saturation and to clarify the terms involved, let us conceptually construct an oil-bearing formation and measure its electrical properties as we do so.

Definition of R_w

Visualize an open-top cubic tank one meter in all dimensions. It has electrically nonconducting sides except for two opposite walls that are metal and serve as electrodes.

First, the tank is filled with water containing about 10% sodium chloride (NaCl) by weight to simulate an average formation water. A low-frequency alternating voltage, V, is applied across the electrodes, and the resulting current I_1 is measured (Fig. 2–1a). The ratio V/I_1 (volts/amperes) is R_w, the resistivity of the formation water, in units of ohm-meters. This resistivity is an intrinsic property of the water and is a function of its salinity and temperature. The higher these two variables, the more conductive the water will be and the lower its resistivity.

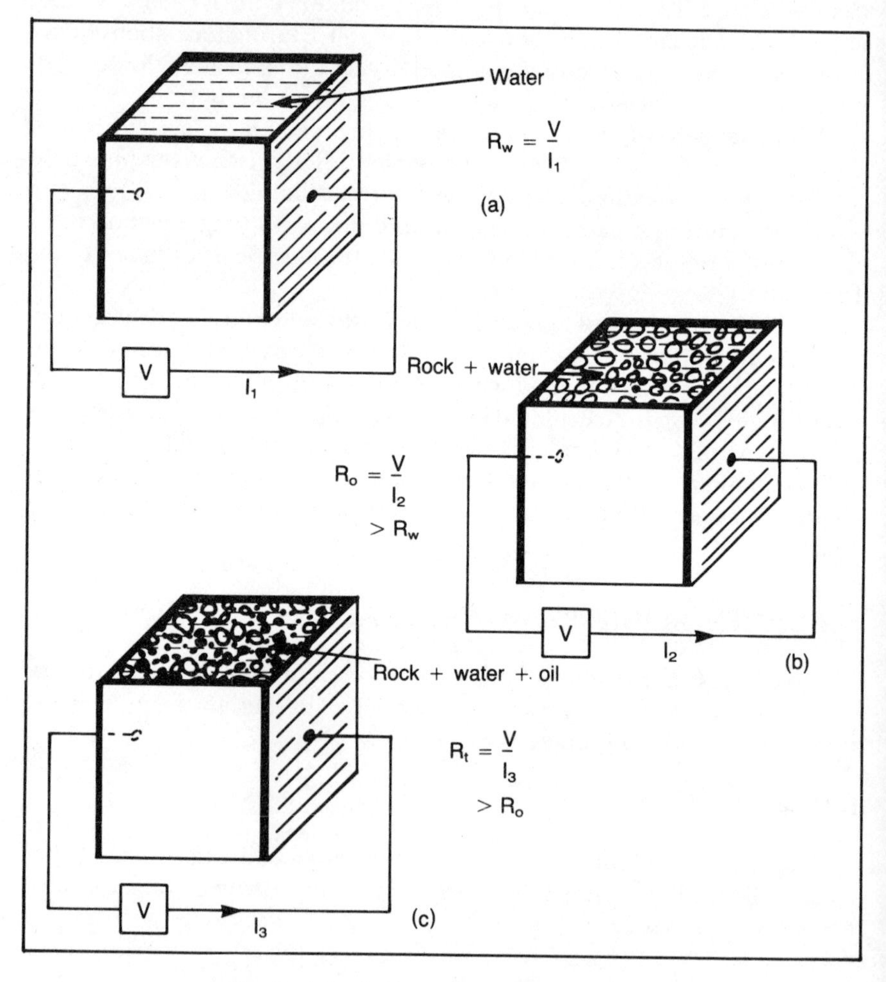

Fig. 2–1 Definition of resistivities

Definition of R_o

Next, sand is poured into the water-filled tank and the volume of water expelled is measured. When the sand is level with the top, the result is a porous, water-bearing formation of one-meter dimensions. About 0.6 cu m of water will have been expelled, so the porosity of the formation will be (1 – 0.6) or 0.4. Again the voltage is applied and a current I_2 is measured (Fig. 2–1b). I_2 will be less than I_1 since there is less water to conduct electricity. The ratio V/I_2 is R_o, the resistivity of the water-bearing formation. It will be larger than R_w.

Formation Factor

The resistivity, R_o, must be proportional to R_w since only the water conducts. Thus

$$R_o = F \cdot R_w \tag{2.1}$$

The proportionality constant F is termed the *formation factor*.

On general principles, formation factor must be related to porosity by a relation of the form

$$F = 1/\phi^m \tag{2.2}$$

because when $\phi = 1$ (all water, no matrix), R_o must equal R_w; and when $\phi = 0$ (no pore water, solid matrix), R_o must be infinite since the rock itself is an insulator. Eq. 2.2 satisfies these conditions regardless of the value of m, which is termed the *cementation exponent*.

The value of m reflects the tortuosity of current flow through the maze of rock pores. If the pore space consisted of cylindrical tubes through an otherwise solid matrix, current flow paths would be straight and m would be 1.0. In the case of porous formations, measurements have shown m to be 2.0 on the average. Accepted relations for the range of porosities encountered in logging are

$$F = 1/\phi^2 \text{ for limestones} \tag{2.3}$$

$$F = 0.81/\phi^2 \text{ or } 0.62/\phi^{2.15} \text{ for sands}^{1} \tag{2.4}$$

We shall use the first of the sand relations; it is less cumbersome than the second. The constant 1, 0.81, or 0.62 is called the *cementation factor* and is designated as *a* in general equations.

Definition of R_t

Now an appreciable fraction of the pore water is replaced by oil, resulting in the situation depicted in Fig. 2–1c. The same voltage, V, is applied, and current I_3 is measured. It will be less than I_2 since even less water is available for conduction. The ratio V/I_3 is R_t, the resistivity of the oil-bearing formation. It will be greater than R_o.

Water Saturation

Knowing R_o and R_t, water saturation, S_w, the fraction of pore space containing water, can be calculated. Again on general principles there must be a relation of the form

$$R_t = R_o/S_w^{\,n} \tag{2.5}$$

because when $S_w = 1$ (all water in the pores), R_t must equal R_o; and when $S_w = 0$ (all oil in the pores, if it were possible), R_t must be infinite, as both oil and rock matrix are insulators. Eq. 2.5 satisifies these conditions regardless of the value of the exponent n.

The constant n is termed the *saturation exponent*. It is close in value to m because the flow of current cannot distinguish between displacement of pore water by sand grains or oil globules of like sizes since neither conducts. Indeed laboratory experiments have shown $n = 2.0$ in the average case. Consequently, water saturation is given by

$$S_w = \sqrt{R_o/R_t} \tag{2.6}$$

This relation can be used directly to calculate the water saturation of a hydrocarbon-bearing zone when an obvious water-bearing zone *of the same porosity and having water of the same salinity is nearby*. An example would be a thick sand with an obvious water-oil contact in the middle.

In general there will not be a nearby water sand to give R_o, so Eq. 2.6 will not apply. Replacing R_o by Eq. 2.1 gives

$$S_w = \sqrt{FR_w/R_t} \tag{2.7}$$

Replacing F by Eq. 2.3 gives

$$\boxed{S_w = c\sqrt{R_w/R_t}/\phi} \tag{2.8}$$

where $c = 1.0$ for carbonates and 0.90 for sands.

This is the basic equation of log interpretation. *It was initially developed by G.E. Archie of Shell and is termed the Archie relation.[2] *The whole well-logging industry is built upon this equation.*

Eq. 2.8 shows that hydrocarbons in place can be evaluated if there are sufficient logs to give interstitial water resistivity (R_w), formation resistivity (R_t), and porosity (ϕ). In practice R_w is obtained either from applying Eq. 2.8 in a nearby water sand ($S_w = 1$) or from the SP log or from catalogs or water sample measurements; R_t is obtained from deep resistivity readings (Induction or Laterolog); and ϕ is obtained from porosity logs (Density, Neutron, or Sonic).

Hydrocarbons in Place

A good estimate of the total quantity of hydrocarbon in situ can be obtained from logs. The product of porosity and hydrocarbon saturation, $\phi \times (1 - S_w)$, is the fraction of the formation by volume that contains hydrocarbons. The thickness of the producing formation, h (ft), can readily be determined from logs. However, the areal extent, A (acres), cannot. Barring other information, it can be taken as the allowed well spacing. With this assumption the total quantity of hydrocarbons in place can be calculated.

For oil, the number of barrels in situ is

$$N = 7{,}758 \cdot \phi(1 - S_w) \cdot h \cdot A \tag{2.9}$$

To obtain stock-tank barrels at the surface, this number is divided by the formation volume factor B, a value slightly greater than unity, which takes into account the shrinkage of oil volume, principally by gas evolution, as it comes to the surface.

For gas, the number of cubic feet in situ is

$$G = 43{,}560 \cdot \phi (1 - S_w) \cdot h \cdot A \tag{2.10}$$

---

*The general equation is $S_w^n = (a/\phi^m)(R_w/R_t)$. Values of a, m, and n can differ from those indicated in specific cases. This is discussed in chapter 6.

This is the amount of gas at reservoir pressure and temperature. To convert it to standard cubic feet at 14.7 psi and 60°F, the number is multiplied by the quantity

$$\frac{P_r}{14.7} \cdot \frac{520}{(460 + T_r)} \cdot \frac{1}{Z} \tag{2.11}$$

where:

P_r = reservoir pressure, psi; if not known it may be taken as 0.46d where d is the vertical depth of the reservoir in feet
T_r = formation temperature, °F, determinable from log information
Z = deviation factor of the gas at formation temperature and pressure, obtainable from charts; it will be close to unity

The amount of hydrocarbons actually recoverable as a fraction of the quantity in place will depend on the reservoir type, the initial hydrocarbon saturation, and the production mechanism. A reasonable estimate for primary production would be 20% for oil and 70% for gas.

THE BASIC INTERPRETATION PROCEDURE

Three basic logs are required for adequate formation evaulation. One is needed to show permeable zones, one to give resistivity of the undisturbed formation and one to record porosity. An idealized set is shown in Fig. 2–2. The permeable zone log is in Track 1, the resistivity log in Track 2, and the porosity log in Track 3. The permeable zone log is either Spontaneous Potential or Gamma Ray, the resistivity is either deep Induction or deep Laterolog, and the porosity log is either Density, Neutron, or Sonic. Given such a set of logs, the problem is to determine

- where are the potential producing zones
- how much hydrocarbon (oil or gas) do they contain

Selection of Productive Zones

The first step is to locate the permeable zones. This is done by scanning the log in Track 1. It has a base line on the right and occasional swings to the

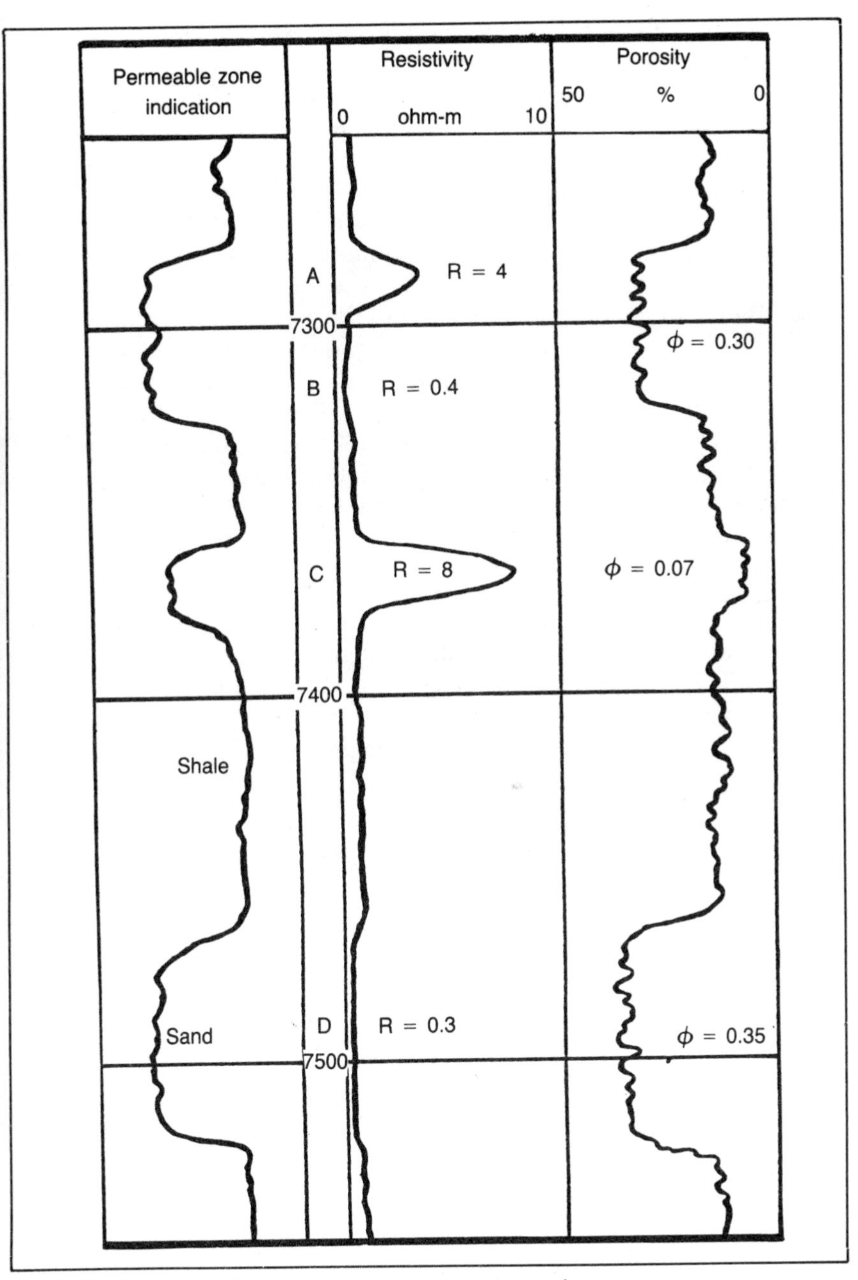

Fig. 2–2 Idealized log set

left over short intervals. The base line indicates shale, which is impermeable and will not produce, and the swings to the left indicate clean zones—generally sand or limestone — that may produce. The interpreter focuses his attention immediately on the zones labeled A, B, C, and D and disregards the rest.

Next, the resistivity log in Track 2 is scanned to see which of the zones of interest gives high resistivity readings. High resistivities reflect either hydrocarbons in the pores, or low porosity (assuming the pore water is reasonably saline). Zones A and C show high resistivity. Zones B and D have low values of resistivity, which can only be a result of the presence of large amounts of salt water in the pores. These are obvious water-bearing intervals.

The question then is whether zones A and C contain hydrocarbons or simply have low porosity. The porosity log in Track 3 shows values of 0.30 and 0.07 for A and C, respectively. At this point one may guess that A contains hydrocarbons and C is simply tight. However, to be sure, it is necessary to apply one of the water saturation relations, Eq. 2.6 or Eq. 2.8, whichever is applicable.

Calculation of Water Saturation

Intervals A and B constitute a formation of uniform porosity with an apparent water-oil contact just above 7,300 ft. Conditions are appropriate for applying Eq. 2.6. The resistivity of the water-bearing section, which by definition is R_o, is 0.4 ohm-m. That of the apparent oil-bearing section, R_t, is 4 ohm-m. Consequently

$$S_w = \sqrt{\frac{0.4}{4}} = 0.32$$

That is, the pore space contains 32% water and 68% hydrocarbons. It should be a productive interval.

Turning now to zone C, we cannot utilize Eq. 2.6 because there is no obvious measurement of R_o for that formation. Eq. 2.8 must be applied. For that, however, a value of water resistivity, R_w, must be known.

The best way to obtain R_w is to apply the same relation, Eq. 2.8, with S_w = 1, in a water-bearing interval that is nearby. This is permissible because water salinity changes only slowly with depth. Zone D is a nearby water-bearing interval. Applying Eq. 2.8 to that zone, assuming the formations are sands

$$1 = \frac{0.9}{0.35} \sqrt{\frac{R_w}{0.30}}$$

which gives

$$R_w = 0.045 \text{ ohm-m}$$

This value of R_w applies throughout the 300 ft interval of interest. For zone C then

$$S_w = \frac{0.9}{0.07}\sqrt{\frac{0.045}{8}} = 0.96$$

Consequently, this zone is simply a tight water-bearing section.

Knowing R_w, the interpretation of zone A can be checked. Applying Eq. 2.8

$$S_w = \frac{0.9}{0.30}\sqrt{\frac{0.045}{4}} = 0.32$$

This verifies the value previously obtained. In fact, it verifies that zone B is 100% water bearing. (In an old producing field with an advancing water drive, a zone such as B could contain residual hydrocarbon.)

The only hydrocarbon to be found is in zone A. The logs of Fig. 2–2 do not indicate whether it is oil or gas, although an appropriate pair of porosity logs could do so. Assume first it is oil. A productive thickness of approximately 10 ft is indicated. With 40-acre spacing the number of reservoir barrels in place is given by Eq. 2.9 as

$$N = 7{,}758 \times 0.30\,(1 - 0.32) \times 10 \times 40 = 633{,}000$$

Assuming a formation volume factor of 1.2 and recovery factor of 0.20, the producible amount would be estimated as 0.2 × 633,000/1.2 or 105,000 stock-tank barrels. This equates to a potential revenue of $2.62 million at a net price of $25/bbl. Completed cost of such a well in the U.S. Gulf Coast would be approximately $700,000. The obvious decision, after log analysis, would be to complete the well and put it on production.

Assume on the other hand the reservoir is gas bearing. From Eq. 2.10 the number of cubic feet in situ, on the basis of 160-acre spacing, is

$$G = 43{,}560 \times 0.30\,(1 - 0.32) \times 10 \times 160 = 14.2 \times 10^6$$

Reservoir pressure would be estimated as $0.46 \times 7{,}300$ or 3,360 psi and reservoir temperature as approximately 150°F. Assuming a gas deviation factor of 1.0, the amount of gas in place in standard cubic feet would be, by Eq. 2.11

$$14.2 \times 10^6 \times (3{,}360/14.7) \times (520/610) = 2{,}770 \times 10^6$$

With a recovery factor of 0.70, the producible amount is estimated at $0.7 \times 2{,}770$ or 1,940 MMscf. A gas price of $3/Mcf would yield a potential revenue of $5.8 million. Again, the well would be definitely worth completing.

IMPACT OF INVASION ON RESISTIVITY MEASUREMENTS

In the foregoing discussion the formation resistivity, R_t, was assumed to be that of the undisturbed reservoir beyond any invasion. The difficulties in measuring that resistivity may be appreciated by considering the disturbance caused by invasion, as illustrated in Fig. 2–3.

Immediately behind the borehole wall is a flushed zone of diameter d_f containing only mud filtrate of resistivity R_{mf} and residual hydrocarbon. The resistivity of that zone is denoted R_{xo} and the water saturation is S_{xo}. The thickness of this zone is of the order of 6 in. but can be much more or much less. Behind the flushed zone is the transition zone of diameter d_j, which may extend several feet. Beyond that is the undisturbed formation with resistivity R_t, interstitial water resistivity R_w, and water saturation S_w.

The existence of invasion has forced the development of resistivity logging tools that measure as deeply as possible in an effort to read R_t uninfluenced by mud filtrate. However, no tool has been developed that can read deeply enough under all circumstances and still maintain good vertical resolution. Consequently, the industry is gradually standardizing on running three resistivity curves at the same time. One is a deep investigation curve, one is a medium curve, and one is a shallow curve. With three curves the reading of the deep one can be corrected for invasion effects to provide an R_t value. As a side benefit the flushed zone resistivity and the diameter of invasion can also be estimated.

Mud Filtrate Resistivity; Fresh Muds and Salt Muds

To appreciate the difference in resistivity readings of shallow, medium, and deep curves, it is necessary to consider the contrast between mud filtrate resistivity, R_{mf}, and interstitial water resistivity, R_w. Mud filtrate resistivity is measured at the wellsite by the logging engineer. He catches a sample of mud, preferably from the mud return line, places it in a mud filter press that forces filtrate through a filter paper and measures the resistivity of the filtrate in a resistivity-measuring cell. The value of R_{mf} along with the temperature at which it is measured are included in the log heading.

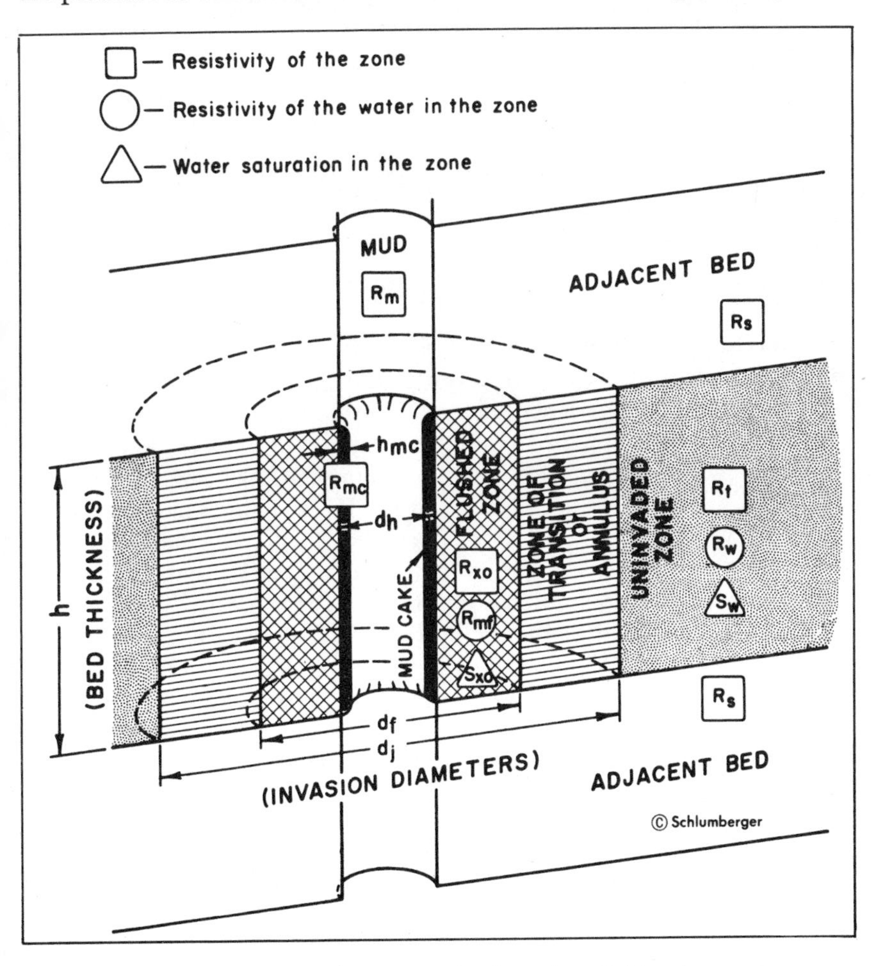

Fig. 2–3 Idealized invasion profile (courtesy Schlumberger)

Most wells are drilled with "fresh" (low-salinity) muds that have R_{mf} values in the range 0.4–2 ohm-m at surface temperature. In these cases R_{mf} will generally be close to R_w in very shallow zones where formation waters are fresh but will be higher by a factor of about 10 in deep zones where formation waters are saline. Typical values at depths of interest are R_w = 0.02–0.1 ohm-m and R_{mf} = 0.2–1.0 ohm-m. This means invaded-zone resistivities will in general be *higher* than undisturbed-zone resistivities in water-bearing intervals.

In certain areas, however, such as the Permian Basin of West Texas, salt beds are penetrated. Saturated salt mud must be used during drilling to prevent the salt beds continually dissolving and caving. R_{mf} values will be 0.1 ohm-m or less at surface temperature. In these cases R_{mf} will be much lower than R_w near the surface and somewhat lower than R_w in deep formations. Invaded-zone resistivities will in general be *lower* than virgin resistivities at depths of interest.

Fig. 2–4 illustrates how invasion manifests itself on the three resistivity curves with different depths of investigation. The LL_8 has a depth of investigation of about 1 ft, the IL_m 2 ft, and the IL_d 5 ft. In the two permeable sands where the curves separate, the LL_8 reads close to R_{xo} and the IL_d reads close to R_t.

Several deductions can be made. First, the sands are obviously water-bearing because of the low IL_d readings relative to that of the intervening shale resistivity (an experience judgment). Second, the well has been drilled with fresh mud since the LL_8 reads much higher than the IL_d in the water sands—in fact, from the ratio of the readings, $R_{mf} \geq 6\ R_w$. Third, the invasion is shallow in the sands because the IL_m reading is close to the IL_d value. If it were deep, the IL_m would be very influenced by the mud filtrate and would read close to LL_8. Thus, the position of the IL_m curve relative to the others is a quick-look indicator of invasion depth. Finally, we can see that the shales have no permeability because they have not invaded; all curves read alike.

In wells drilled with saturated salt muds, the positions of the three resistivity curves in water-bearing intervals will reverse. The shallow curve will read lowest and the deep curve will read highest.

Invasion Profiles

Fig. 2–5 shows three different resistivity profiles proceeding from the flushed to the undisturbed zone for the fresh mud case where $R_{xo} > R_t$. The assumption made in deriving a correction for R_t with three resistivity curves is a step profile—an abrupt change from R_{xo} to R_t at an equivalent diameter

d_i. In actuality the resistivity will change gradually from the R_{xo} value at diameter d_f to the R_t value at diameter d_j, as shown by the dashed curve of Fig. 2–5, but field and laboratory tests have shown that the exact shape of the resistivity profile in the transition zone is not too important. Adequate correction for R_t usually can be made with the step profile assumption.

Occasionally, however, a condition occurs in high-porosity formations with high hydrocarbon saturation wherein invading filtrate displaces hydrocarbon faster than the interstitial water. This creates an annulus or bank of formation water where the resistivity is temporarily lower than either R_{xo} or R_t (Fig. 2–5, dotted line). It is a transient phenomenon, lasting only a few days.[3]

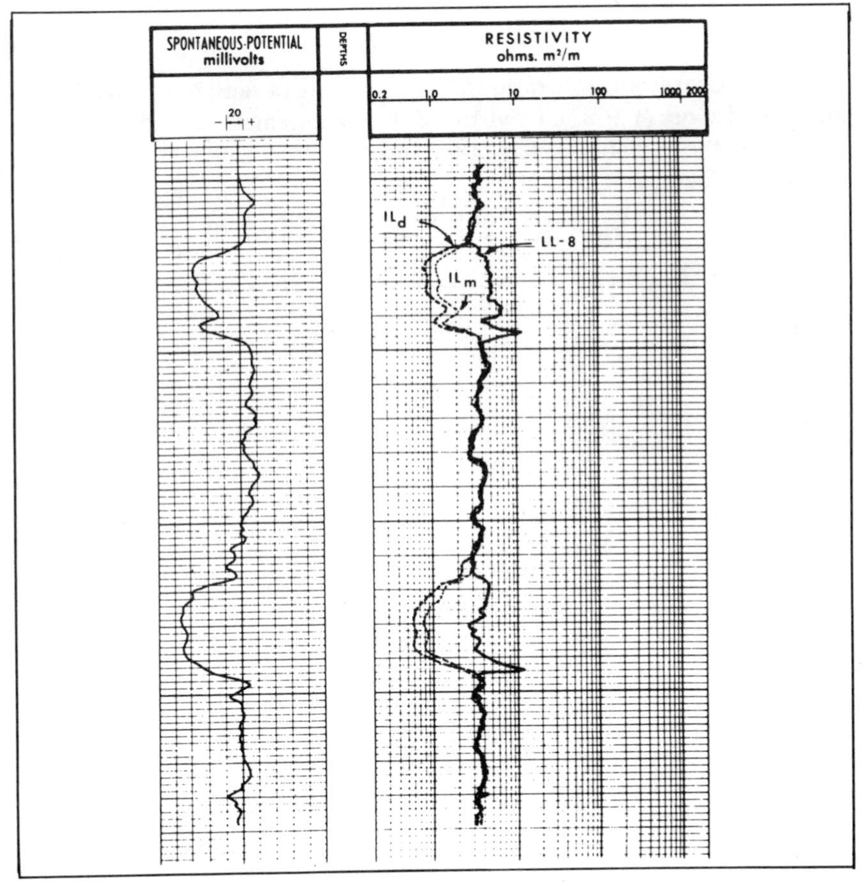

Fig. 2–4 Effect of invasion on resistivity measurements (courtesy Schlumberger)

Movable Oil Calculation

Invasion has one redeeming feature: It can provide information on hydrocarbon producibility through comparison of hydrocarbon saturation in the flushed and undisturbed zones. The difference between the two saturations represents hydrocarbon that has been pushed back by invading fluid and therefore should be recoverable on production, at least by a water drive simulating invasion in reverse.

The water saturation relations already developed can be applied to the flushed zone provided appropriate resistivity values are used for the fluid-filled rock, R_{xo}, and the pore fluid, R_{mf}. Following Eq. 2.8, the water saturation in the flushed zone is

$$S_{xo} = c\sqrt{R_{mf}/R_{xo}}/\phi \tag{2.12}$$

The hydrocarbon saturation $(1 - S_{xo})$ will be less than that in the uninvaded zone, $(1 - S_w)$, by virtue of the displacement caused by filtrate. Therefore, the movable oil saturation is the difference, which is $(S_{xo} - S_w)$. By Eqs. 2.12 and 2.8

$$S_{xo} - S_w = c\left(\sqrt{R_{mf}/R_{xo}} - \sqrt{R_w/R_t}\right)/\phi \tag{2.13}$$

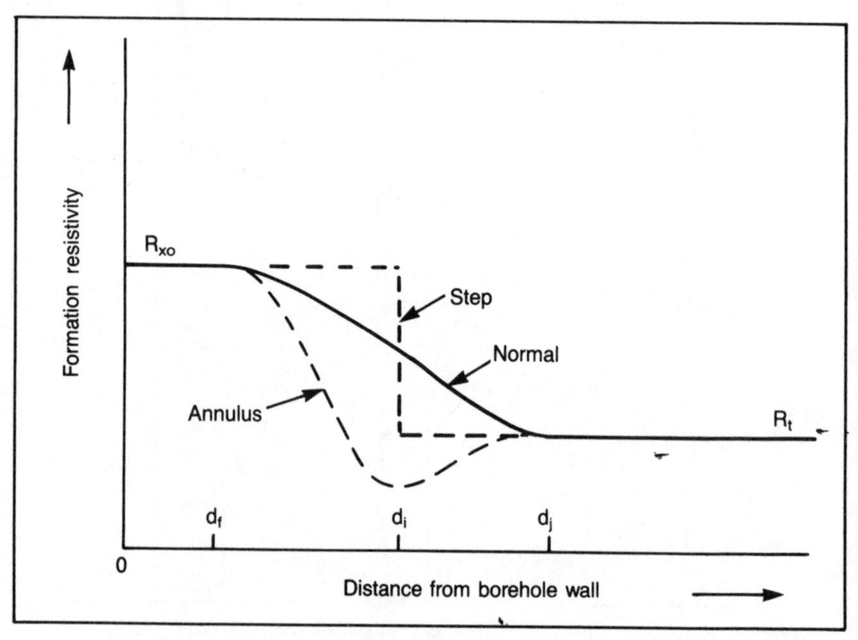

Fig. 2–5 Resistivity profiles—step, gradual, and annulus

This is the fraction of pore fluid that constitutes moved oil. It represents an upper limit of what might be recovered on production.

As an example, assume a limestone formation with $\phi = 0.18$, $R_w = 0.04$, $R_{mf} = 0.50$, $R_t = 10$, and $R_{xo} = 25$ ohm-m

$$S_{xo} - S_w = 1.0\,(\sqrt{0.5/25} - \sqrt{0.04/10})/0.18 = 0.43$$

That is, 43% of the reservoir pore space constitutes movable oil. Alternatively, the bulk volume fraction of movable oil is $\phi\,(S_{xo} - S_w)$ or 7.7%.

Eq. 2.13 works well in salt-mud situations but tends to overestimate the amount of movable oil in fresh-mud conditions. Recent investigation has led to the conclusion that all of the connate water in the hydrocarbon-bearing zone may not be replaced by mud filtrate; some may be shielded by the residual oil and left in place.[4] This incomplete replacement does not matter when $R_{mf} \approx R_w$, as in the case of salt mud. However for fresh mud where $R_{mf} \gg R_w$, R_{mf} is too high a value to use for flushed-zone water resistivity. The error can be large when the oil is heavy and residual oil saturation is high. The same overestimation occurs if the invasion is so shallow that the R_{xo} reading is influenced by the formation water.

Consequently, the preferred method of estimating movable oil in fresh mud is with the electromagnetic propagation technique described in chapter 5.

SUMMARY

ARCHIE RELATIONS FOR WATER SATURATION

$$\text{General: } S_w = c \cdot \sqrt{R_w/R_t}\,/\phi \qquad \text{(a)}$$

R_t = deep resistivity, ohm-m
R_w = interstitial water resistivity, ohm-m
ϕ = porosity, fraction
c = 1.0 for carbonates, 0.9 for sands

$$\text{Specific: } S_w = \sqrt{R_o/R_t} \qquad \text{(b)}$$

R_o = resistivity of water-bearing formation of same porosity and R_w as for R_t formation

RECOVERABLE OIL, STOCK-TANK BBL

$$N = 7758 \cdot r \cdot \phi \cdot (1 - S_w) \cdot h \cdot A/B \quad \text{(c)}$$

r = recovery factor, ≈ 0.2
h = thickness of producing formation, ft
A = well spacing, acres
B = volume factor of oil, ≈ 1.2

RECOVERABLE GAS, MMscf

$$G = 1.54 \cdot r \cdot \phi \cdot (1 - S_w) \cdot h \cdot A \cdot P_r/[Z \cdot (460 + T_r)] \quad \text{(d)}$$

r = recovery factor, ≈ 0.7
P_r = reservoir pressure, psi
T_r = reservoir temperature, °F
Z = gas deviation factor, 0.8–1.2

BASIC INTERPRETATION STEPS

1. Scan lithology logs in Track 1 (SP and/or GR); pick out permeable zones; discard shales.
2. Scan deep-reading resistivity curve in Track 2 (usually Induction) in permeable zones, looking for high values. These must be due either to hydrocarbons in pores or to low porosity.
3. Read porosities in zones of interest (Track 3 or separate log).
4. If there is a nearby water sand of the same porosity, apply Eq. *b* to obtain water saturation.
5. If step 4 does not apply, determine R_w by applying Eq. *a* to the nearby water sand with $S_w = 1$. Then apply Eq. *a* again to the zone of interest to obtain water saturation.
6. Calculate recoverable hydrocarbon by Eqs. *c* or *d*.

INVASION

- Can correct R_{deep} to R_t with three resistivity curves.
- Drilling mud may be fresh ($R_{mf} > R_w$) or saline ($R_{mf} < R_w$).
- Movable oil can be calculated

$$S_{xo} - S_w = c\left(\sqrt{R_{mf}/R_{xo}} - \sqrt{R_w/R_t}\right)/\phi \quad \text{(e)}$$

R_{mf} = mud filtrate resistivity at level of interest, ohm-m

R_{xo} = flushed zone resistivity, ohm-m.

The calculation is more reliable with salt mud than fresh mud.

REFERENCES

[1]W.O. Winsauer, H.M. Shearin Jr., P.H. Masson and M. Williams, "Resistivity of Brine-Saturated Sands in Relation to Pore Geometry," *AAPG Bull.*, Vol. 36 (February 1952), pp. 253–277.

[2]G.E. Archie, "The Electrical Resistivity Log as an Aid in Determining Some Reservoir Characteristics," *SPE-AIME Transactions*, Vol. 146 (1942), pp. 54–62.

[3]M. Gondouin and A. Heim, "Experimentally Determined Resistivity Profiles in Invaded Water and Oil Sands for Linear Flows," *SPE paper 712* presented October 6–9, 1963.

[4]C. Boyeldieu and A. Sibbit, "A More Accurate Water Saturation Evaluation in The Invaded Zone," *SPWLA Logging Symposium Transactions* (June 1981).

Chapter 3

PERMEABLE ZONE LOGS

The first step in analyzing a set of logs, as outlined previously, is to pick out the permeable zones, which may be sands or carbonates, and discard the impermeable shales. The logs used for this purpose are the Spontaneous Potential (SP) and the Gamma Ray (GR). They are always recorded in Track 1.

The two logs distinguish shales from nonshales by quite different mechanisms. The SP is an electrical measurement and the GR is a nuclear measurement. Sometimes the logs are virtually identical; sometimes they are vastly different. Fortunately, when one is poor, the other is usually good.

Fig. 3–1 compares SP and GR logs in a typical soft rock sand-and-shale sequence. Both curves are good in this case and clearly distinguish the shales on the right from the permeable sands on the left. In soft rock the SP generally gives a more black-and-white distinction between the shales and the sands than does the GR. The latter shows more variability in both shale and sand readings.

By contrast, in hard limestone formations the SP may be a lazy, poorly developed curve that hardly resolves permeable and impermeable zones. The GR is superior under these conditions, giving good shale-carbonate distinction and bed resolution.

Both curves are used to indicate the shale content of a permeable zone for shaly formation interpretation (chapter 7). The GR is more quantitative than the SP in this respect. On the other hand the SP may be used to give the formation water resistivity required for saturation calculations.

SPONTANEOUS POTENTIAL (SP) LOG

The SP log is a recording of the difference in electrical potential between a fixed electrode at the surface and a movable electrode in the borehole. The hole must be filled with conductive mud. No SP can be measured in oil-base mud, empty holes, or cased holes. The scale of the SP log is in millivolts. There is no absolute zero; only changes in potential are recorded.

Measurement of the SP is simple, but understanding the log is not so easy. The log is the one curve among modern logs whose boundary response and bed definition vary widely with formation and mud properties. The SP curve is used to

- select permeable zones
- provide R_w values
- estimate the degree of shaliness of reservoir rock

SOURCE OF SPONTANEOUS POTENTIAL

The potential sensed is a combination of four electrical potentials set in motion when the drillhole penetrates the formations. These are shown in Fig. 3–2 with polarities appropriate for the usual fresh-mud case where

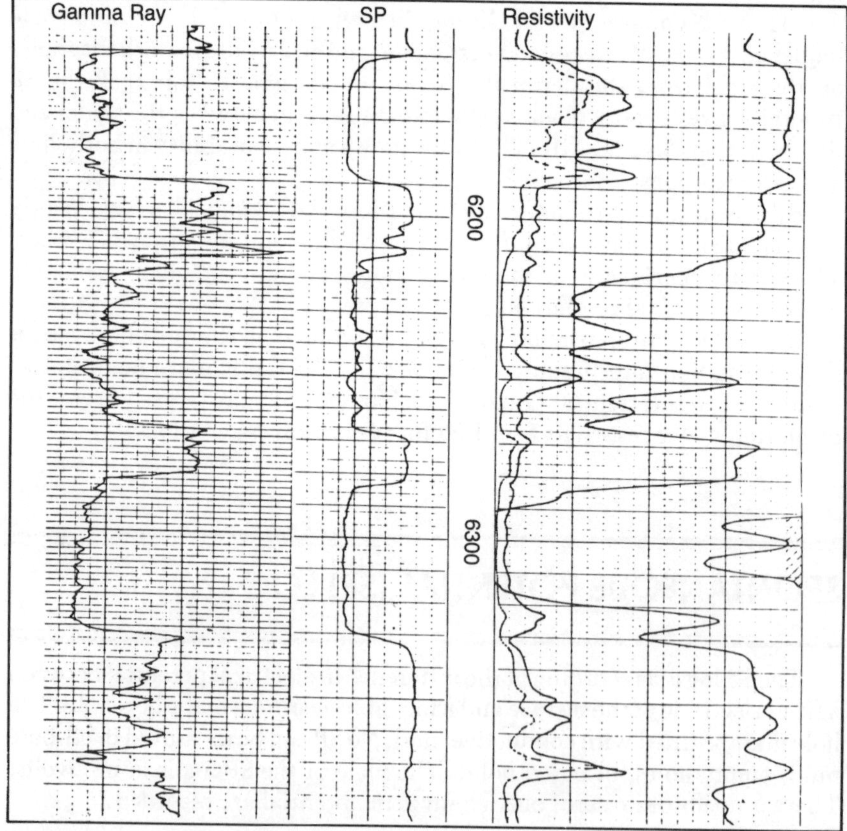

Fig. 3–1 Comparison of SP and GR logs, soft formations (from *Applied Openhole Log Interpretation*, courtesy D.W. Hilchie)

$R_{mf} > R_w$. They are, in order of importance

- an electrochemical potential, E_{sh}, existing across the impermeable shale between its horizontal interface with a permeable zone and its vertical interface with the borehole
- an electrochemical potential, E_d, existing across the transition between invaded and noninvaded zones in the permeable bed
- an electrokinetic potential, E_{mc}, existing across the mud cake
- an electrokinetic potential, E_{sb}, existing across a thin layer of shale next to the borehole

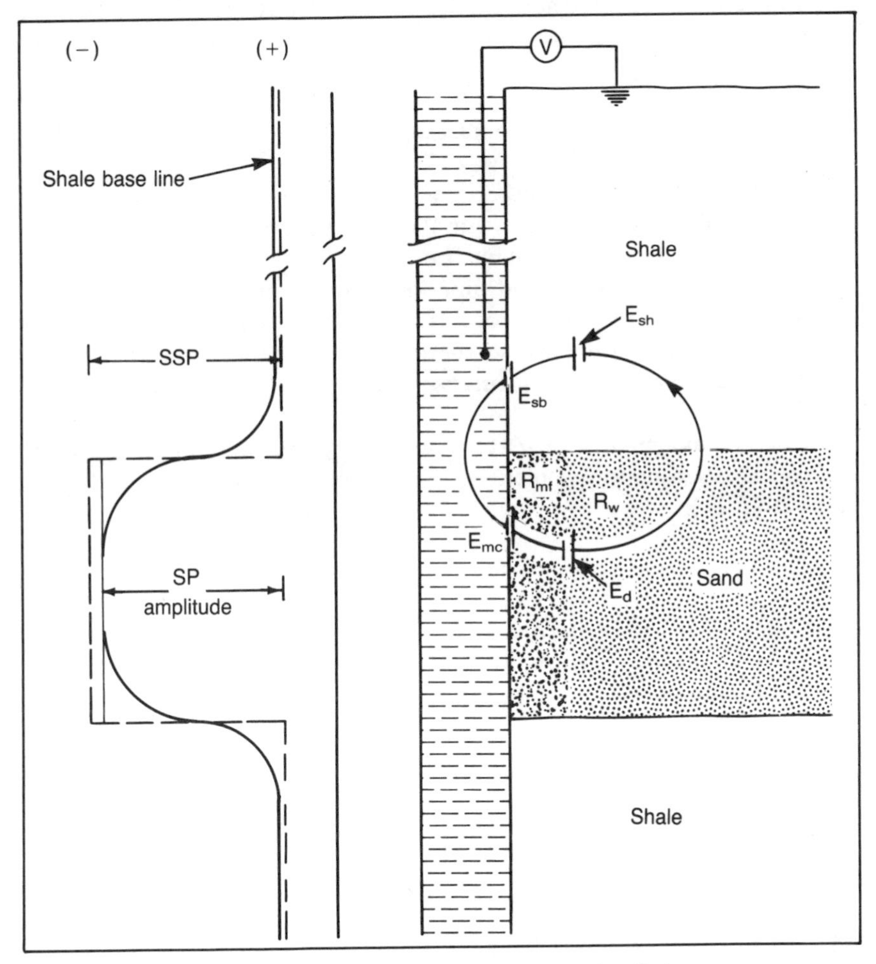

Fig. 3–2 Origin of spontaneous potential

Electrochemical Potentials[1,2,3]

The potential E_{sh} is a membrane potential associated with selective passage of ions in the shale. Because of layered clay structure and charges on the layers, shales pass Na+ ions but retain Cl− ions. When shale separates sodium chloride solution of different concentrations, the Na+ ions move through the shale from the more-concentrated solution (normally the formation water) to the less-concentrated solution (normally the mud). This constitutes a current flow. The magnitude of the potential causing this flow is a function of the ionic activities of the two solutions; these in turn are related to their resistivities. For sodium chloride solutions at 77°F, the potential in millivolts is

$$E_{sh} = -59.1 \log (R_{mfe}/R_{we}) \tag{3.1}$$

R_{mfe} and R_{we} are equivalent resistivities closely related to actual values of R_{mf} and R_w. The negative sign means the potential is negative relative to the value observed in a thick shale far from a boundary.

The potential E_d is a liquid junction or diffusion potential that exists across the interface of solutions of different salinity. Cl− ions have greater mobility than do Na+ ions, so there is a net flow of negative charges from the more-concentrated formation water to the less-concentrated mud filtrate.

This is equivalent to a current flow in the opposite direction. The magnitude of the corresponding potential, in millivolts, for sodium chloride solutions at 77°F is

$$E_d = -11.5 \log (R_{mfe}/R_{we}) \tag{3.2}$$

This potential has the same dependency as E_{sh} but is only one-fifth as large. It is of such polarity as to add to E_{sh}.

Combining the two, the total electrochemical potential is

$$E_{ec} = -K \log (R_{mfe}/R_{we}) \tag{3.3}$$

K, which is proportional to absolute temperature, is given by

$$K = 61 + 0.13T \tag{3.4}$$

where T is in degrees Fahrenheit.

Typically, E_{ec} is 70–100 mv for fresh mud and saline formation waters.

Electrofiltration Potentials [4-8]

The potential E_{mc} across the mud cake is of a different nature. It is an electrofiltration or streaming potential that is produced by a flow of electrolyte, the mud filtrate, through a porous medium, the mud cake. The magnitude cannot be predicted with any accuracy. However, an approximate relation for its value, in mv, has been given as

$$E_{mc} \approx 0.04 \cdot \Delta P \sqrt{R_{mc} \cdot t_{mc} \cdot f} \quad (3.5)$$

where

ΔP = pressure differential between borehole and formation, psi
R_{mc} = mud cake resistivity, ohm-m
t_{mc} = mud cake thickness, in
f = API water loss of the mud, cc/30 min

With $\Delta P = 200$, $R_{mc} = 1$, $t_{mc} = 0.25$, and $f = 4$, which are fairly normal conditions, Eq. 3.5 gives $E_{mc} = 8$ mv.

Typically E_{mc} is a few millivolts and adds to E_{ec} for the normal case of $R_{mf} > R_w$. However, it is partially counterbalanced by a similar potential, E_{sb}, of opposite polarity at the shale-borehole boundary where the shale itself acts as a thick mud cake with very low water loss. Consequently, the net electrofiltration potential is generally small. The principal exceptions are where permeable formations have been pressure-depleted by previous production or where very heavy muds are opposite normally pressured formations. In such cases the net electrokinetic potential can be quite large — many tens of millivolts — if mud resistivity and filtration rates are high.

The Total Potential

Normally the net electrokinetic potential is negligible so that the total SP, denoted static SP (SSP), is the electrochemical contribution

$$SSP = -(61 + 0.13T) \log (R_{mfe}/R_{we}) \quad (3.6)$$

where SSP is in millivolts and T is in °F.

This is the fundamental SP equation. It is an important relation as it is used to derive R_w from measured values of SSP and R_{mf}.

SP BEHAVIOR OVER A LONG LOG

As indicated by Eq. 3–6, the SP becomes zero when $R_{mfe} = R_{we}$ and reverses when $R_{mfe} < R_{we}$. In a typical fresh-mud well, the following behavior is observed. At very shallow depths where formation water is fresh, the SP in sands is reversed, i.e., has positive polarity relative to the shale value. Slightly deeper, perhaps at about 1,000 ft, the SP goes to zero. As depth progresses the formation water gradually becomes more saline and the SP increases in magnitude (negatively). Values of 70–100 mv at depths of 8,000 ft are typical. At greater depths the connate water sometimes decreases again in salinity, especially when overpressured formations are encountered. In such cases the SP correspondingly reduces in magnitude. In unusual cases it may even reverse again, as Fig. 3–3 illustrates. The SP in the permeable bed at 6,300 ft is normal; the SP at 9,100 ft is reversed.

In typical salt muds the SP is often useless because the SP magnitudes at depths of interest are small ($R_{mf} \approx R_w$) and because boundary definition with low-resistivity mud and high-resistivity formations is extremely poor. This is explained in the following.

SHAPE OF THE SP CURVE

The measured SP is really the potential change that occurs in the borehole as a result of SP-generated currents flowing through the resistive borehole fluid. This is illustrated in Fig. 3–4. Opposite a shale far from a permeable zone boundary, no current flows, so the potential is constant. As the boundary of a permeable zone is approached, current is encountered, which causes the potential to go negative with respect to the shale.

Directly opposite the boundary, current flow is maximum so the potential change per unit length of hole, which is the slope of the SP curve, is greatest. Beyond the boundary the current density decreases and gradually goes to zero. If the bed is thick enough the potential becomes constant very close to the SSP value. This value is always measured relative to the shale base line. As the other boundary with the shale is approached, the reverse situation is encountered and the potential returns gradually to the shale value.

The sharpness of the SP curve at a boundary, and therefore the vertical bed resolution, depends on the pattern of current flow at the boundary. This is a complicated function of the relative resistivities of the mud, permeable zone (invaded and noninvaded portions), and shale and to a lesser extent is a function of hole diameter and depth of invasion. The overriding principle is that currents seek the lowest resistance path in a loop. In a given formation they will spread out to the point that the formation resistance, which decreases as 1/area of flow, is negligible with respect to the borehole resistance, which increases linearly with length of current path in the mud.

Consequently, when the ratio of formation to mud resistivity is high, the currents will spread widely. There will be long flow paths in the borehole, and bed boundaries will be poorly defined. Conversely, when that ratio is

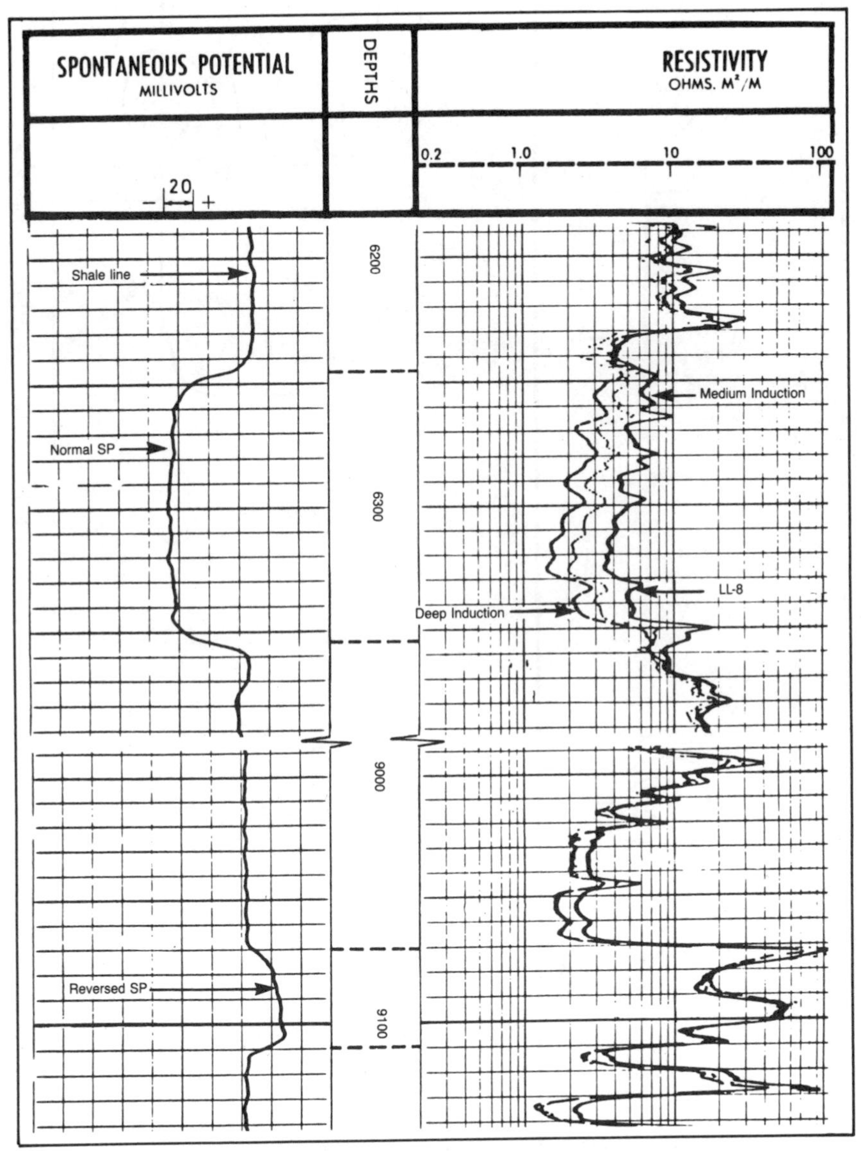

Fig. 3–3 Example of normal and reversed SP deflections

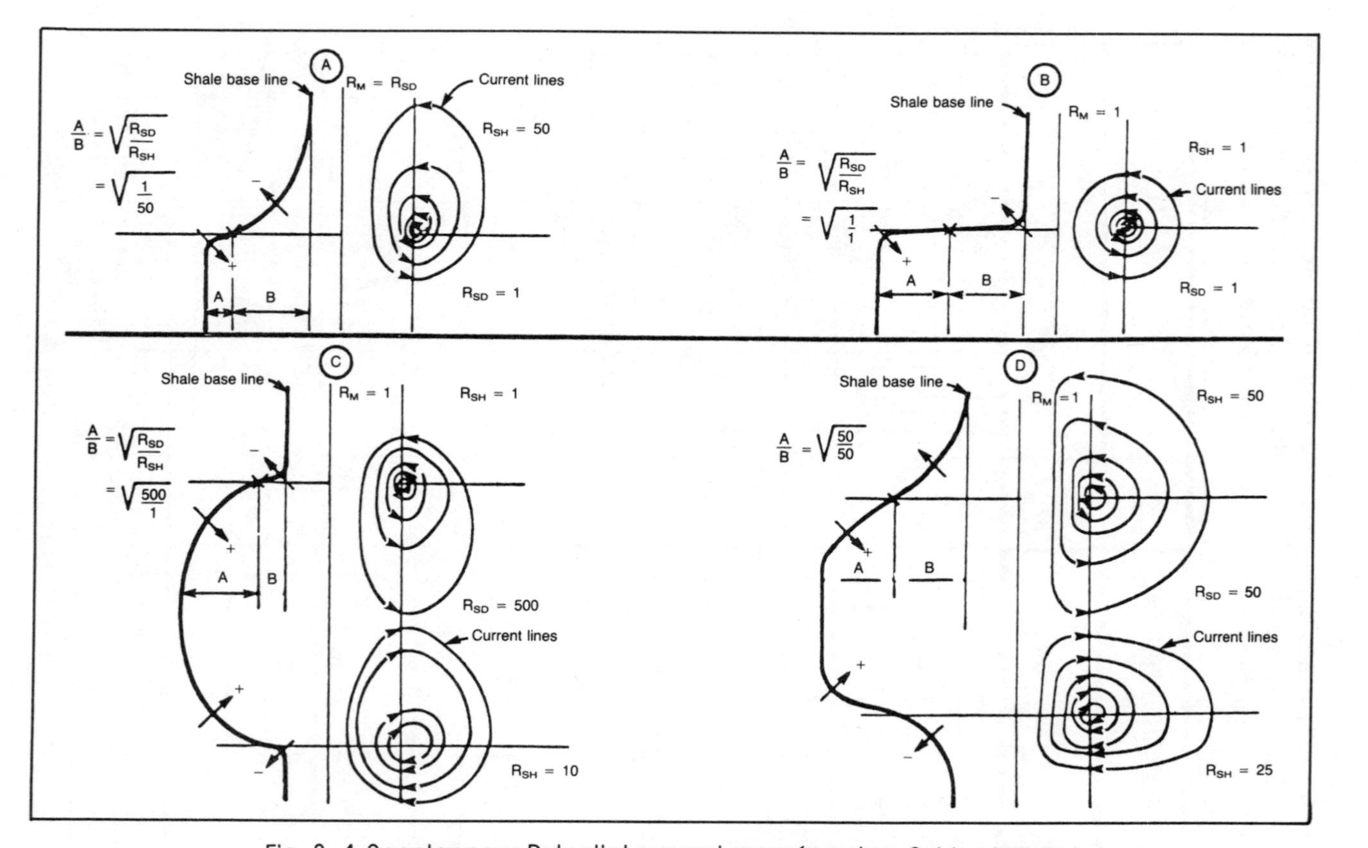

Fig. 3–4 Spontaneous Potential curve shapes (courtesy Schlumberger)

low, boundaries will be sharply defined. Note that the boundaries are opposite the points of inflection (maximum slope) of the SP curve, not necessarily opposite the halfway point between shale and permeable-zone SP levels.

The more gradual the boundary transition on the SP, the poorer the bed resolution will be. A rule of thumb is that the bed resolution h (ft), defined as that thickness required for the SP to reach 80% of the thick-bed reading (the SSP), is given by

$$h = \sqrt{R_i/R_m} \tag{3.7}$$

where R_i is the resistivity of the invaded zone (the shallow log reading) and R_m is the mud resistivity. An even coarser approximation is

$$h = 1/\phi \tag{3.8}$$

where ϕ is the porosity (fractional). This indicates that permeable bed resolution varies from about 3 ft at 30% porosity to 30 ft at 3% porosity.

SP logs therefore delineate permeable beds quite well in porous sand-and-shale sequences but resolve beds very poorly in tight formations. In hard rock areas there may be massive low-porosity limestone or anhydrites of very high resistivity. These show up as long, straight lines (no current at all entering or leaving the hole opposite these formations) on the SP and make interbedded permeable zones recognizable only by changes in the slope of the lazy SP curve. Fig. 3–5 is an illustration.

Service company charts are available to correct the SP response in a thin bed,[9,10] such as zone B of Fig. 3–6, to what it would be in a thick bed if there are no thick beds to indicate the SSP directly. However, these charts are not too precise. A correction factor greater than 1.3 has dubious accuracy.

COMPUTATION OF R_W FROM THE SP

Eq. 3.6 is used extensively to determine the formation water resistivity that is required for water saturation calculations. First, the SSP is read from the log as the difference in millivolts between the shale level and the thick-clean-water-sand level near the zone of interest. This is illustrated in Fig. 3–6. The shale line is taken as the maximum SP excursion to the right. The sand line is taken as the maximum deflection to the left in zone A, which is a water-bearing sand, as shown by the low reading (1 ohm-m) of the deep resistivity (dashed) curve. The SSP is read as −68 mv, the scale being 10 mv per chart division as indicated in the log heading.

Next, the R_{mf} value is read from the log heading and converted to the temperature at the zone of interest. The temperature is obtained from linear interpolation between surface temperature and bottomhole temperature. That is

$$T_d = ST + (BHT - ST) \cdot d/D \tag{3.9}$$

where

T_d = temperature, °F, at depth of interest d, ft
ST = surface temperature, assumed in the 60–80 ° F range
BHT = bottom-hole temperature, °F, given on the log heading
D = bottom-hole depth from the log heading, ft
d = depth at zone of interest, ft

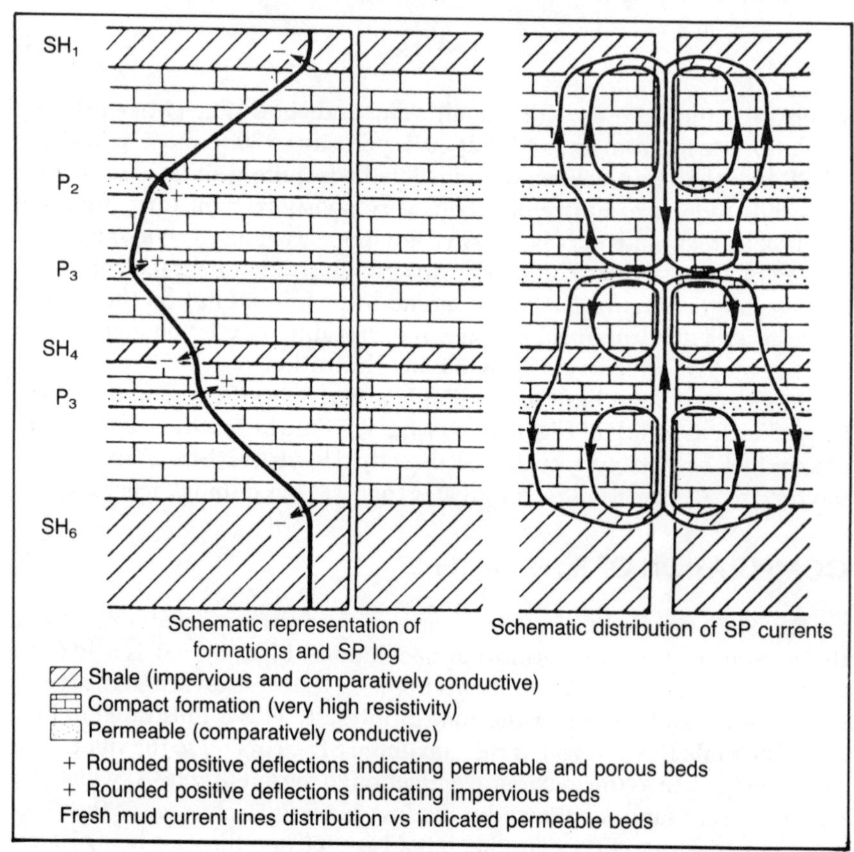

Fig. 3–5 SP in hard rock (courtesy Schlumberger)

R_{mf} at depth d is then given by

$$(R_{mf})_d = (R_{mf})_m \cdot (T_m + 7)/(T_d + 7) \qquad (3.10)$$

where

$(R_{mf})_m$ = measured R_{mf} value from the log heading, ohm-m
T_m = temperature of R_{mf} measurement from the log heading, °F

For the example of Fig. 3–6, assume the log heading gives

BHT = 196°F at total depth D = 9,400 ft
$(R_{mf})_m$ = 0.71 ohm-m at T_m = 68°F

Assuming a surface temperature of 75°F, the temperature at 4,170 ft (middle of zone A) is

$$T_d = 75 + (196 - 75) \cdot 4{,}170/9{,}400 = 129°F$$

The R_{mf} value at 4,170 ft is then

$$R_{mf} = 0.71 \cdot (68 + 7)/(129 + 7) = 0.39 \text{ ohm-m}$$

Charts are available for determining T_d and $(R_{mf})_d$, but it is simpler to use Eqs. 3.9 and 3.10 directly.

From this point Figs. 3–7 and 3–8 are used to obtain R_w. The procedure is as follows:

1. Multiply R_{mf} by 0.85 to obtain R_{mfe}. Mark the value on stem 3 of Fig. 3–7.
2. Mark the SSP value on stem 1 of Fig. 3–7 and project a line from that point through the formation temperature to read R_{mfe}/R_{we} on stem 2. From this point, project a line through the value of R_{mfe} on stem 3 to read R_{we} on stem 4.
3. Enter the value of R_{we} on the vertical axis of Fig. 3–8, project horizontally to the formation temperature, and then project vertically to the horizontal scale to read R_w.

Following this procedure with the example of Fig. 3–6, R_{mfe} = 0.85 × 0.39 = 0.33, R_{we} = 0.048 from Fig. 3–7, and R_w = 0.063 from Fig. 3–8.

Considering all of the uncertainties involved, the probable error in R_w values obtained from the SP in normal formations is at least ±10% and perhaps as much as ±20%. Errors can be much larger, however, where the formation water contains an appreciable fraction of salts other than sodium chloride. In particular calcium and magnesium ions, which are divalent, create excessively large SP deflections, which lead to unduly low values of R_w if not taken into account.[11] The situation is most likely to occur in shallow formations with fairly fresh water.

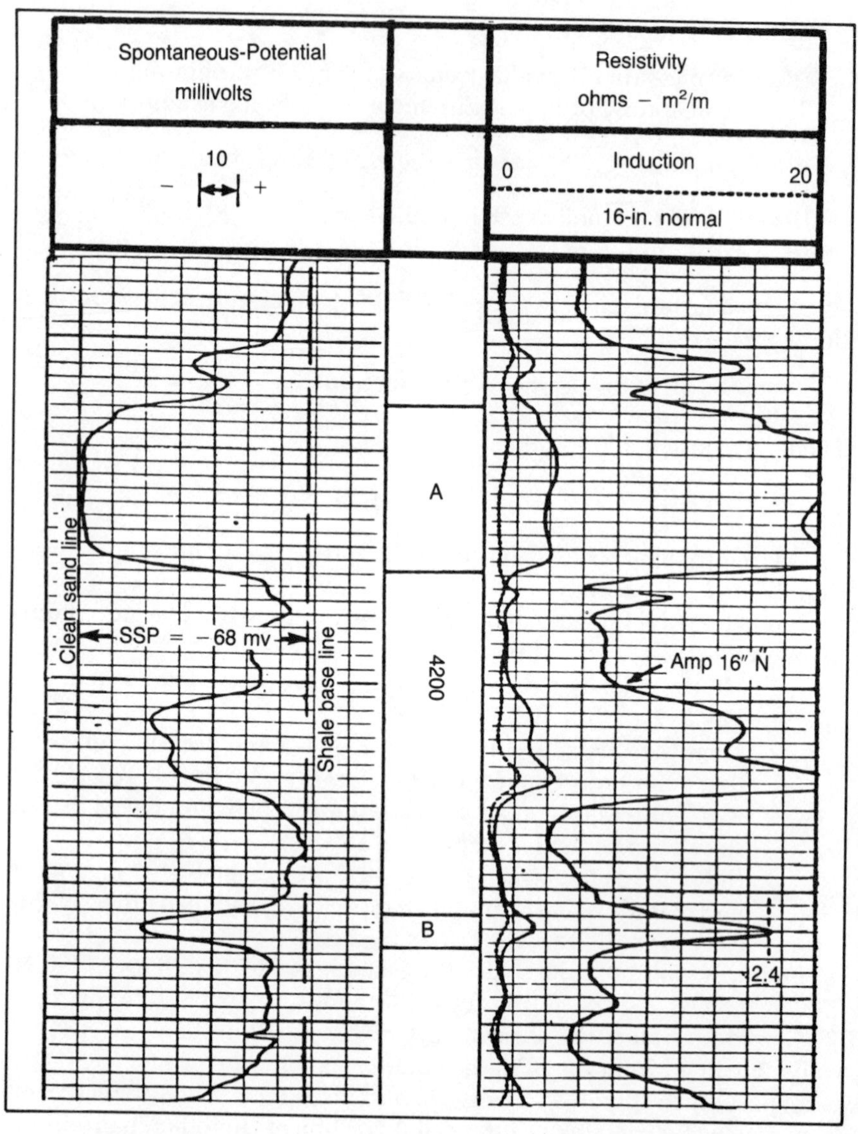

Fig. 3–6 Location of shale and clean sand lines on the SP

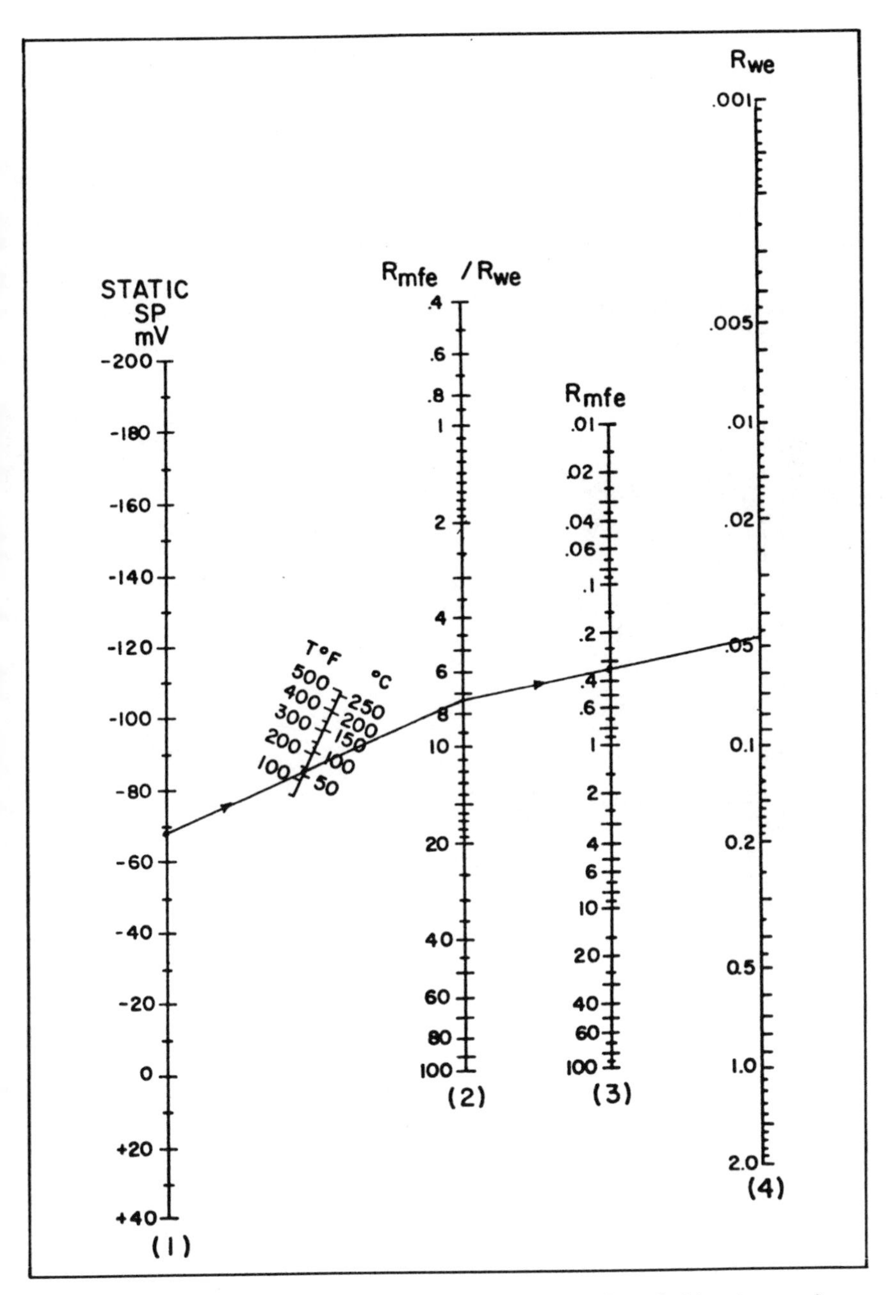

Fig. 3–7 Determination of R_{we} from the SP (courtesy Schlumberger)

In such cases the dashed lines in the lower half of Fig. 3–8 should be used to transform R_{we} to R_w. These lines are for average fresh formation waters; corrections can be large. For example, $R_{weq} = 0.7$ at 100°F leads to $R_w = 1.4$, a factor of two increase.

Whenever possible the R_w value obtained from the SP should be checked against that obtained by applying Eq. 2.8 with $S_w = 1$ in the same or nearby water sands. If there is good agreement, water saturation in hydrocarbon zones can be calculated with confidence. If there is substantial disagreement, the value derived from Eq. 2.8 is preferred. However, the reason for

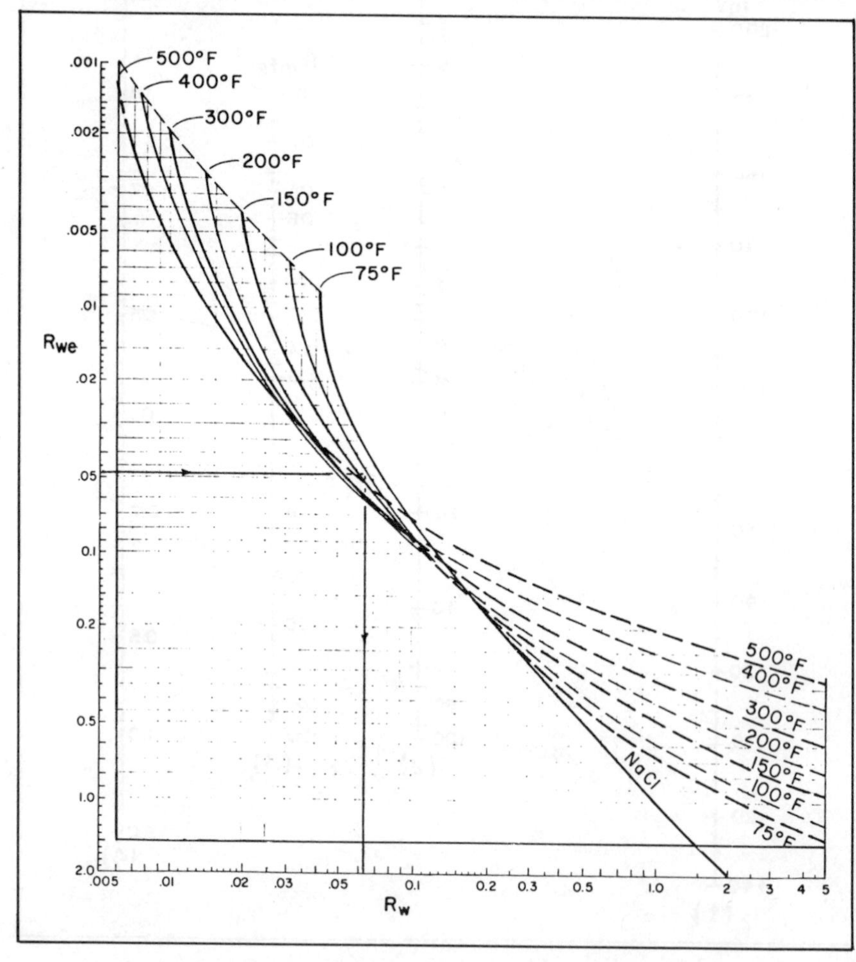

Fig. 3–8 Determination of R_w from R_{we} (courtesy Schlumberger)

the discrepancy should be sought, particularly by checking the R_{mf} measurement and if possible obtaining a sample of formation water for analysis.

SP LOG IN SHALY SANDS[12, 13]

In shaly sands whether finely laminated or containing dispersed clay, the shale layers or clay particles create internal membrane potentials that, when added together, constitute a potential opposing the normal electrochemical potential in the adjacent shale. This reduces the SSP to a pseudostatic value called PSP. Under ideal conditions where the shale laminations have the same resistivity as the sand laminations (both invaded and univaded portions), the percent reduction in SSP equals the percentage of shale by volume.

In the case where the sand is substantially more resistive than the shale, the percent reduction in SSP is much greater than the percentage of shale. Consequently, the SP deflection will be less opposite an oil-bearing portion of a shaly sand than opposite the water-bearing portion. This is frequently observed even when the sand does not contain much shale. If the sand were clean, the only effect would be a more rounded transition of the SP into the oil-bearing zone by virtue of its higher resistivity.

The SP log is used as one of the prime indicators of shale fraction, V_{sh}, in a shaly sand. Computation of V_{sh} is described in chapter 7. The value obtained from the SP tends to be an upper limit for the reasons just indicated.

SP ANOMALIES DUE TO VERTICAL MIGRATION OF FILTRATE[14]

When a very permeable sand containing salt water is invaded by fresh-mud filtrate, the lighter mud filtrate will float up toward the upper boundary of the sand. After a few days, invasion will be quite deep just below the upper boundary and very shallow near the lower. The SSP will be significantly reduced at the bottom because the diffusion potential, E_d, disappears (with no invasion). Further, it is replaced by an electrochemical potential across the mud cake that now directly separates the borehole mud and the formation water that has reclaimed the originally invaded pores. This potential opposes the normal shale potential E_{sh}. Both effects reduce the normal SP deflection at the bottom of the sand. At the top of the sand, the SP deflection will be rounded due to the higher resistivity of the deeply invaded fresh filtrate.

Further, if such a sand bed contains thin, interbedded shale streaks, the SP will show negative and positive undulations from the normal level above and below each streak.

SP ANOMALIES DUE TO NOISE

Spurious DC currents in the ground and in the borehole can be caused by telluric currents, bimetallic potentials, cathodic protection devices, leaky rig power sources, welding machines, etc. Other sources of noise are magnetized cable drums and intermittent contacts between casing and logging cable. Occasionally these may disturb the normal SP, but usually they can be eliminated with appropriate measures.

THE GAMMA RAY (GR) LOG

The basic GR log is simple and easy to record and to understand. It was routinely recorded for many years with only minor improvements in instrumentation and with limited quantitative use. In the past few years, however, the introduction of the spectral GR has given the recording of earth radioactivity a new lease on life. It is exciting to see new uses for the technique unfold.

BASIC GR LOGS

The basic GR log is a recording of the natural radioactivity of formations. The radioactivity arises from uranium (U), thorium (Th), and potassium (K) present in the rock. These three elements continuously emit gamma rays, which are short bursts of high-energy radiation similar to X-rays. The gamma rays are capable of penetrating a few inches of rock. A fraction of those that originate close to the borehole traverse the hole and can be detected by a suitable gamma-ray sensor. Typically, this is a scintillation detector, 8–12 in. in active length. The detector gives a discrete electrical pulse for each gamma ray detected. The parameter logged is the number of pulses recorded per unit of time by the detector.

GR logs are scaled in API units (APIU). An APIU is 1/200 of the response generated by a calibration standard, which is an artificial formation containing precisely known quantities of uranium, thorium, and potassium maintained by the American Petroleum Institute (API) in Houston.[15] The response generated by this formation is defined as 200 APIU. By design the calibration standard has twice the activity of an average shale, considered to contain 6 ppm (parts per million) uranium, 12 ppm thorium, and 2% potassium. Consequently, shales read in the vicinity of 100 APIU on GR logs.

Response to Different Formations

Gamma Ray logs are effective in distinguishing permeable zones by virtue of the fact that the radioactive elements tend to be concentrated in the shales, which are impermeable, and are much less concentrated in carbonates and sands, which are generally permeable. Fig. 3–9 shows typical responses. Limestones and anhydrites have the lowest reading, 15–20 APIU; dolomites and clean sands have slightly higher values, about 20–30 APIU.

Shales average about 100 APIU but can vary from 75 to 150. A few very radioactive shales—the Woodford, for example—may read 200–300 APIU. Normally, therefore, the GR log separates clean sands and carbonates from shale quite nicely. However, there are localized areas where sands and dolomites, even though fairly free of clay, are radioactive enough that distinguishing them from shales is difficult. Among the less commonly encountered formations, coal, salt, and gypsum give quite low readings; volcanic ash and potash beds give high readings.

Depth of Penetration and Vertical Resolution

The depth of penetration of the GR log is 6–12 in., being somewhat higher at low formation density (high porosity) than at high density. Vertical bed resolution is about 3 ft. It is dependent on logging speed, as explained later.

Borehole Effects

The GR log is calibrated under conditions of 8-in. hole, 10-lb mud, with the logging tool (3⅝-in. diameter) excentered in the hole. No corrections are required for these conditions. With larger hole sizes and heavier muds or with the logging tool centralized, there is more gamma-ray-absorbing material between the formation and tool, and the response drops. Conversely, the response is increased in smaller or empty holes. Correction curves may be found in service company chart books.

Correction factors are normally modest, in the range of 1.0–1.3. They can be ignored in hand interpretation except in the combination of circumstances where the GR is being used for determination of shale content, the shales are badly caved relative to the sands, and the mud weight is very high. The corrections are also important in the infrequent cases where the GR log is used to assay potash or uranium deposits.

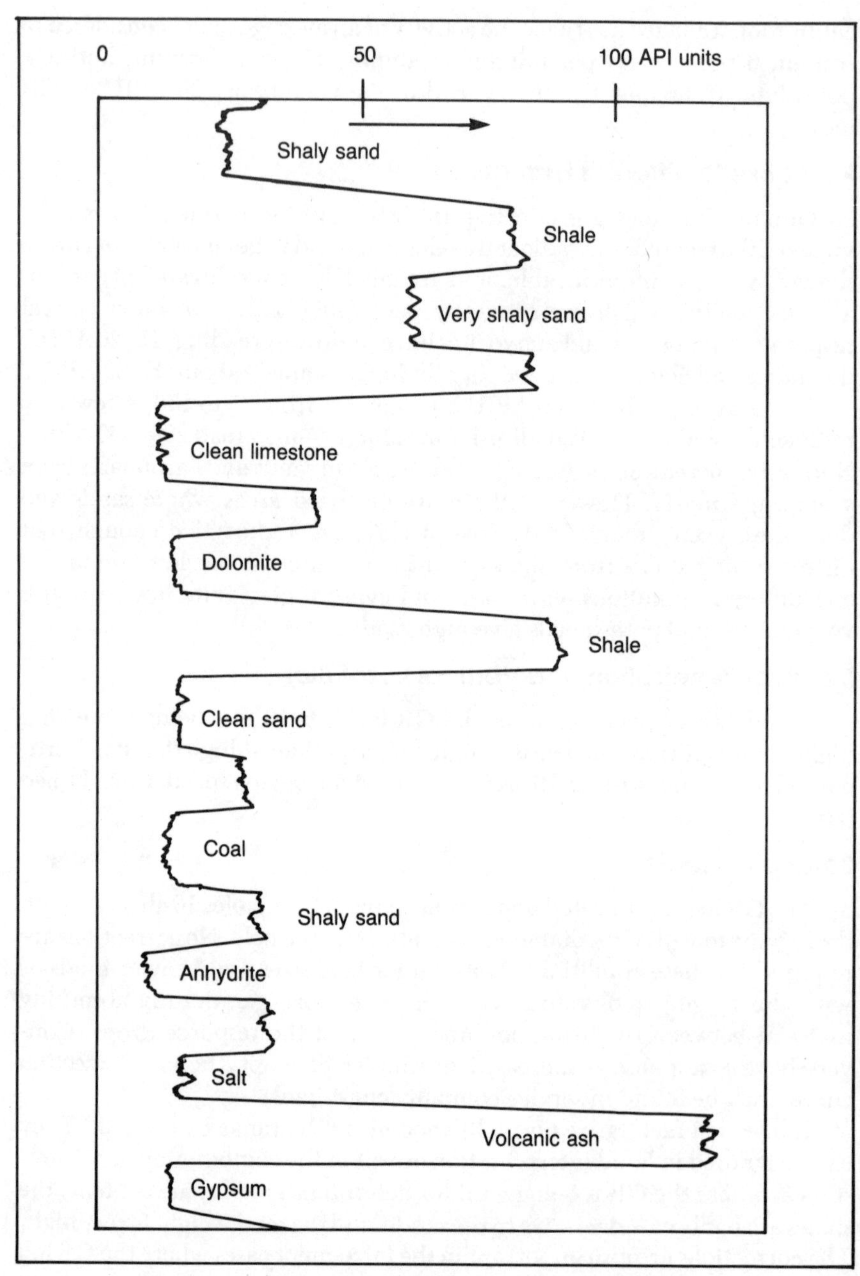

Fig. 3–9 Gamma Ray response in typical formations

Occasionally the mud in a well contains excessive amounts of potassium or uranium, either because potassium chloride has been added (to prevent shale swelling) or because potash or uranium beds have been penetrated. The potassium or uranium in the mud will contribute to the GR log, giving an abnormally high background level (that will vary with hole diameter) on which the normal formation response is superimposed. An empirical method of correction for this effect has been published.[16]

Shale Determination

Because uranium, thorium, and potassium are largely concentrated in clay minerals, the GR log is used extensively in shaly sand interpretation to estimate the fraction of shale by volume, V_{sh}, in the sand.[17] This procedure is described in chapter 7. Basically, it is a matter of estimating the clean sand and 100% shale levels on the log and interpolating between them to determine V_{sh} in a partially shaly interval. It is not a very precise technique, so other shale indicators are used as well.

SPECTRAL GR LOGS[18,19]

The radioactive elements uranium, thorium, and potassium emit gamma rays of different energies, as shown in Fig. 3–10. Potassium has a single energy at 1.46 mev (million electron volts). Thorium and uranium emit gamma rays of various energies, the major distinction being a prominent thorium energy at 2.62 mev and a predominant uranium energy at about 0.6 mev. In principle it is therefore possible to distinguish the three emitters by analyzing the energies of detected gamma rays.

As gamma rays progress from the point of origin in the formation to the detector in the borehole, their energies suffer severe degradation. Further smearing takes place in the detector itself. Nevertheless, it is possible with appropriate instrumentation and careful analysis of the spectrum of pulse amplitudes from the detector to break down the total GR log into its uranium, thorium, and potassium components and to generate spectral GR logs showing directly the concentrations of each of these elements in the logged formations.

Fig. 3–11 is an example. Uranium and thorium are scaled in parts per million (ppm) and potassium is scaled in percent by weight (wt%). The three elements contribute roughly equally to the total counting rate even though potassium is present in much greater concentration. The spectral log shows large differences in the shales above and below the clean zone from

4,120–4,220 ft. The upper shale is low in thorium and potassium and high in uranium. It is really a shaly carbonate that produces in some areas. The lower bed is a true shale, high in thorium and potassium.

Application of Spectral GR Logs

Uses of spectral GR logs are evolving at present.[20,21,22] In some areas it has been established that the thorium-potassium components with uranium

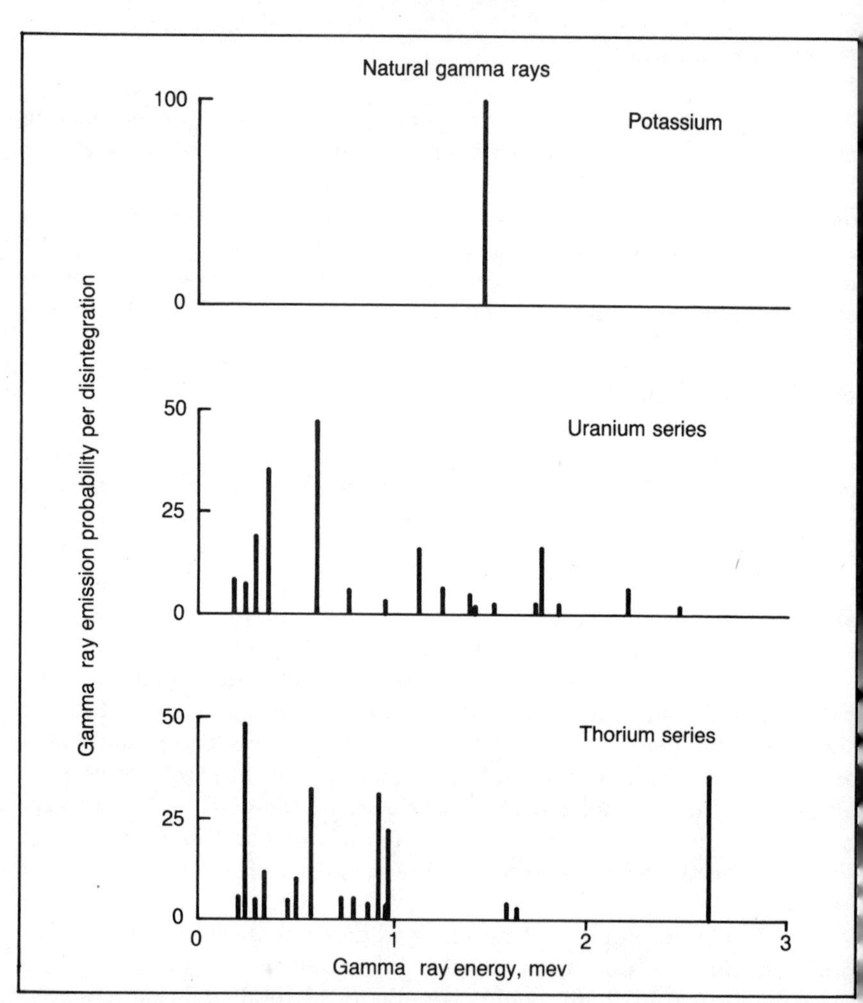

Fig. 3–10 Gamma Ray spectra of potassium, uranium, and thorium (courtesy Schlumberger)

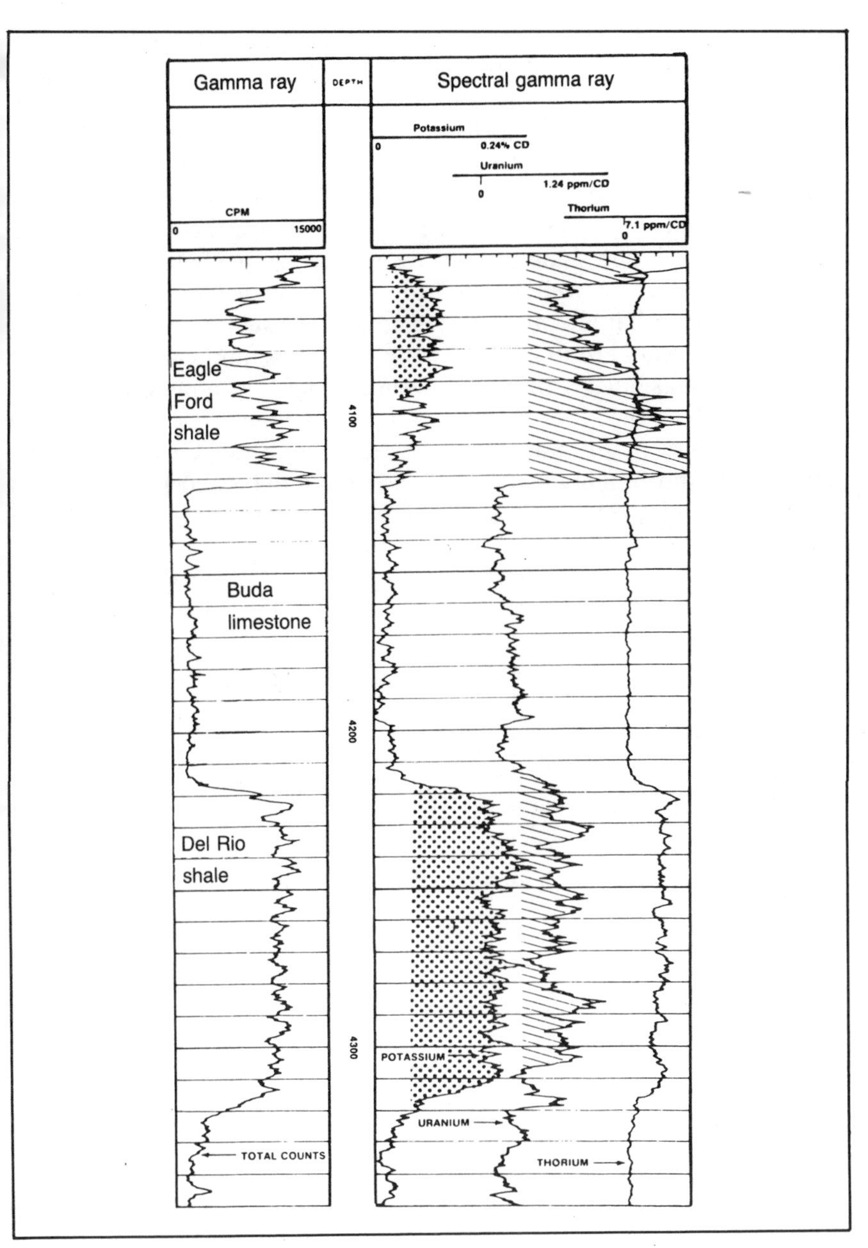

Fig. 3–11 Spectral GR log in Austin chalk sequence (Texas) (courtesy Dresser)

eliminated are better indicators of shale content than the total GR activity. This is because uranium salts are soluble and can be transported by liquid movement after primary deposition. They may be picked up in one location and precipitated elsewhere, particularly where a pressure drop occurs. Fig. 3–12 shows a case where the standard GR log indicates interval A to be three separate intervals separated by shale streaks at 1,600 and 1,638 ft. However, the thorium and potassium spectral curves clearly show the reservoir to be a continuous unit. The shale streaks are simply permeable zones high in uranium content. New wells in old fields often show such response opposite permeable zones that are producing in adjacent wells.

In other areas where abnormal amounts of potassium-bearing micas and feldspars are associated with the sands (the North Sea, granite wash areas of the southwestern U.S.), the thorium component alone is the best indicator of shale content. In any case, whichever shale indicator is used, thorium-potassium or thorium alone, the shale fraction is derived from the desired spectral curve in exactly the same fashion as described in chapter 7 for the total GR curve.

Efforts are underway to determine from spectral GR logs the type as well as the amount of clay encountered in formations. Fig. 3–13 is a proposed

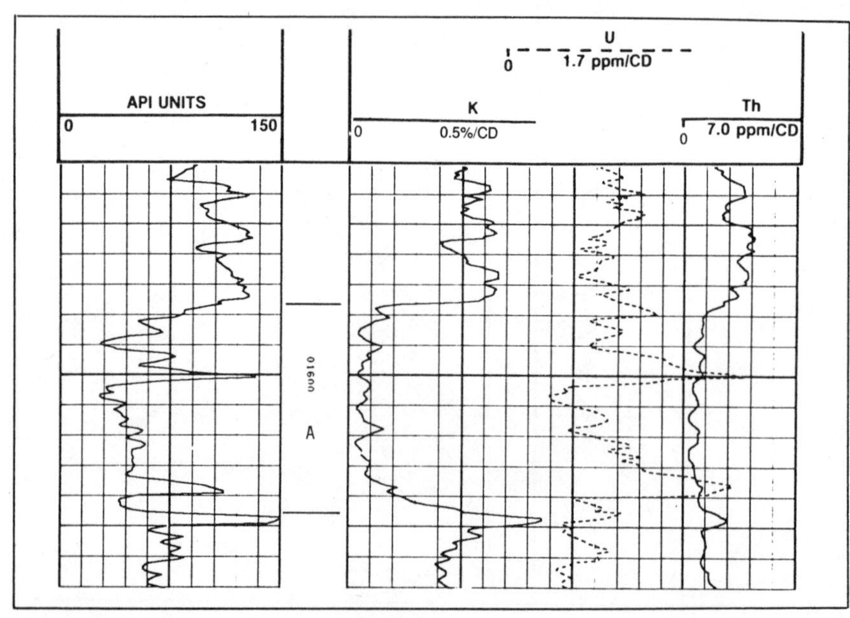

Fig. 3–12 Spectral GR log showing single reservoir instead of three in Interval A (Alberta) (courtesy Dresser)

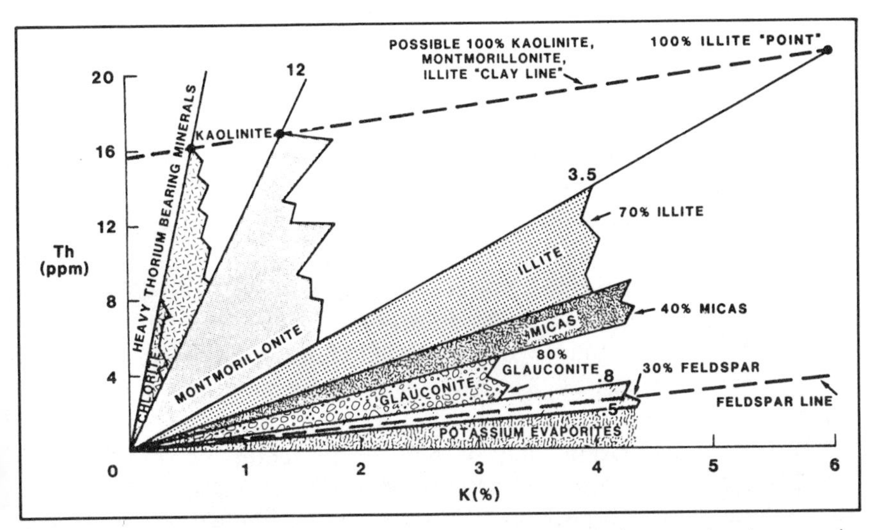

Fig. 3–13 Identification of clays and other minerals from potassium and thorium responses (courtesy Schlumberger, © SPE)

chart for identification of clay type from thorium and potassium log responses.[23] Approaches such as this will assist in determing formation productivity, since different types of clay have quite different surface-to-volume ratios and therefore water-retention and permeability-reduction capabilities.

In another application the uranium component of the spectral curves has been used to indicate fractured zones in tight carbonates.[24] High uranium content is taken as an indication that fissures in the zones were permeable enough to flow fluids in the past and therefore may still be able to conduct oil and gas from remote porosity.

STATISTICAL FLUCTUATIONS

Gamma Ray logs never repeat exactly. The same is true for all types of nuclear logs. Some of the small wiggles on the logs are statistical fluctuations that do not reproduce and do not represent formation variations. In reading the logs, averages over about 3–4 ft should be used. The exception is a bed that is less than 3 ft thick, in which case the peak reading should be taken.

The source of statistical fluctuation is the random nature of nuclear processes. The emission of any one gamma ray in a formation is in no way time related to any other. Consequently, the pulses from the gamma ray detector appear as a random sequence. The number occurring in any given time interval will differ from that in a successive but identical time interval, even though the detector is stationary. Percentage-wise this difference will be small if the time interval is long enough that a large number of pulses is counted. Conversely, it will be large if the averaging time is small. A measure of the percentage fluctuation is $100/\sqrt{N}$, where N is the average number of pulses in the measuring interval.

Typical pulse rates for the basic GR log are 50–300 pulses per second. The usual averaging time is 2 sec. This means the log will show fluctuations of about 4% above and below the mean reading in shales ($N \approx 600$) and 10% either side of the mean in clean sands or carbonates ($N \approx 100$). Absolute magnitudes will be in the range of ±5–10 API units in shales and ±2–4 API units in clean formations. The best way to appreciate this is to overlay repeat logs.

Statistical fluctuations may be reduced by increasing the averaging time, but the value employed sets a limit on the logging speed. Fig. 3–14 shows the approximate response (without statistical fluctuation) to a 4 ft bed with a detector of 8–12 in. active length for a 2-sec averaging time and for logging speeds of 600, 1,800, and 5,400 ft/hr. At 600 ft/hr the boundary response is primarily determined by the length of the counter; resolution will not improve at slower speeds. At 1,800 ft/hr the bed is adequately resolved, although it appears to be shifted about 6 in. in the direction of logging; compensation is made in the recording process for this shift. At 5,400 ft/hr the bed is quite distorted and its amplitude is reduced as well as being shifted about 18 in.

If the averaging time is doubled, for example, logging speed must be halved to maintain the same bed resolution. A compromise must always be made between log accuracy and logging speed. A reasonable logging speed is that which moves the detector its effective length (1 ft) in the averaging interval. For 2-sec averaging this is 1,800 ft/hr; most nuclear logs are run at this speed. If the speed varies, bed resolution will change but the magnitudes of statistical fluctuations will remain constant.

With logging-truck computers, averaging over a fixed depth interval rather than time interval may be employed. Typically this is set at 1 ft for the GR log. With logging speed of 1,800 ft/hr, effective averaging time is again 2 sec. However, if logging speed is doubled, averaging time is reduced by a factor of 2. Bed resolution remains fixed but statistical variations increase by a factor of $\sqrt{2}$.

The problem of statistical uncertainty is aggravated with spectral GR logs because the counting rates in the uranium, thorium, and potassium channels are 3–10 times lower than that of the total GR, depending on the

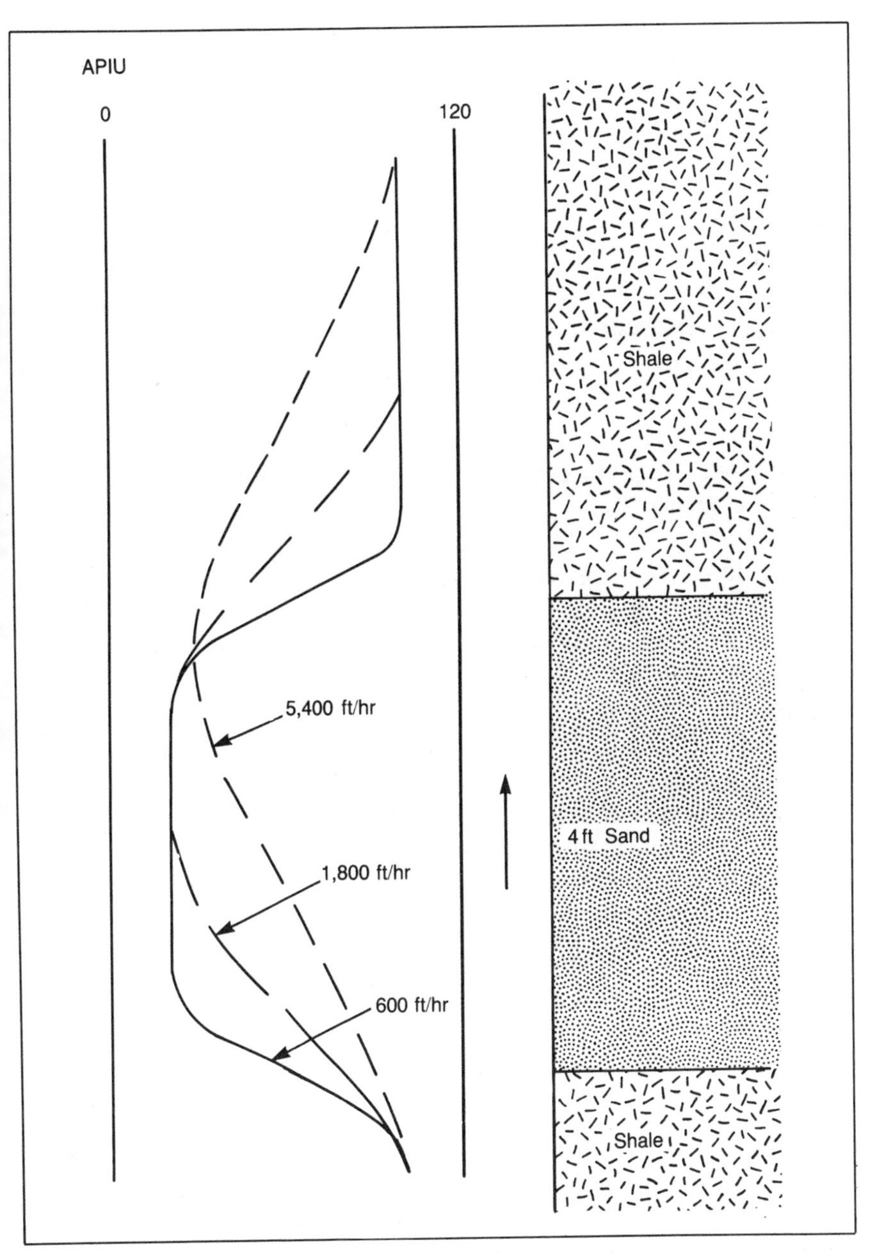

Fig. 3–14 Effects of GR averaging time and logging speed on bed resolution

particular instrumentation. This means averaging times must be increased and logging speed must be slowed. Typical values are 4–6 sec and 900–600 ft/hr. These are painfully slow logging speeds, so the best technique is to first log the well at normal speed while recording a total GR curve and then rerun the short sections of vital interest recording the spectral curves.

SUMMARY

SPONTANEOUS POTENTIAL LOG

- A recording of voltage generated by electrochemical and electrokinetic action at the junction of a permeable zone and shale.
- Distinguishes impermeable shales from permeable sands or carbonates.
- SP can develop with only a fraction of a millidarcy, but there is no correlation between magnitude and absolute permeability or porosity.
- Shales appear as excursions to the right, permeable formations to the left.
- Bed boundaries are at points of inflection, not halfway points.
- Bed resolution is good at high porosity, poor at low porosity.
- Magnitude depends on contrast between R_{mf} and R_w. Thus, SP delineates permeable zones well in fresh mud where $R_{mf} \gg R_w$ but poorly in salt mud where $R_{mf} \approx R_w$.
- R_w obtainable from SP, but values must be chosen carefully, especially at depths at less than 3,000 ft.
- Shaly sands reduce SP especially when hydrocarbon bearing. Maximum value of shale content can be computed from log.
- Abnormal readings obtained in pressure-depleted zones and in high vertical-permeability zones.

GAMMA RAY LOG

- A measurement of gamma-ray intensity in the borehole due to natural disintegration of U, Th, and K.
- U, Th, and K tend to concentrate in shales and occur least in clean sands and carbonates.

- Shales appear as excursions to right; clean formation to the left.
- GR curve is good in hard rock regions where SP is deficient.
- A prime indicator of degree of shaliness of formation. Shaliness estimated by interpolating between clean formation line and shale line.
- Spectral GR logs separate total response into U, Th, and K contributions. Improves reservoir delineation and shale estimation.
- Do not repeat exactly. Statistical fluctuations limit speed to 1,800 ft/hr (vs 5,000 ft/hr for Sonic or Electric). Also holds true for all nuclear logs. Do not read sharp peaks. Averages over 3–4 ft should be taken. Repeat run overlays indicate magnitude of statistical fluctuations.
- Depth of penetration—6 in.; vertical resolution — 3 ft.

REFERENCES

[1] H.G. Doll, "The SP Log: Theoretical Analysis and Principles of Interpretation," *Trans AIME*, Vol. 179 (1948).

[2] M.R.J. Wyllie, "A Quantitative Analysis of the Electro-Chemical Component of the SP Curve," *Jour. Pet. Tech.*, Vol. 1 (1949), p. 17.

[3] M. Gondouin, H.J. Hill, and M.H. Waxman, "An Investigation of the Electro-Chemical Component of the SP Curve," *Jour. Pet. Tech.* (March 1962).

[4] M.R.J. Wyllie, "An Investigation of the Electrokinetic Component of the Self-Potential Curve," *Trans. AIME*, Vol. 192 (1951), p.1.

[5] M. Gondouin and C. Scala, "Streaming Potential and the SP Log," *Jour. Pet. Tech.* (August 1958).

[6] M.R.J. Wyllie, A.J. deWitte, and J.E. Warren, "On the Streaming Potential Problem," *Trans. AIME*, Vol. 213 (1958), p. 409.

[7] H.J. Hill and A.E. Anderson, "Streaming Potential Phenomena in SP Log Interpretation," *Trans. AIME*, Vol. 216 (1959), p.203.

[8] Voy E. Althaus, "Electrokinetic Potentials in South Louisiana Tertiary Sediments," *The Log Analyst* (May-July 1967), p.29.

[9] Schlumberger, *Log Interpretation Charts* (1979).

[10] Dresser Atlas, *Log Interpretation Charts* (1979).

[11] M. Gondouin, M.P. Tixier, and G.L. Simard, "An Experimental Study on the Influence of the Chemical Composition of Electrolytes on the SP Curve," *Jour. Pet. Tech.* (February 1957).

[12] H.G. Doll, "The SP Log in Shaly Sands," *Trans. AIME*, Vol. 189 (1950).

[13] L.J.M. Smits, "SP Log Interpretation in Shaly Sands," *Soc. of Pet. Eng. J.*, Vol. 8 (1968), pp. 123–136.

[14] F. Segesman and M.P. Tixier, "Some Effects of Invasion on the SP Curve," *Jour. Pet. Tech.* (June 1959).

[15] W.B. Belknap, J.T. Dewan, C.V. Kirkpatrick, W.E. Mott, A.J. Pearson, and W.R. Rabson, "API Calibration Facility for Nuclear Logs," *Drill. and Prod. Prac.* (Houston: API, 1959).

[16] J.W. Cox and L.L. Raymer, "The Effect of Potassium-Salt Muds on Gamma Ray and Spontaneous Potential Measurements," *SPWLA Logging Symposium Transactions* (June 1976).

[17] A. Heslop, "Gamma Ray Log Response of Shaly Sandstones." *The Log Analyst* (September-October 1974).

[18] G.A. Lock and W.A. Hoyer, "Natural Gamma-Ray Spectral Logging," *The Log Analyst* (September-October 1971).

[19] Dresser Atlas, "Spectralog" (1980).

[20] M. Hassan, A. Hossin, and A. Combaz. "Fundamentals of the Differential Gamma Ray Log-Interpretation Technique," *SPWLA Logging Symposium Transactions* (June 1976).

[21] O. Serra, J.L. Baldwin, and J.A. Quirein, "Theory and Practical Applications of Natural Gamma Ray Spectroscopy," *SPWLA Logging Symposium Transactions* (July 1980).

[22] W.H. Fertl, and E. Frost, Jr., "Experiences with Natural Gamma Ray Spectral Logging in North America," *SPE 11145*, New Orleans (September 1982).

[23] J.A. Quirein, J.S. Gardner, and J.T. Watson, "Combined Natural Gamma Ray Spectral/Litho-Density Measurements Applied to Complex Lithologies," *SPE 11143*, New Orleans (September 1982).

[24] Dresser Atlas, "Spectralog," *ibid.*

Chapter 4

RESISTIVITY LOGS

The basic interpretation relation in well logging, as developed in chapter 2, is the water saturation relation

$$S_w = c\sqrt{R_w/R_t} / \phi \tag{4.1}$$

The most important input to this equation (since its value can never be guessed) is the resistivity, R_t, of the uninvaded region of the formation in question.

No resistivity-measuring tool has yet been designed that can reach deep enough to guarantee reading R_t under all possible invasion conditions while retaining good bed resolution. Therefore, from early days resistivity logs have consisted of three curves: deep, medium, and shallow investigation. With these three measurements and the assumption of a step invasion profile, correction can be made to the deep reading to obtain R_t.

Nevertheless, many logs have been run with only two curves, deep and shallow reading. These clearly show invasion effects but do not permit a correction to the deep reading, which must be assumed equal to R_t. The assumption is reasonable in high-porosity areas where invasion is shallow but can lead to significant errors in low-porosity regions where invasion may be deep.

Over the years there has been a continual succession of resistivity tools with improved designs replacing older ones. It would be convenient to forget the obsolete versions, but we cannot. Company files and log libraries still abound with old logs that are continually being reviewed for new drilling or production prospects.

CLASSIFICATION AND APPLICATION

Table 4–1 is a classification of the major resistivity tools that have been used or are in use. The curves are categorized by their radii of investigation, i.e., deep (3+ ft), medium (1.5–3 ft), shallow (0.5–1.5 ft), and flushed zone

TABLE 4-1

CLASSIFICATION OF RESISTIVITY TOOLS

	Flushed Zone 1–6 in.	Shallow 0.5–1.5 ft	Medium 1.5–3 ft	Deep 3 + ft	Years	Designations	Comments
		16″ Normal	64″ Normal	18′ Lateral	up to 1955	ES, EL	obsolete
Fresh mud $R_{mf} > 2R_w$ or $R_t < 200$	Microlog (ML) Minilog Contact)	16″ Normal		Induction (6FF40)	1955–80	IES, IEL	obsolete
		Spherically Focused		Induction (6FF40)	1970–85	ISF	phasing out
	Proximity (PL)	LL8/short Guard	Medium Induction (IL_m)	Deep Induction (IL_d)	1965–	DIL-LL8, DIFL, DISG	current
		Spherically Focused	Medium Induction (IL_m)	Deep Induction (IL_d)	1975–	DIL-SFL or DISF	current
Salt mud $R_{mf} < 2R_w$ or $R_t > 200$	Microlaterolog (MLL) (FoRxo)		Laterolog-7 Laterolog-3/Guard		1955–80	LL-7 LL-3, guard	obsolete
	MLL or FoRxo	Shallow Laterolog (LL_s)		Deep Laterolog (LL_d)		DLL-MLL (or FoRxo)	current
	Micro Spherically Focused	Shallow Laterolog (LL_s)		Deep Laterolog (LL_d)	1972–	DLL-MSFL	current

Note: Numbers are 50% radius of investigation measured from borehole wall.

(1–6 in.). The number given for any curve represents the 50% response depth measured from the borehole wall, meaning that 50% of the response of the tool comes from the formation contained within the indicated depth and 50% comes from beyond that depth. As an average the 80% response depth is approximately twice the 50% depth. The figures apply to homogeneous formations of average resistivity.

All deep, medium, and shallow curves are obtained with electrodes or coils mounted on cylindrical mandrels that are run more or less centralized in the hole. In contrast the flushed zone (microresistivity) curves are obtained with pad-mounted electrodes forced to ride one side of the hole. The combinations circled on Table 4–1 represent those curves that can be obtained simultaneously on a single pass in the hole.

ELECTRICAL SURVEY (ES) TOOLS[1,2]

Prior to 1950, all resistivity measurements were made with simple electrode arrays of the type shown in Figs. 4–1 (the Normal) and 4–2 (the Lateral). A constant survey current I was emitted from electrode A and returned to electrode B. The voltage V between electrodes M and N was measured. The ratio V/I, multiplied by a constant dependent on the electrode spacing, gave the resistivity. After some period of experimentation, the ES tool configuration settled down in soft rock areas to a short Normal with 16-in. spacing, a long Normal with 64-in. spacing, and a Lateral with 18-ft, 8-in. spacing, all curves being obtained simultaneously. Other spacings continued to be used in hard rock areas. The greater the spacing, the greater the depth of investigation.

The ES logs were difficult, sometimes almost impossible, to interpret. Extensive charts are required to correct for borehole, bed thickness, and adjacent-bed resistivity effects.[3] In particular the curves were relatively useless for bed thicknesses less than about 1.5 times the spacing, i.e., 28 ft for the Lateral and 8 ft for the long Normal. The short Normal curve was the most usable, but it was severely affected by invasion. The basic problem with the ES logs was that the direction of the survey current was not controlled. It took the path of least resistance, favoring conductive mud and conductive shoulder beds over resistive beds at the level of the tool.

As a result the long Normal and Lateral curves were replaced in the 1950s by focused logs in which the path of the survey current was controlled. The focusing minimized borehole and adjacent bed effects and provided simultaneously both deep penetration and good bed resolution. Two types of focused tools were introduced. One was the Induction, which works best in fresh-mud, medium-to-high-porosity conditions. The other was the Laterolog, which is best suited to salt-mud, low-porosity conditions.

FRESH MUD TOOLS

The first major system to replace to ES combination for fresh mud logging was the deep Induction combined with the short Normal.[4] This was called IES or IEL, depending on the service company. (The deep Induction

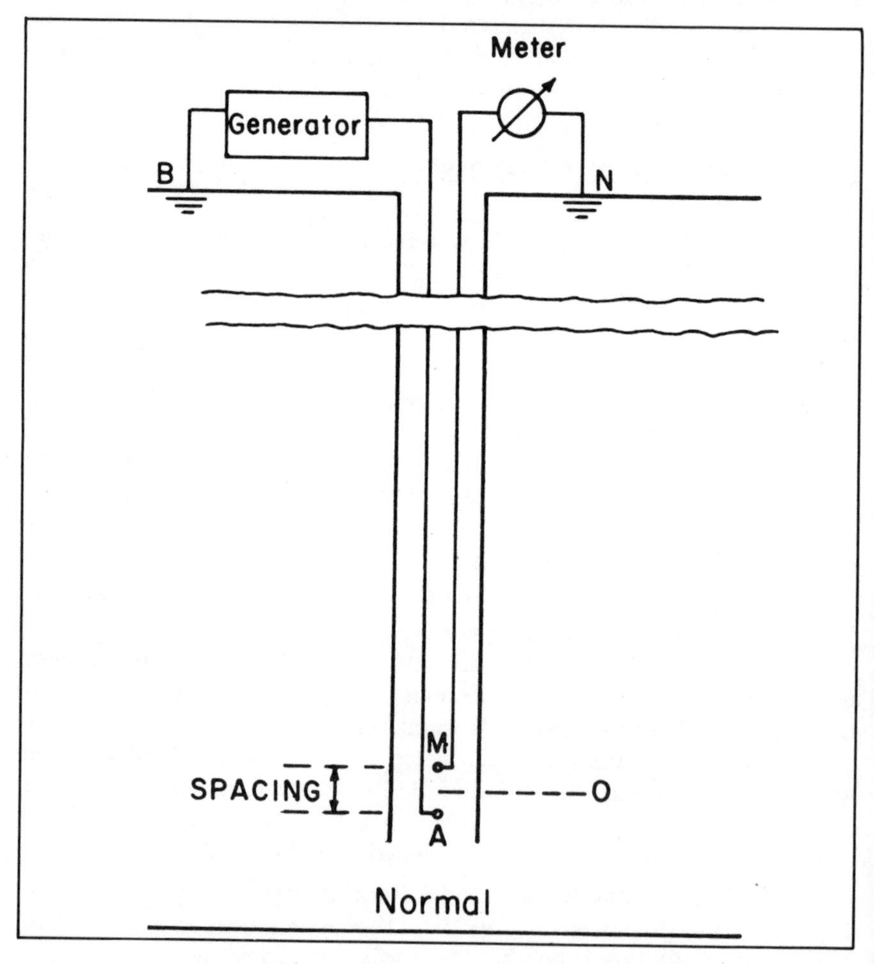

Fig. 4–1 Normal array (courtesy Schlumberger)

was preceded for a few years by a medium Induction that did not last). Later Schlumberger replaced the short Normal by a shallow focused log called the Spherically Focused.[5] Thousands of logs have been run and are still being run with these IES/IEL/ISF combinations.

In the mid 1960s the Dual Induction tool was introduced.[6] It comprised a deep-reading Induction, a medium-reading Induction, and a shallow focused-electrode array. The combination was designated DIL-LL8, DIFL, or DISG, depending on the service company. An improved version

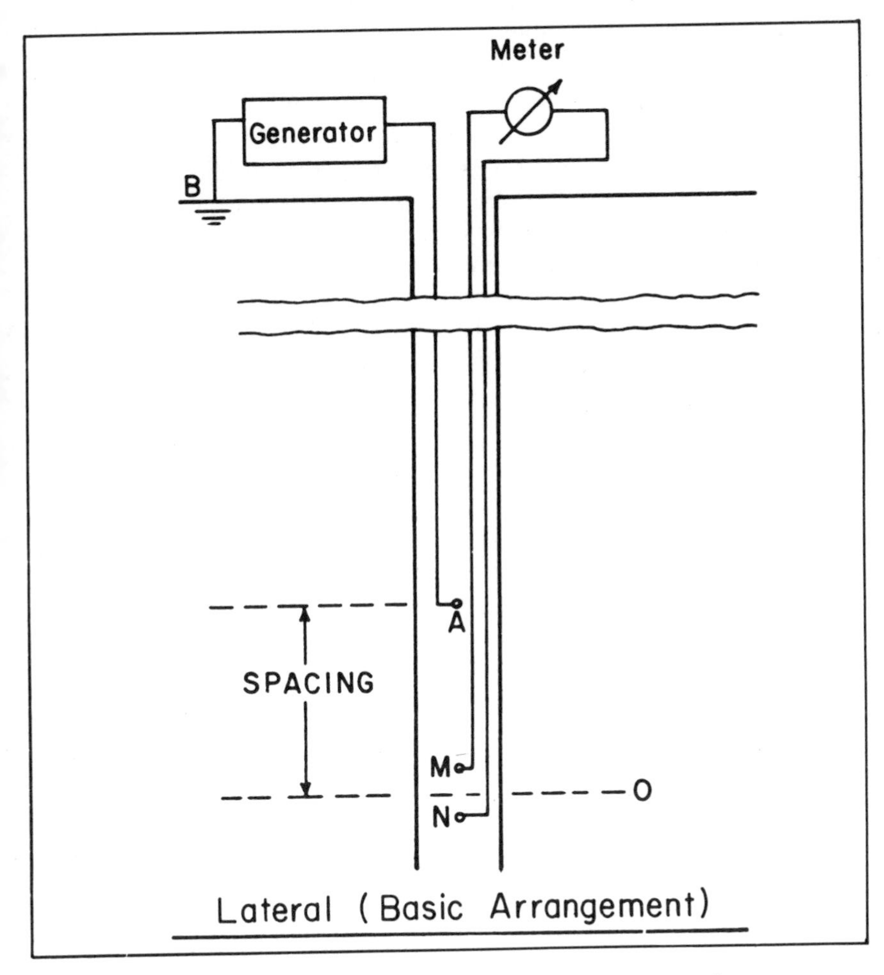

Fig. 4–2 Lateral array (courtesy Schlumberger)

with the Spherically Focused shallow log, the DISF, was introduced by Schlumberger in the mid 1970s. Gradually the Dual Induction tools are replacing the single Induction systems because of their invasion-correcting capabilities.

Two flushed-zone tools are shown in the fresh mud category. The Microlog, also called Minilog and Contact log, was introduced about 1950.[7] It is a nonfocused device with very shallow depth of investigation. It primarily indicates the presence or absence of mud cake in great vertical detail. Therefore, it is an excellent permeable-zone or sand-count indicator—the best available. Under favorable circumstances it can give the flushed zone resistivity, R_{xo}, but is not really designed for that purpose. The Microlog is discussed in chapter 5 along with the Electromagnetic Propagation Log, with which it has been recently combined.

The Proximity log is a focused curve that measures flushed zone resistivity, R_{xo}, in the presence of thick mud cakes that can occur with fresh mud. The R_{xo}value can be used for movable oil calculation or for deriving additional information about the invasion profile. The tool must be run separately from the Induction logs but can be run simultaneously with the Microlog. It has been superseded by the MicroSFL, which can be run simultaneously with the Induction.

SALT MUD TOOLS

Two medium-investigation focused tools were introduced in the 1950's for salt mud surveying. They were the Laterolog-7 and the Laterolog-3, also called the Guard log.[8] At the same time a flushed zone tool called the Microlaterolog or FoRxo, was introduced.[9] It could provide good R_{xo} values for mud cakes up to ⅜ in. thick. The medium and flushed-zone curves had to be run separately and could only provide good water saturation and movable oil values if invasion was not too deep.

As a consequence the separate tools were succeeded in the 1970s by Dual Laterolog systems comprising deep and shallow curves run simultaneously with a flushed zone log.[10] Designations for Dresser and Welex are DLL-MLL and Dual Guard-FoRxo. In the case of Schlumberger, the Dual Laterolog is combined with a Microspherically-Focused curve that can read R_{xo} accurately over a wider range of conditions than can the MLL/FoRxo curve. The combination is denoted DLL-MSFL. Once again the single Laterolog tools are being phased out in favor of the dual systems with their invasion-correcting capability.

RANGES OF APPLICATION OF INDUCTION LOGS AND LATEROLOGS

The Induction combination is the one suitable in the majority of cases. It applies wherever the mud is reasonably fresh and the resistivity is not too high. Fig. 4–3 is more specific. It shows that the Induction should be run whenever R_{mf}/R_w is greater than approximately 2 and the formation resistivity, R_t, does not exceed about 200 ohm-m; the log is not accurate at higher resistivities. Resistivity increases as porosity decreases, so there is a low-

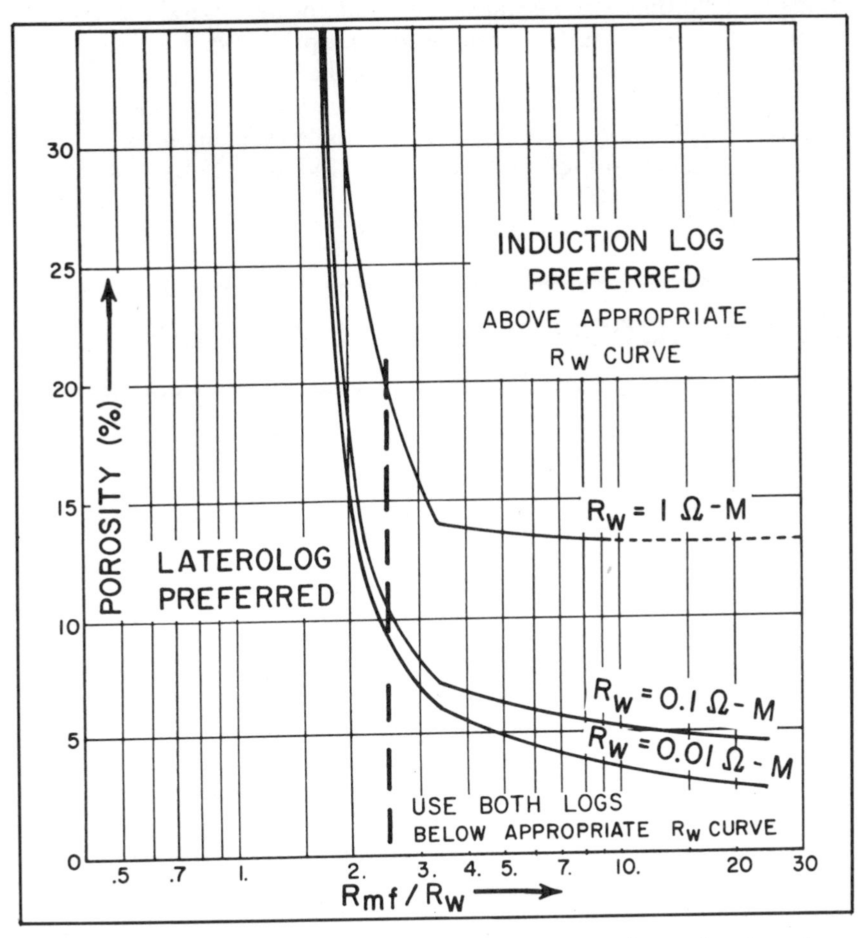

Fig. 4–3 Ranges of application of Induction log and Laterolog (courtesy Schlumberger)

porosity limit. The cutoff corresponding to 200 ohm-m is dependent on the R_w value, as indicated. Typically, R_w is about 0.05 and R_{mf}/R_w is approximately 10 in fresh mud, in which case the low-porosity cutoff is 5%.

If the mud is conductive relative to formation water, i.e., $R_{mf}/R_w < 2$, the Laterolog should be run. It is also the best in fresh mud when resistivities above 200 ohms are encountered because it is accurate at high resistivities. In borderline cases, large boreholes (>12 in.) and deep invasion (>40 in.) favor the Laterolog. This commonly occurs in the Eastern Hemisphere.[11]

The reasons why the Induction and Laterolog are preferred under the indicated conditions will become apparent in the next two sections where the principles and interpretation of the tools are presented. The discussion is confined to the Schlumberger combinations, the DIL-SFL on the one hand and the DLL-MSFL on the other, since they are the most advanced tools in each category. If these are understood, there is no difficulty in interpreting logs run with older or single-curve versions of Induction or Laterolog tools.

DUAL INDUCTION—SPHERICALLY FOCUSED LOGS

The Induction log inherently senses the conductivity of the formation, which is the inverse of its resistivity. In commonly used units

$$\text{conductivity in mmho/m} = 1{,}000/\text{resistivity, ohm-m}$$

The principle of the Induction system is illustrated in Fig. 4–4.[12,13] A constant current of 20 kHz frequency is fed to a transmitter coil. This generates an alternating magnetic field that causes a circular current (Foucault or eddy current) to flow in the surrounding medium. This current in turn creates a magnetic field that induces a voltage in the receiver coil. The induced voltage is approximately proportional to the surrounding conductivity. From this voltage the formation conductivity and thence its resistivity is derived for presentation on the log.

With the single-transmitter, single-receiver system shown, contributions from the borehole and invaded zone as well as from neighboring beds above and below the coil pair would constitute a significant portion of the receiver signal. Practical Induction tools therefore utilize an array of auxiliary transmitter and receiver coils, spaced above and below the main ones, to minimize these contributions and to maximize depth of penetration and vertical resolution. Typically, six or more coils with approximately 40-in. spacing between the main transmitter-receiver pair are used (Fig. 4–5) to obtain the deepest reading curve, denoted IL_d. Fewer coils are used to provide the medium reading curve, denoted IL_m.

The Induction log requires no conductive fluid in the borehole for its operation. It works very well—in fact, best—in holes filled with air or gas or with oil-base mud. No other resistivity tool can be used under these circumstances.

Depth of Penetration of Induction Logs

Fig. 4–6 shows the depth of penetration of the deep, IL_d, and medium, IL_m, arrays. The *integrated geometrical factor* plotted is the relative weight that the tool assigns to a cylindrical shell of the surrounding medium extending from the surface of the sonde to any particular diameter. Each shell contributes to the total conductivity signal in accordance with the product of its conductivity and its relative weight. For example, if C_{xo} is the

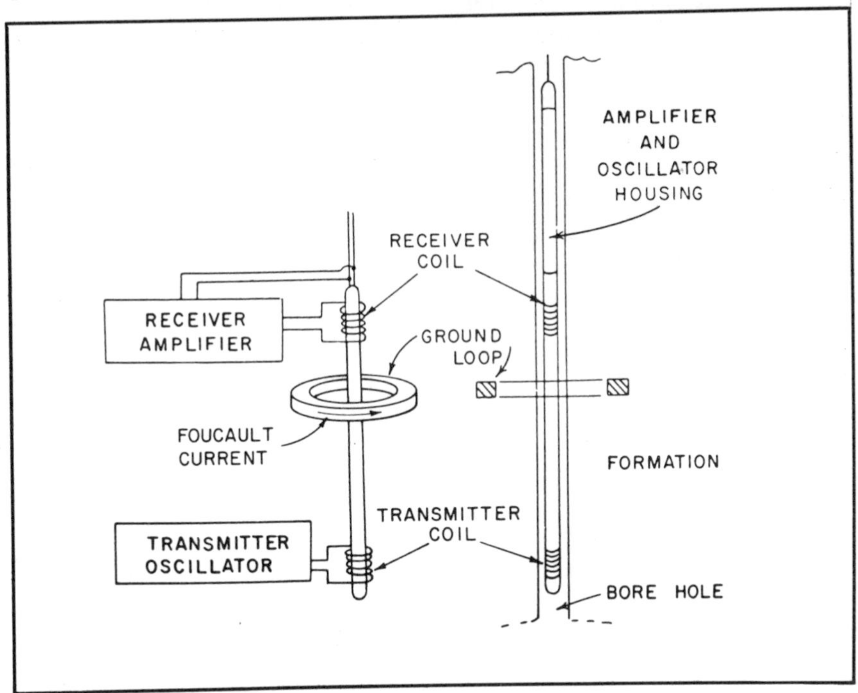

Fig. 4–4 Induction principle (courtesy Schlumberger)

conductivity of the invaded zone, d_i its diameter, and C_t the conductivity of the undisturbed formation, then the apparent conductivity, C_a, read by the tool (neglecting any borehole signal) is

$$C_a = C_{xo}G_{di} + C_t(1 - G_{di}) \tag{4.2}$$

In terms of resistivity

$$1/R_a = G_{di}/R_{xo} + (1 - G_{di})/R_t \tag{4.3}$$

For homogenous formations with no invasion, Fig. 4–6 (solid line) gives relative signal contributions from different regions. For example, 50% of the signal comes from within and 50% comes from beyond a diameter of 11 ft for the IL_d. The corresponding figure for the IL_m is 5 ft, so it has only half the depth of investigation as does the IL_d. Note that the IL_d assigns greatest weight to that part of the formation situated at about 60 in. diameter; the IL_m assigns it to that part at 40 in. diameter, as shown by the maximum-slope points on their respective curves.

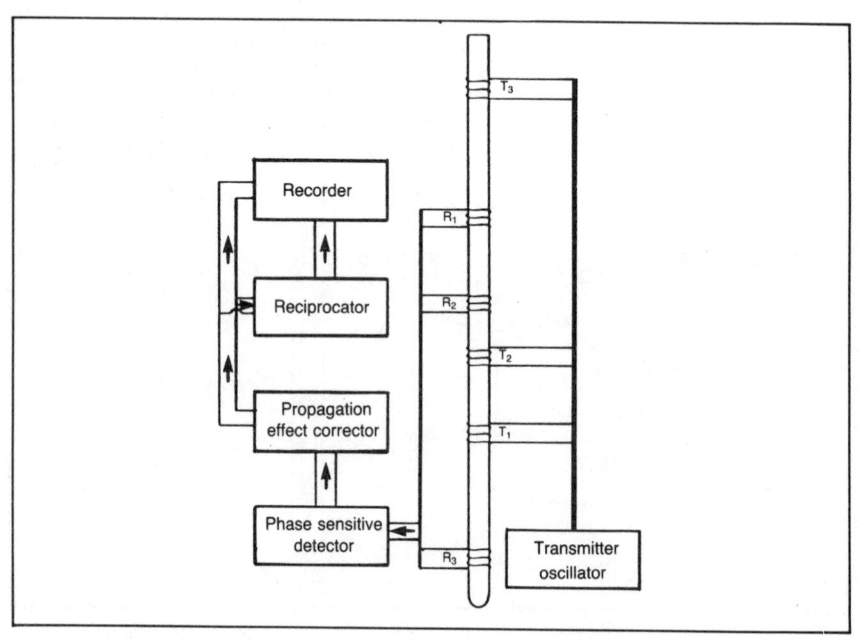

Fig. 4–5 Practical Induction array (courtesy Dresser)

For the usual case of invaded formations, the above equations may be used to illustrate why the Induction works well in fresh mud and poorly in salt mud.

Consider a typical fresh mud case where $R_m = 1$, $R_t = 10$, $R_{xo} = 20$, and $d_i = 65$ in. The IL_d geometrical factor for 65 in. is 0.2. Eq. 4.3 gives

$$1/R_a = 0.2/20 + (1 - 0.2)/10$$
$$R_a = 11 \text{ ohm-m}$$

The IL_d therefore reads only 10% in error. Applying a small correction factor obtainable from the IL_m and SFL readings is reasonable.

The situation with salt mud is quite different. The same formation, drilled with salt mud of $R_m = 0.05$ ohm-m, would have $R_{xo} = 1.0$. The IL_d reading would then be

$$1/R_a = 0.2/1.0 + (1 - 0.2)/10$$
$$R_a = 3.6 \text{ ohm-m}$$

In this case the IL_d reads far from the correct value. It would be impossible to obtain a sufficiently accurate correction factor from the IL_m and SFL curves.

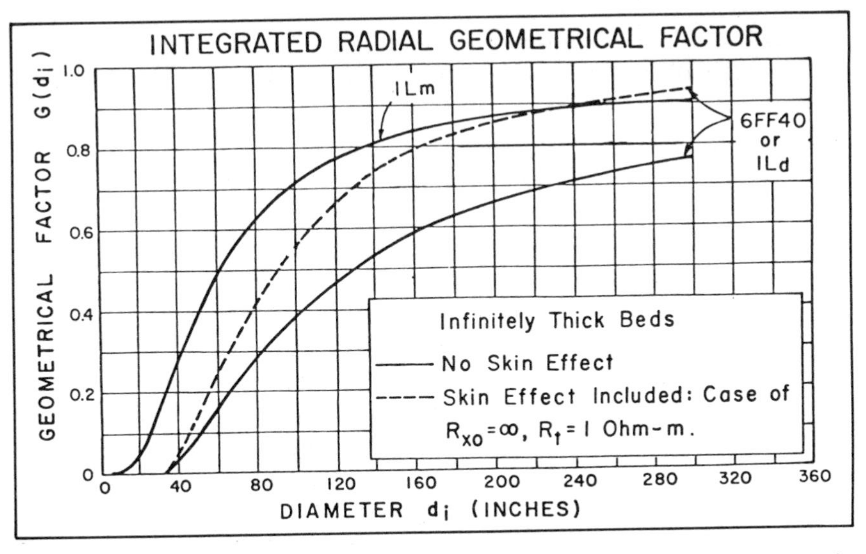

Fig. 4-6 Depth of penetration of Induction log (courtesy Schlumberger)

In reality the error in salt mud conditions is even greater than illustrated because the geometrical factor curves of Fig. 4–6 that apply to quite resistive formations shrink to smaller diameters as resistivity decreases, giving greater weight to the invaded zone (dashed line). This phenomenon is called *skin* or *propagation effect*.

The important point is that the Induction log will be adversely affected by an invaded zone more conductive than the undisturbed formation. It prefers invaded zones less conductive (more resistive) than the uninvaded formation.

Borehole Effects

Borehole fluid contributes an undesired signal to the Induction response. It is inconsequential when the mud is fresh, the hole size is small, or the formation conductivity is high. However, it can be significant under the reverse conditions.

Borehole correction is made by means of Fig. 4–7. For the example illustrated, IL_d with 1.5-in. standoff in a 14.6-in. hole, the borehole geometrical factor is 0.002. With R_m = 0.35 ohm-m (2,857 mmho/m), the hole signal is (0.002 × 2,857) or 5.7 mmho/m, as indicated by the nomograph. This signal must be subtracted from the IL_d conductivity reading to obtain the corrected formation conductivity. The correction is negligible if the IL_d reading is less than 10 ohm-m (i.e., greater than 100 mmho/m) but significant at higher resistivities. For example, if the IL_d read 50 ohm-m (20 mmho/m), the corrected resistivity would be 1,000/(20 − 5.7) or 70 ohm-m.

The situation is aggravated when the mud is more conductive. For the same conditions as cited, except with R_m = 0.15 ohm-m, the hole signal is 13.3 mmho/m. An IL_d reading of 50 would correct to 149 ohm-m, an excessively large adjustment. In general, a correction of more than 50% to the indicated resistivity is unreliable because there is some uncertainty in the hole geometrical factor, the hole may be out-of-round, the standoff may be affected by mud cake or hole rugosity, and there may be some uncertainty in mud resistivity.

Fig. 4–7 shows two additional important points. First, borehole corrections are much greater with the Induction tool against the wall than stood off from the wall. A minimum standoff of 1.5 in. should always be used. This is normal procedure for service companies but should be checked at the wellsite. Standoff devices are sometimes omitted when there is difficulty in getting the tool downhole.

Second, the hole corrections are much larger for the medium-reading IL_m than for the IL_d. In fact, corrections become excessive with hole diameters larger than 12 in., even with fresh mud. This limits the use of the dual Induction tool in some areas.

Bed Thickness Effects

The Induction log has a vertical bed resolution of approximately 4 ft, as determined primarily by the main coil spacing.

There are, however, contributions from beds well above and below the 4-ft section directly opposite the tool. These contributions are negligible in soft rock but are significant in hard rock when the shoulders are much more conductive than the bed of interest, even when the latter is much thicker

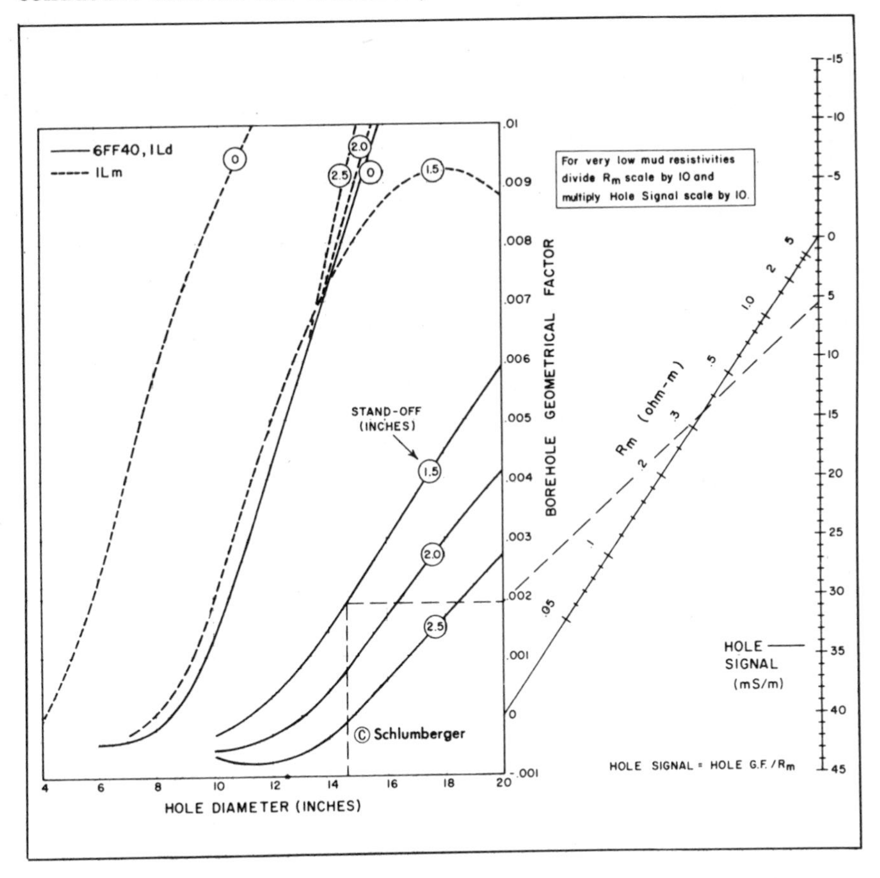

Fig. 4–7 Borehole corrections for ILd and ILm (courtesy Schlumberger)

than 4 ft. Such could be the case for example of a 10–ft hydrocarbon-bearing low-porosity limestone sandwiched between two shales. The Induction could read too low by a factor of two in a worst case.

Charts exist in service-company manuals for bed thickness correction. However, they presume thick homogeneous shoulder beds, a condition not often satisfied in hard rock. For the most part, then, Induction log bed thickness corrections are ignored, either as unnecessary or inapplicable to the conditions at hand.

Sonde Error

If the Induction sonde is lifted high in the air with no conductive material in its vicinity, there can still be a small signal on the order of a few mmho/m due to residual coupling between transmitter and receiver coils or to imbalance in the receiver circuits. This is termed *sonde error*. Normally the error is balanced out before the log is run. The log heading should so indicate. There is no guarantee however that sonde error remains the same in the borehole, under high temperature and pressure conditions. This, along with borehole effect, is the reason the Induction log is not accurate in measuring formation conductivities less than 5 mmho/m (resistivities above 200 ohm-m).

THE SPHERICALLY FOCUSED LOG (SFL)

The SFL is the shallow-reading resistivity curve of the DIL-SFL combination. It is obtained with a set of separate electrodes mounted on the Induction sonde. Fig. 4–8 illustrates the principle. Survey current i_o, flows from the center electrode, A_o. A variable focusing current, i_a, flows between A_o and the auxiliary electrode pair, A_1 and A_1', connected together. By appropriate adjustment, the focusing current forces the survey current to enter the formation in the same manner as it would if there were no borehole (with spherical equipotential lines such as B and C, hence the name). With this system, conflicting requirements of shallow formation penetration and independence of borehole size and salinity over a wide range are met.

The depth of penetration of the SFL is shown in Fig. 4–9. It is significantly shallower than that of the predecessor curves, the LL-8 (or short Guard), and 16-in. Normal. This means it gives greater weight to the invaded zone, which is desired, but in general it still reads too deep to give an accurate measurement of flushed zone resistivity, R_{xo}.

The vertical bed resolution of the SFL, LL-8, and short Guard is about 1 ft. Bed thickness corrections are not required.

Borehole effects for the SFL are normally negligible. All shallow resistivity curves tend to read resistivities too low if the borehole becomes quite large and invaded zone resistivity becomes high relative to mud resistivity (meaning low porosity). For borehole diameters of 6–12 in., SFL corrections are negligible up to $R_{SFL}/R_m = 2{,}000$. LL-8 or short Guard correction becomes significant at $R_{LL8}/R_m > 100$ and 16-in. Normal corrections significant if $R_{16}/R_m > 30$. Correction charts are found in service company chart books but are not often necessary.

LOG PRESENTATION

Fig. 4–10 shows a typical presentation of the 5-in./100-ft DIL-SFL log when it is run in combination with the Sonic log. The SP curve, which is obtained simultaneously, is recorded in Track 1 on a linear scale. Also shown is an R_{wa} curve; this is described in chapter 9. The three resistivity curves are recorded in Track 2 and half of Track 3 on a logarithmic scale

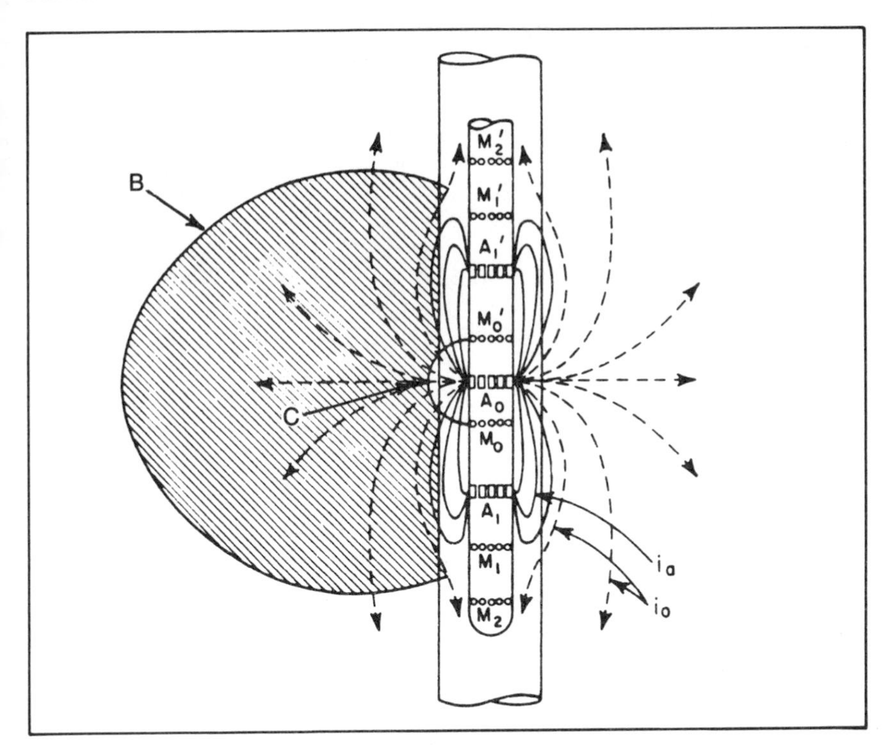

Fig. 4–8 Principle of the spherically focused log (courtesy Schlumberger)

covering 0.2–200 ohm-m. The deep curve is always heavy dashed, the medium curve is light dashed, and the shallow curve is solid; the latter is unaveraged (SFLU) in this case. In the center of Track 3 is the caliper curve (dotted) with the bit size (7⅞ in.) shown dashed. The Sonic log, to be discussed later, is recorded in the right-hand side of Track 3. Note the breaks in the right and left edges of the grid. These occur at 1-min intervals and serve to indicate logging speed. In this case it was 86 ft/min or 5,200 ft/hr.

On the 2-in./100-ft log the resistivities are presented on a linear scale to facilitate correlation with older linear scale logs. The SFL or LL8 is averaged over 3 ft to reduce its detail to that of the Induction curves. Further, the deep Induction conductivity reading is presented in Track 3 on a scale from right to left. This facilitates reading very low IL_d values, such as may occur in salt-water sands. For example, an IL_d value of 0.55 ohm-m, difficult to read accurately on the resistivity scale, can be read easily as 1,820 mmho/m on the conductivity scale.

In Fig. 4–10, which is a log of the Travis Peak formation in East Texas, the SP curve indicates permeable zones at A, C, E, and G and relatively

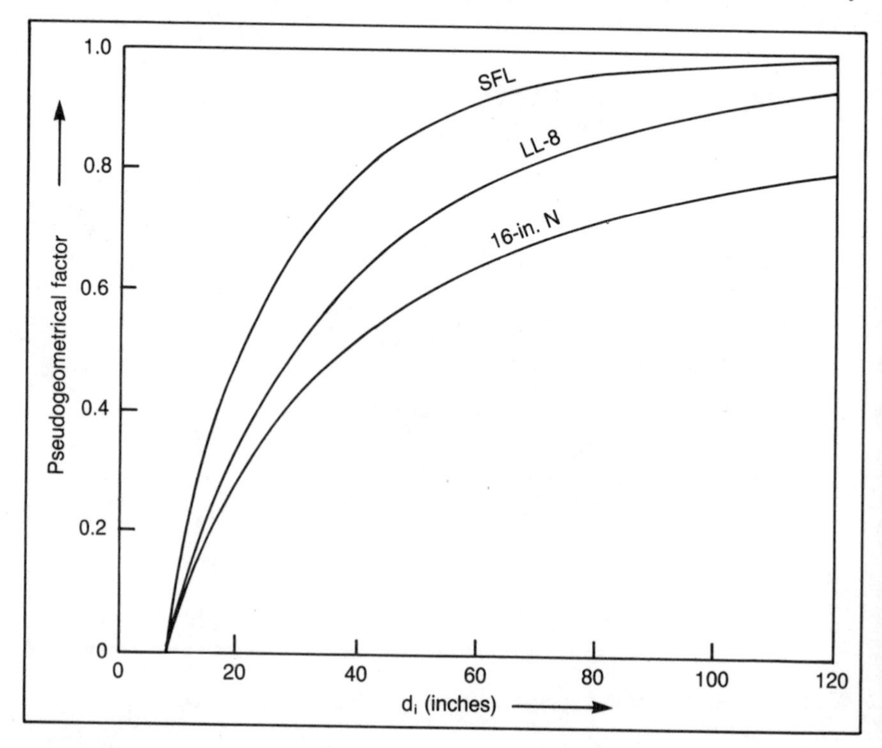

Fig. 4–9 Depths of penetration of the shallow resistivity curves (courtesy Schlumberger)

impermeable intervals at B, D, and F. In the permeable zones the three resistivity curves show wide separation indicating deep invasion. This is supported by the caliper curve where mud cakes up to about ½ in. in thickness are indicated (shaded sections). Zones D and F show no curve separation, indicating that they are formations, probably shales, with zero permeability. On the other hand interval B shows some separation of the SFL and Induction curves but none between the two Induction logs. This

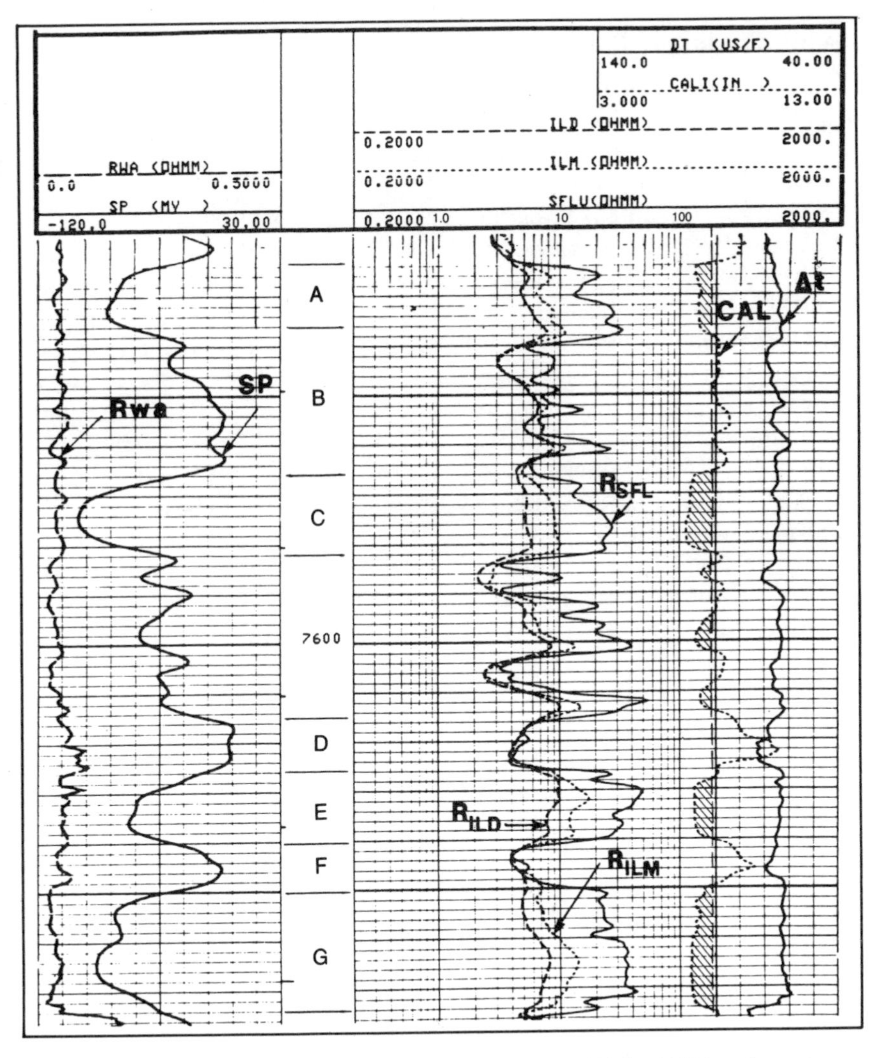

Fig. 4–10 Typical presentation of Dual Induction-SFL (or LL8) log (courtesy Schlumberger)

implies low but finite permeability with shallow invasion. The section is a shaly sand or carbonate, not a pure shale.

It is apparent from both the negative SP values and the $R_{SFL} > R_{ILD}$ separation in the permeable zones that the well was drilled with fresh mud, with $R_{mf} \gg R_w$. However, it is not evident from the resistivity logs alone that hydrocarbons are present. There is no obvious water-bearing interval for reference. We could calculate water saturations with porosities derived from the Sonic log but will reserve this until later.

Determination of R_t

The object of the interpretation procedure is to utilize the three resistivity values obtained in a given zone to correct the deep reading for invasion and thus to arrive at the best possible R_t value. This is done by means of the tornado chart in Fig. 4–11.

The IL_d, IL_m, and SFL readings are first corrected for borehole and bed thickness effects, if necessary, to give corrected R_{ID}, R_{IM}, and R_{SFL} values. Then the ratios R_{IM}/R_{ID} and R_{SFL}/R_{ID} are entered on the chart and their point of intersection is established. From the location of this point, the value of R_t/R_{ID} is found by interpolating between the dotted curves. (R_{xo}/R_t and invasion diameter, d_i, can also be found from the other sets of curves, but their values are of academic interest.) For example, in the middle of zone C where $R_{ID} = 5.2$, $R_{IM} = 9$, and $R_{SFL} = 26$, R_{IM}/R_{ID} of 1.7 and R_{SFL}/R_{ID} of 5.0 gives R_t/R_{ID} of 0.65, which means the correct R_t is 0.65×5.2 or 3.4 ohm-m. Invasion diameter is approximately 100 in. R_{xo} is found to be 9×3.4 or 31 ohm-m.

Several points are noteworthy. First, the R_t correction factor is between 1.0 and 0.6, although some charts go to 0.5. This means that even with deep invasion the IL_d reads close to R_t. It is for this reason that single Induction tools were successful for many years. A 20% error in R_t translates to only a 10% error in water saturation, for example, when the saturation equation, Eq. 4.1, is applied.

Second, the depth of invasion is reflected by the relative values of the two Induction readings. A ratio R_{IM}/R_{ID} greater than 1.5 indicates quite deep invasion ($d_i > 70$ in.) and a value less than 1.2 indicates shallow invasion ($d_i < 40$ in.). These ratios convert to constant separations between the curves on a logarithmic scale so the depth of invasion can be eyeballed easily on the logs.

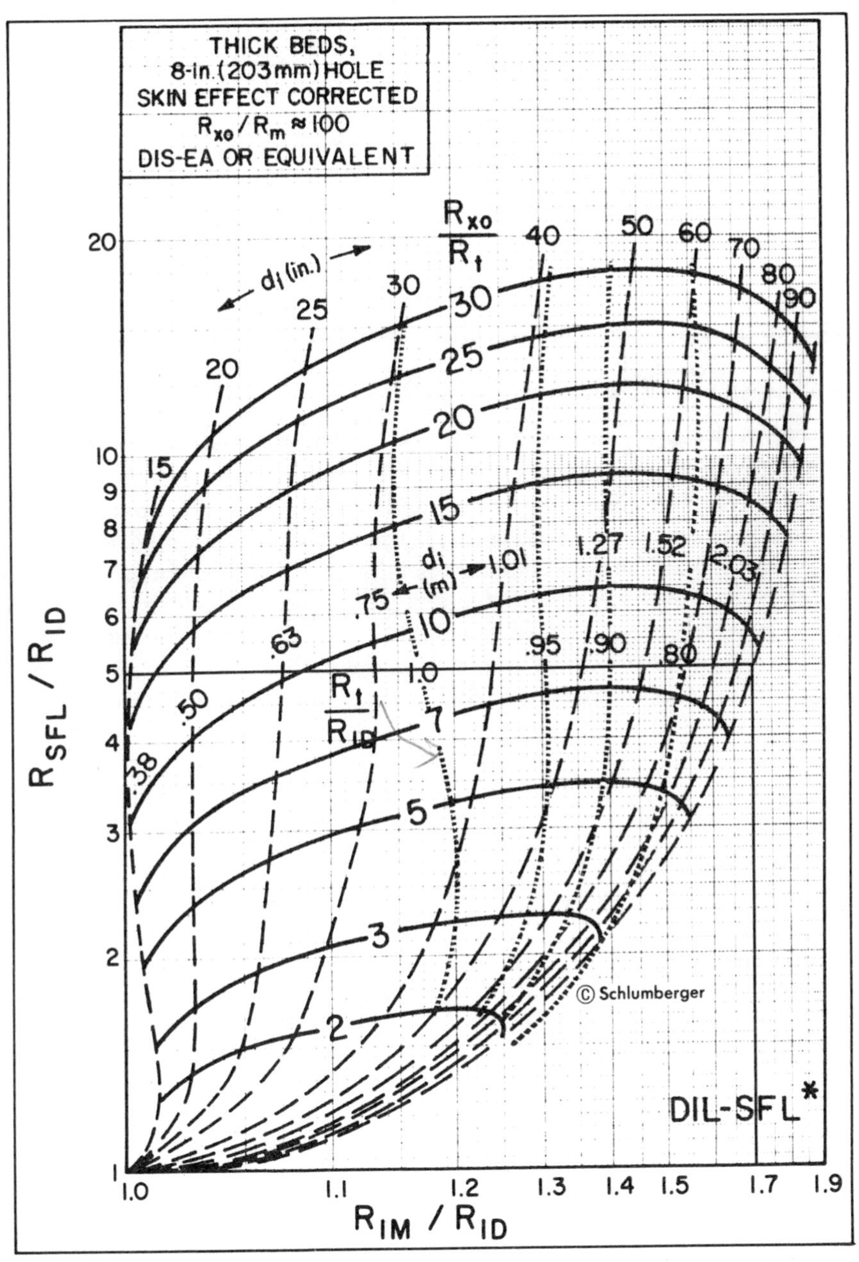

Fig. 4–11 Invasion correction chart for Dual Induction-SFL (courtesy Schlumberger)

Third, the value of R_{xo} derived from the chart, generally 1.5 to 2 times the SFL reading, is not sufficiently accurate for movable oil calculation because of approximations and uncertainties in the chart. In any case movable oil calculation with fresh mud is suspect, for reasons given in chapter 2.

DUAL LATEROLOG—R_{xo} LOGS

Laterolog systems utilize a multiple electrode array to force survey current to travel laterally across the mud and into the adjacent formation. The advantages that accrue are the ability to operate in very salty mud, excellent bed definition, and independence to neighboring bed resistivities.

There are two basic types of focused-electrode Laterolog arrays. One is the 3-electrode system, commonly called Guard log or LL3, and the other is the 7 to 9 electrode system, with designations such as LL7, LL8, LL_D, and LL_S. Both systems operate on much the same principle, as illustrated in Fig. 4–12.

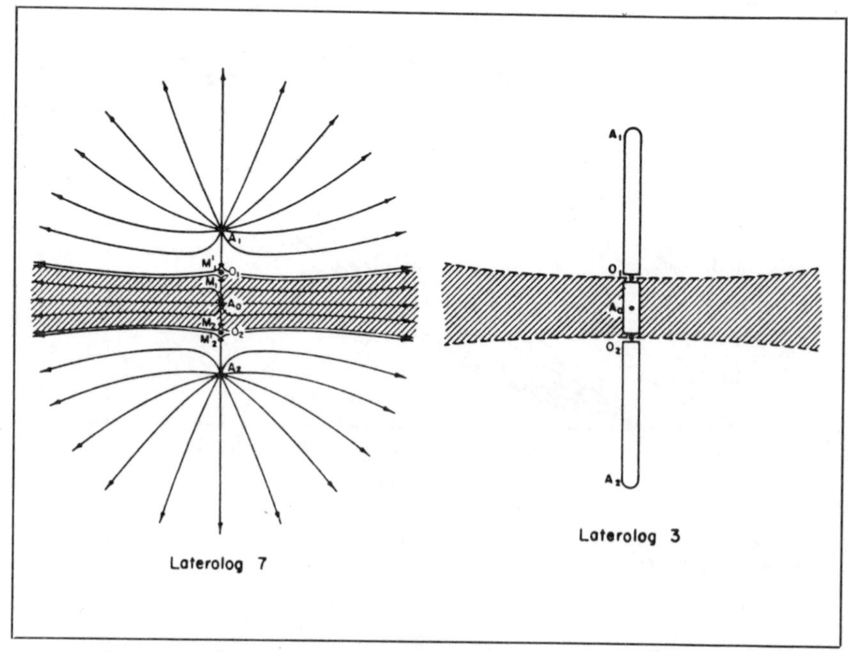

Fig. 4–12 Basic Laterolog arrays (courtesy Schlumberger)

Considering first the LL3, survey current i_o is sent out from the center electrode A and bucking current i_a is sent from the guard electrodes A_1 and A_2, which are connected together. The bucking current is adjusted to maintain zero voltage between A_0 and the A_1-A_2 pair. There is then no current flowing up or down the hole, which means the survey current is forced to flow in a lateral sheet into the formation. The width of the A_0 electrode, commonly 1 to 12 in., determines the vertical resolution of the tool. The length of the guard electrodes, usually 2.5 to 5 ft, and the proximity of the survey current return point determine the depth of penetration. The longer the guards and the more remote the return point, the deeper the penetration. Formation resistivity is proportional to V/i_o where V is the common electrode potential relative to a far electrode.

In the case of multielectrode systems, illustrated by the Laterolog 7, survey current flows from the A_o electrode and bucking current flows from the A_1 and A_2 electrodes, which are connected together. The bucking current is adjusted to maintain zero voltage across the monitor electrodes M_1 and M_1' (which are connected to M_2 and M_2', respectively). The net result is the same as for the guard system.

THE DLL-MSFL TOOL

Fig. 4–13 is a schematic of this system. The main part of the tool constitutes a 9-electrode array that provides deep (LL_d) and shallow (LL_s) resistivity curves. On the bottom of the tool is a pad-mounted, Microspherically Focused array that provides a flushed zone (R_{xo}) resistivity curve. The four-arm linkage that supports the MSFL pad provides a caliper curve and also centralizes the bottom of the tool. The top of the tool is centered with another centralizer.

Fig. 4–14 shows how the same set of electrodes is used to obtain the deep and shallow curves by using currents at two different frequencies. The deep measurement is made at 35 Hz and the shallow one is made at 280 Hz. The LL_d achieves deep penetration by having a long electrode array (28 ft) and returning the current to a surface electrode. With the LL_s, current is returned to a nearby electrode that gives it shallow penetration. Beam width and therefore vertical bed resolution is 24 in. for both curves.

Fig. 4–15 shows the MSFL array. Five rectangular electrodes are mounted on an insulating pad that is forced to ride the side of the hole. Survey current i_o flows from A_o and bucking current i_a flows between A_o and A_1. The latter current is adjusted to maintain zero voltage between the monitor electrodes indicated. This forces the survey current directly into the

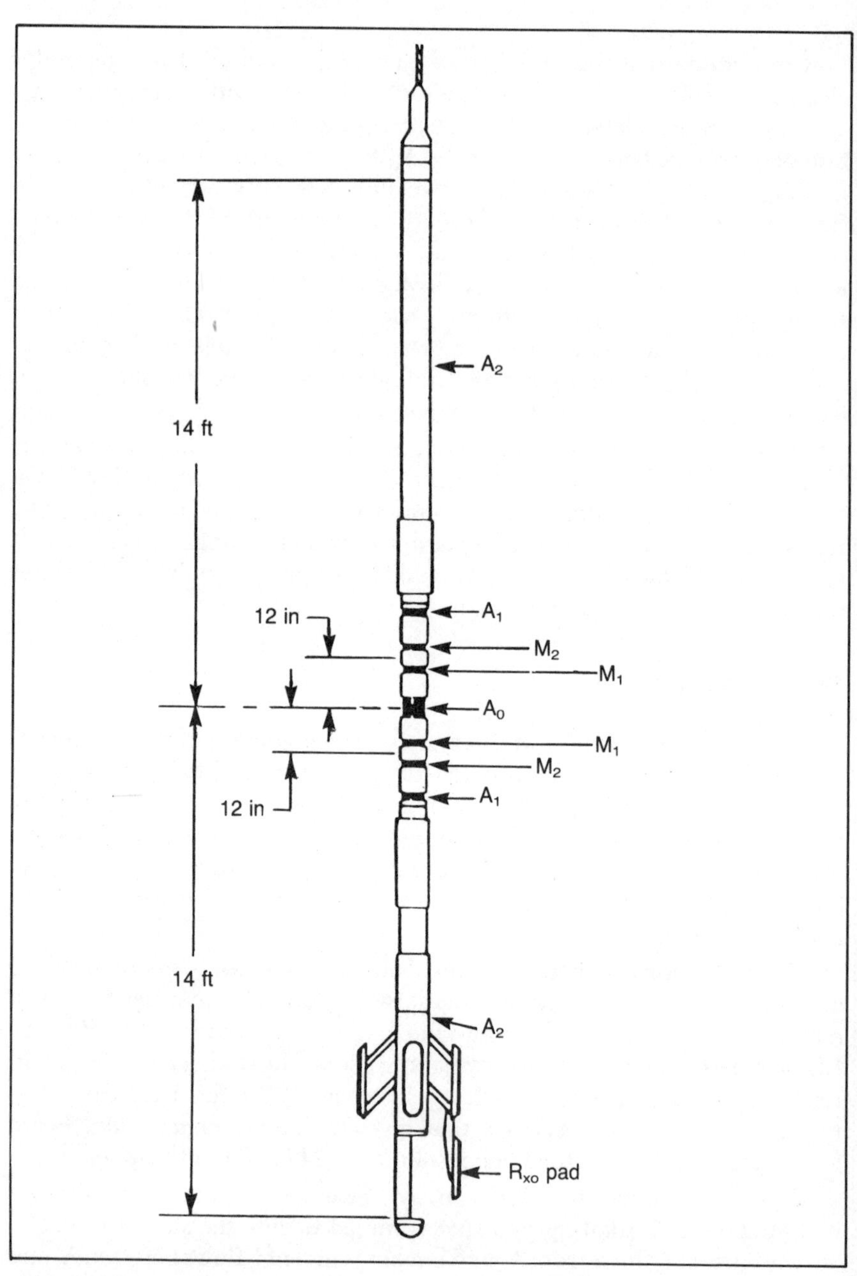

Fig. 4–13 Schematic of Dual Laterolog-Rxo tool (courtesy Schlumberger, © SPE-AIME)

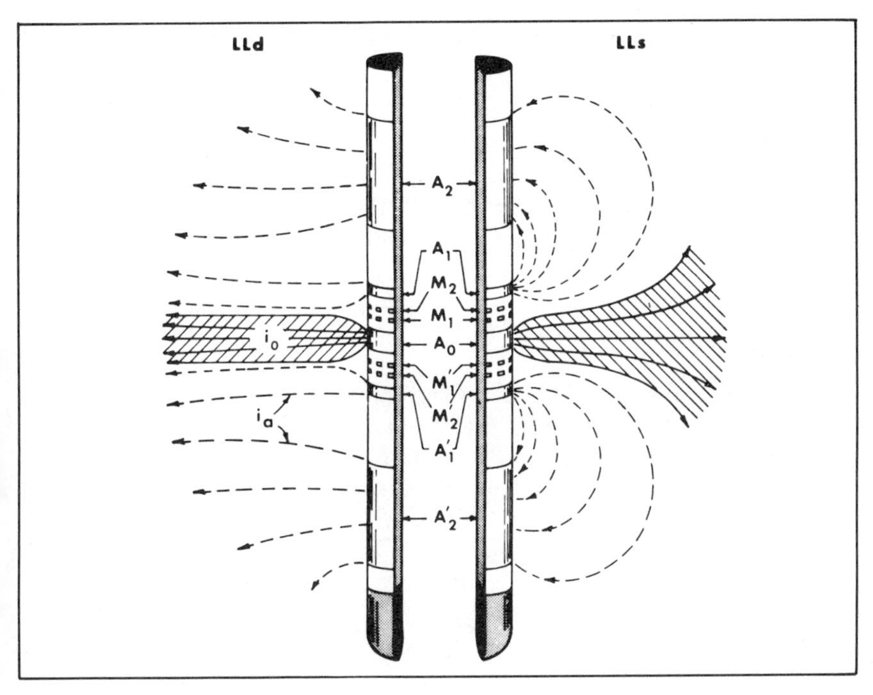

Fig. 4–14 Current flow patterns for LLd and LLs (courtesy Schlumberger, © SPE-AIME)

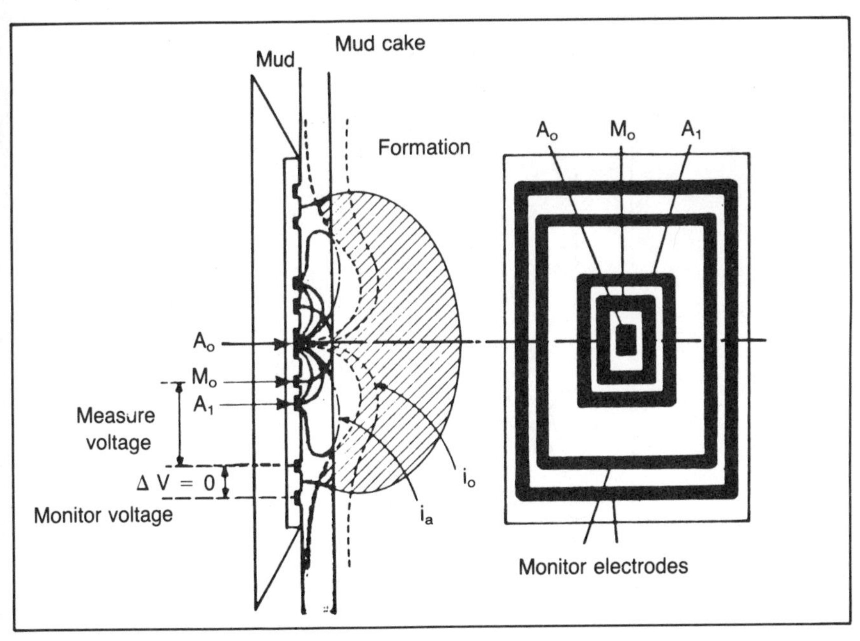

Fig. 4–15 MicroSFL array (courtesy Schlumberger, © SPE-AIME)

formation, where it bells out quickly and returns to a nearby electrode. The voltage V between electrode M_o and the monitor electrodes is measured. Resistivity is proportional to V/i_o. With this system the MicroSFL has sufficiently shallow penetration to read flushed-zone resistivity, R_{xo}, directly, even in the presence of mud cakes up to ¾ in. thick.

DEPTH OF INVESTIGATION

Fig 4–16 shows the depth of investigation of the LL_d and LL_s curves along with that of the older LL3 and LL7 tools, now obsolete. The pseudogeometrical factor, J, is the relative weight that the particular array assigns to the invaded zone. The term *pseudo-* is used because the weight

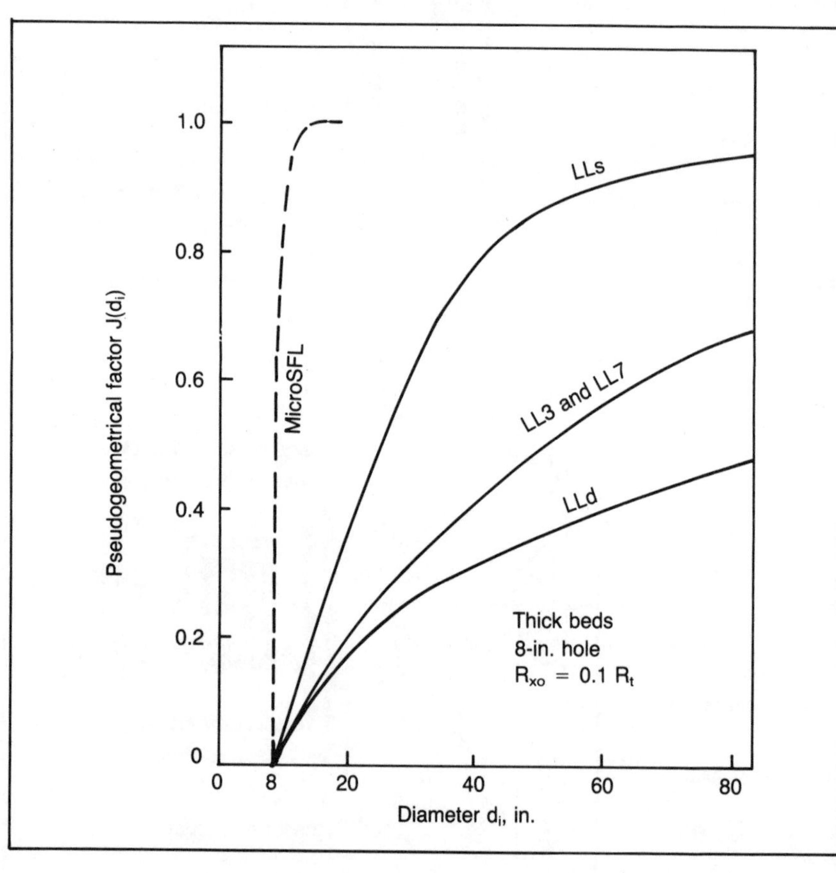

Fig. 4–16 Depths of Investigation of DLL-Rxo curves (courtesy Schlumberger, © SPE-AIME)

depends on the relative resistivities of invaded and uninvaded zones as well as the invasion diameter.

The invaded and uninvaded zones contribute to the total resistivity read by the tool in accordance with the product of their resistivities and weights. That is

$$R_a = R_{xo}J_{di} + R_t(1 - J_{di}) \tag{4.4}$$

This equation illustrates why the Laterolog is better than the Induction log in salt mud. Consider the same salt-mud situation as for the induction case: $R_m = 0.05$, $R_{xo} = 1$, $R_t = 10$, $d_i = 65$ in. The LL_d geometrical factor is 0.42. Consequently, the LL_d would read an apparent resistivity

$$R_a = 1 \times 0.42 + 10\ (1 - 0.42) = 6.2 \text{ ohm-m}$$

This is considerably closer to the true value than the Induction reading, which would be less than 3.8 ohm-m. Nevertheless, a significant correction factor is required, which illustrates the need for a three-curve combination.

For the same fresh-mud conditions as previously cited, $R_m = 1.0$, $R_{xo} = 20$, $R_t = 10$, $d_i = 65$ in., the Laterolog would give an approximate value

$$R_a = 20 \times 0.42 + 10\ (1 - 0.42) = 14.2 \text{ ohm-m}$$

The actual value would be higher because J (at 65 in.) would be greater than 0.42 with the more resistive invaded zone. The Induction log would read 11 ohm-m under these conditions; it would be the better choice. This illustrates that the Laterolog tool is adversely affected by an invaded zone more resistive than the undisturbed formation, which is opposite to the Induction case.

Vertical Resolution

As mentioned, the vertical resolution of the LL_d and LL_s curves is 2 ft. Adjacent bed effect is small.[14] No bed thickness corrections are required.

Borehole Effects

Borehole effects on the LL_d and LL_s are insignificant under most conditions. For hole diameters in the range 6–12 in., the corrections are less than 15% and therefore may usually be neglected. For larger diameters the LL_s may require correction.

CHARACTERISTICS OF THE MicroSFL (MSFL)

The MicroSFL succeeded the Microlaterolog and Proximity log as a device to measure flushed-zone resistivity. The 50% depths of penetration of the three tools for homogeneous formations are approximately 1, 2, and 3 in., respectively; 80% depths of penetration are about twice those values. Therefore, the MSFL has the advantage when invasion is shallow.

Bed resolution of the MSFL and other R_{xo} tools is extremely good—on the order of 6 in. In fact, there is so much detail the curve is often averaged over 2 ft during recording to make it more compatible with the LL_d and LL_s curves.

Borehole (i.e., mud cake) corrections for the MSFL are negligible (<15%) for mud-cake thicknesses between ⅛ and ¾ in. On the other hand, the Microlaterolog requires significant correction for mud cakes greater than ⅜ in. thick. Since mud-cake thickness is rarely known with any accuracy, the MicroSFL is the preferred curve.

LOG PRESENTATION

Fig. 4–17 is an example of the DLL-MSFL log in salt mud. A GR curve, which can be run simultaneously, is presented in Track 1 since the SP is poor in salt mud. These three resistivity curves are recorded in Tracks 2 and 3 on the several-decade logarithmic scale. Normal presentation is LL_d heavy-dashed, LL_s light-dashed, and MSFL solid. With salt mud the shallowest curve reads lowest resistivity and the deepest curve reads highest in water-bearing zones, which is the reverse of the fresh-mud situation.

Determination of R_t

Correction of the LL_d to obtain R_t is made by means of Fig. 4–18. The three resistivity curves are read at the point of interest and the chart is entered with the ratio R_{LLd}/R_{LLs} on the horizontal scale and R_{LLd}/R_{xo} (where $R_{xo} = R_{MSFL}$) on the vertical scale. From the point of intersection the correction factor R_t/R_{LLd} is read from the dotted lines trending northeasterly. For level A of Fig. 4–17, $R_{LLd}/R_{LLs} = 2.2$ and $R_{LLd}/R_{xo} = 3$, which gives $R_t/R_{LLd} = 1.6$. R_t is then 6.3×1.6 or 10 ohm-m. Invasion diameter, as read from the dashed lines, is 80 in., which is quite large. Notice that the invasion near the bottom of the section is much smaller because the LL_s curve falls closer to the LL_d curve. For level B the chart indicates an invasion diameter of only 40 in.

QUICK-LOOK HYDROCARBON INDICATION

Fig. 4–19 is an example from the Middle East where the DLL-R_{xo} tool was run, not because of salt mud, but because hole size was large, and because in the hydrocarbon-bearing zones (denoted 2 and 3) $R_t > R_{xo}$ and $R_t > 100$ ohm-m. Correcting for invasion by means of Fig. 4–18 gives $R_t = 650$ ohm-m in the center of Zone 2 and $R_t = 140$ ohm-m in Zone 3.

With the DLL-R_{xo} combination an approximate water saturation can be estimated from the resistivity logs only. Combining the saturation equations

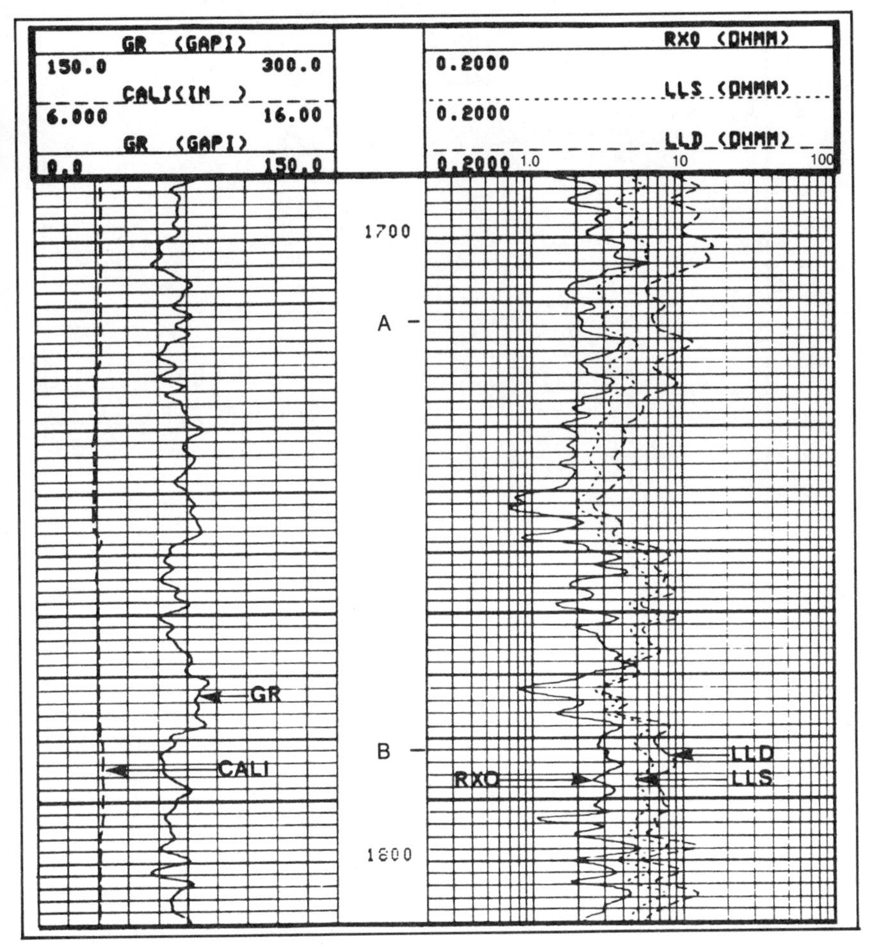

Fig. 4–17 Dual Laterolog-Rxo log in salt mud (courtesy Schlumberger)

for the undisturbed and flushed zones, as given by Eqs. 2.8 and 2.12

$$\frac{S_w}{S_{xo}} = \left[\frac{R_w/R_t}{R_{mf}/R_{xo}} \right]^{1/2} \tag{4.5}$$

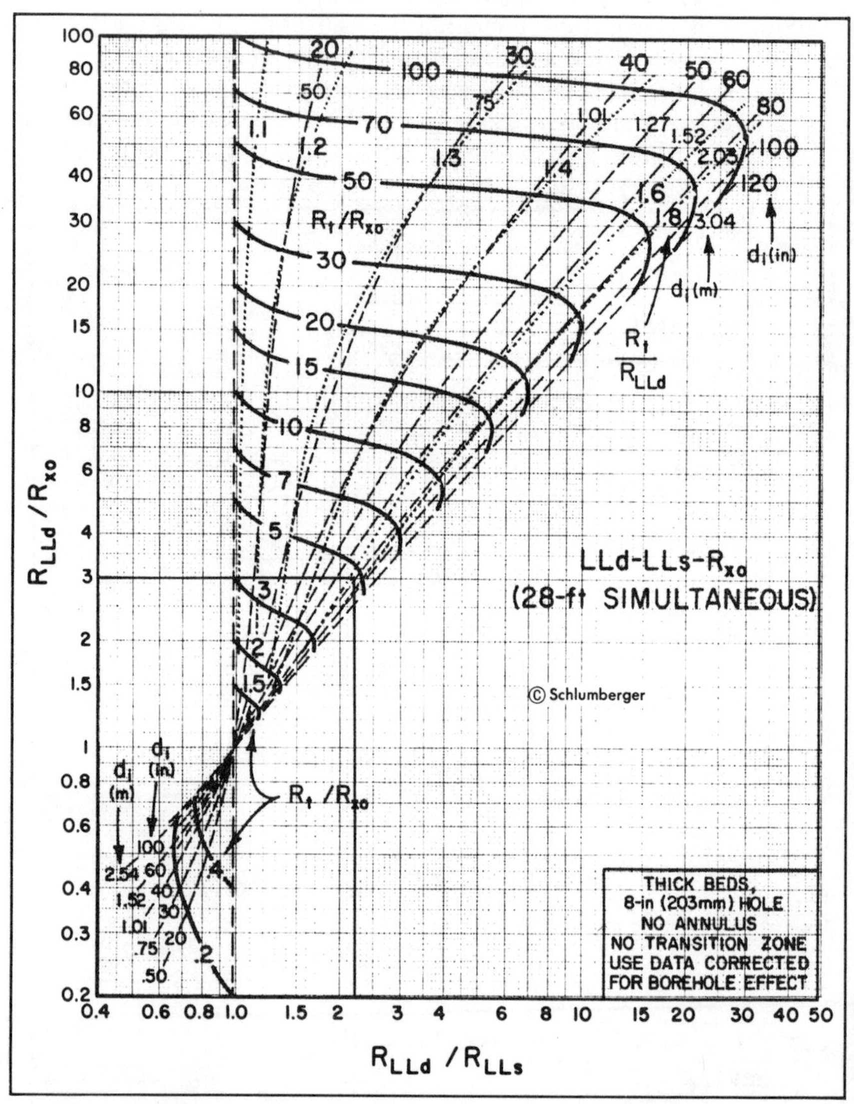

Fig. 4–18 Invasion correction chart for Dual Laterolog-Rxo (courtesy Schlumberger)

Experience in the industry has indicated that for average oil conditions

$$S_{xo} = S_w^{0.2} \tag{4.6}$$

Combining the two equations

$$S_w = \left[\frac{R_{xo}/R_t}{R_{mf}/R_w} \right]^{5/8} \tag{4.7}$$

Consequently, if the ratio R_{mf}/R_w is known, water saturation can be estimated directly.

Normally R_{mf} must be obtained from mud filtrate measurement and R_w needs to be known for the interval in question. However, even that is not necessary if there is an obvious water-bearing interval. In the case of Fig. 4–19, Zone 1 is water bearing, as indicated by the low value of R_{LLd}. Also no invasion correction is necessary since $R_{LLd} = R_{LLs}$. Eq. 4.5 then gives for this zone, since $S_w = S_{xo} = 1$

$$R_{mf}/R_w = R_{xo}/R_t = 0.45/0.3 = 1.5$$

Consequently, applying Eq. 4.7 for the center of Zone 2

$$S_w = [(2/650)/1.5]^{5/8} = 0.02$$

For Zone 3

$$S_w = [(7/140)/1.5]^{5/8} = 0.12$$

These values are only approximate but it is clear the two zones in question are hydrocarbon-saturated.

Use of this technique provides a little extra time for completion planning at the wellsite while the porosity logs are being run. More accurate saturation values should be computed using the Archie equation 2.8 as soon as the porosity data is obtained.

The ratio S_w/S_{xo} as obtained from Eq. 4.5 is also useful as a movable hydrocarbon indicator. If the ratio approaches unity, the hydrocarbon is immovable, regardless of the absolute values of S_w and S_{xo}; this can happen with very heavy oil. Conversely, if it is low, whatever hydrocarbon exists in the reservoir is certainly producible. Empirical guidelines which have been established for the Permian Basin of West Texas are[15]

S_w/S_{xo} less than 0.6—hydrocarbon production
greater than 0.8—water production
between 0.6 and 0.8—production test required

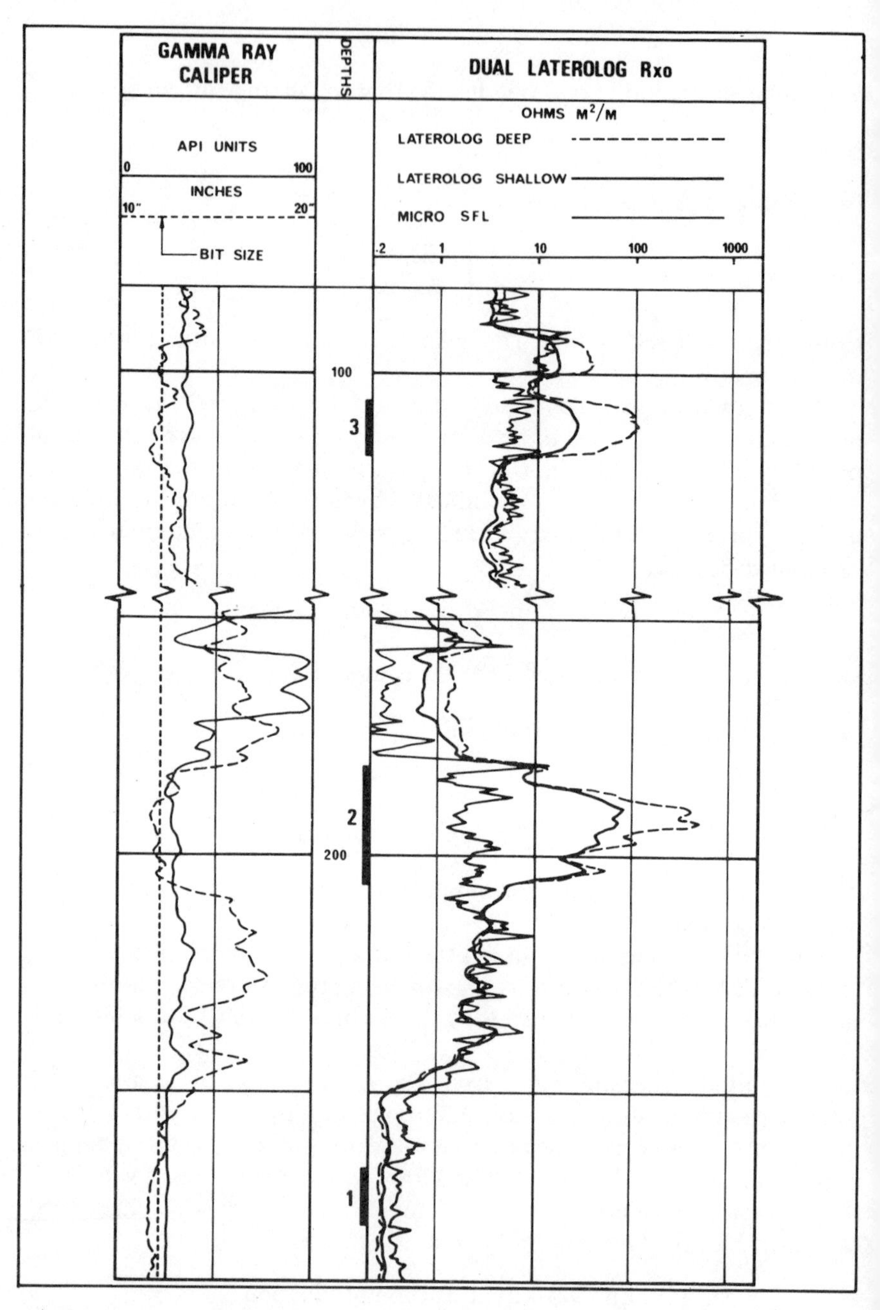

Fig. 4–19 Dual Laterolog—R_{xo} log in large hole, high resistivity conditions (courtesy Schlumberger)

For Zones 1 and 3 of Fig. 4–19, Eq. 4.5 gives S_w/S_{xo} values of 0.04 and 0.18, respectively. Therefore, excellent production would be expected from these zones.

In cases of old wells where resistivity but not porosity logs are available from log libraries, the resistivity ratio method is the only technique available for estimating water saturation. For best results a true R_{xo} curve, from MLL, PL, or MSFL, is required along with a deep resistivity curve. However, under certain circumstances a variation of the ratio method, wherein the SFL, LL-8 or 16 in. Normal curves are used in lieu of an R_{xo} log, has been used with success.[16,17]

Local experience may dictate an exponent other than 0.2 in Eq. 4.6; it may vary up to 0.5 in the case of gas-bearing or viscous oil-bearing formations where flushing action is poor. Sometimes, rather than using Eq. 4.6, an educated guess of S_{xo} is inserted directly in Eq. 4.5. Values from 0.8 for medium gravity oil to 0.6 for gas or heavy oil are typical.

SUMMARY

Modern resistivity combinations are Dual Induction-SFL in fresh mud with medium to low resistivity and Dual Laterolog-R_{xo} for salt mud or high formation resistivity.

DUAL INDUCTION LOGS

- Run when $R_{mf} > 2R_w$ and $R_t < 200$ ohm-m.
- IL_d reading normally close to R_t. Correction for invasion usually 0.7 to 1.0.
- Vertical resolution—4 ft for Induction, 1 ft for SFL. Logging speed—5,000–6,000 ft/hr. Should be run with 1.5 in. of standoff to minimize borehole effects.
- Can be recorded in oil-base (nonconductive) mud or in empty hole. No shallow curve obtained.

DUAL LATEROLOGS

- Run when $R_{mf} < 2R_w$ or when $R_t > 200$ ohm-m.
- LL_d curve should be corrected for invasion to obtain R_t, even for wellsite interpretation; correction factors can be as high as 2.
- Vertical resolution—2 ft for Laterolog, < 1 ft for MSFL. Logging speed—5,000–6,000 ft/hr.
- Quick-look hydrocarbon saturation and movable oil indication can be obtained.

REFERENCES

[1] S.J. Pirson, *Handbook of Well Log Analysis* (Englewood Cliffs, NJ: Prentice-Hall Inc., 1963).

[2] SPWLA, Houston Chapter, *The Art of Ancient Log Analysis*, (Houston: SPWLA, 1979).

[3] D.W. Hilchie, *Old Electrical Log Interpretation*, (Golden, CO: D.W. Hilchie Inc., 1979).

[4] M.P. Tixier, R.P. Alger, and D.R. Tanguy, "New Developments in Induction and Sonic Logging," *SPE 1300-G*, Dallas (October 1959).

[5] N.A. Schuster, J.P. Badon, and E.R. Robbins, "Application of the ISF/Sonic Combination Tool to Gulf Coast Formations," *Transactions of Gulf Coast Association of Geological Societies* (1971).

[6] R.P. Alger, W.P. Biggs, and B.N. Carpenter, "Dual Induction-Laterolog: A New Tool for Resistivity Analysis," *SPE 713* New Orleans (October 1963).

[7] H.G. Doll, "The Microlog," *Trans. AIME*, Vol. 189 (1950).

[8] H.G. Doll, "The Laterolog," *Jour. Pet. Tech.* (November 1951).

[9] H.G. Doll, "The Microlaterolog ," *Jour. Pet. Tech.* (January 1953).

[10] J. Suau, P. Grimaldi, A. Poupon, and P. Souhaite, "The Dual Laterolog — R_{xo} Tool," *SPE 4018*, San Antonio (October 1972).

[11] P. Souhaite, A. Misk, and A. Poupon, "R_t Determination in the Eastern Hemisphere," *SPWLA Logging Symposium Trans.* (June 1975).

[12] H.G. Doll, "Introduction to Induction Logging," *Jour. Pet. Tech.* (June 1949).

[13] J.H. Moran and K.S. Kunz, "Basic Theory of Induction Logging," *Geophysics* (December 1962).

[14] Schlumberger, *Log Interpretation Principles* (1972).

[15] G. Horst and L. Creagar, "Progress Report on the Interpretation of the Dual Laterolog—R_{xo} Tool in the Permian Basin," *SPWLA Logging Symposium Transactions* (June 1974).

[16] M.P. Tixier, "Electric Log Analysis in the Rocky Mountains," *Oil & Gas Journal* (June 23, 1949).

[17] Schlumberger, *Log Interpretation Charts* (1979).

Chapter 5

POROSITY LOGS

Returning to the basic log interpretation equation

$$S_w = c\sqrt{R_w/R_t}/\phi \tag{5.1}$$

we see that porosity is the third and final input needed to calculate water saturation. An accurate value of porosity is required since any error in its value will translate to the same percentage error in water saturation. The error will be magnified in calculating hydrocarbon volume, $\phi(1 - S_w)$. For example, if ϕ is too low by 10%, S_w will be too high by 10% and hydrocarbon volume will be too low by 20% (for a water saturation of 50%).

There are three porosity-measuring tools in common use at the present time — Density, Neutron, and Sonic. Why three when only one value is needed? It is because all three tools respond not only to porosity but also to the type of rock matrix and to the make-up of fluid filling the pore space — principally whether it contains gas. When the matrix is known and pore fluid is all liquid, one measurement may suffice. In other cases all three measurements are needed to sort out the parameters.

THE CURRENT TREND IN POROSITY LOGGING

For many years the Sonic was the popular porosity tool. It was less sensitive to borehole and mud-cake variations than early Density and Neutron tools. It could be run in combination with the Induction, giving both R_t and ϕ values for Eq. 5.1. In shaly sands R_t would be abnormally low and ϕ would be abnormally high, providing a compensating effect such that Eq. 5.1 would give reasonable water saturation values even under these conditions. Porosity values, however, would be optimistic.

In recent years the Density-Neutron combination has become the primary source of porosity information, displacing the Sonic. There are several reasons:

- Porosity can be determined without precise knowledge of rock matrix.
- There is no need for the compaction correction required with Sonic porosity.
- Overlay of Density-Neutron curves is an excellent gas indicator.

- Transitions from one type of rock matrix to another can often be distinguished.
- Shale effects are more evident and can be accounted for more precisely.

Consequently, the Sonic is becoming a backup porosity tool to be used where the hole is very irregular, where secondary porosity is important, or where heavy minerals such as pyrite adversely affect the Density. It is also required when a synthetic seismogram will be generated (from Sonic and Density logs) for depth calibration of seismic sections. This is, in fact, the application for which the Sonic tool was originally designed.

RECENT DEVELOPMENTS

The porosity tools in common use are the Compensated Density, the Compensated Neutron, and the Compensated Sonic instruments. These represent second-generation (third, in the case of the Neutron) tools much less sensitive to borehole and mud-cake effects than their uncompensated predecessors. They have been in use for the last 10 years.

In each case, however, there is a new generation of tool coming into use: the Litho-Density, the Dual Porosity Neutron, and the Long-Spaced Sonic. These tools provide new and significant information about the formation. The Litho-Density indicates the type of rock matrix — whether sand, limestone, or dolomite, for example. The Dual Porosity neutron gives better gas indication in shaly formations and probably will provide more accurate porosity determination in tight formations. The Long-Spaced Sonic can yield shear wave velocity, which is important in determining mechanical properties of the formation; it is also the preferred Sonic device when holes are large and altered shales exist.

In addition, a new and different porosity curve is entering the picture. It is the Electromagnetic Propagation log. In essence it gives water-filled porosity in the formation very close to the borehole. Comparison with total fluid porosity derived from the other porosity curves yields water saturation in the flushed zone without the benefit of resistivity logs. Consequently, this promises to be an excellent device for oil detection in freshwater areas where electrical logs are not definitive. It also can provide, along with standard logs, movable oil indication in fresh mud areas.

The principles and responses of these new tools, as well as the current ones, are discussed in the following sections. It appears that the Litho-

Density will eventually displace the Compensated Density, and that the Long-Spaced Sonic and Electromagnetic Propagation log will find good use. However, it is a little early to speculate on the impact of the Dual Porosity Neutron.

COMPENSATED DENSITY AND LITHO-DENSITY LOGS

The Density tool senses formation density by measuring the attentuation of gamma rays between a source and a detector.[1] Fig. 5–1 shows the arrangement of the Compensated Density tool. A source and two detectors are situated on a pad about 3 ft long that is forced against the side of the borehole with a backup arm.

Gamma rays emitted continuously by the source (typically 0.66 mev energy, from Cs^{137}) are channeled into the formation. There they undergo multiple collisions with electrons that cause them to lose energy and scatter in all directions — a mechanism called *Compton scattering.* When their energies drop below about 0.1 mev, the gamma rays die by a process called *photoelectric absorption.* Compton scattering depends only on the electron density of the formation (the number of electrons per cc), which is closely related to bulk density. This is the basis of the standard density measurement. On the other hand, photoelectric absorption depends on both electron density and the average atomic number of the material making up the formation. This mechanism is utilized by the Litho-Density tool to indicate rock type.

We may visualize at any instant a gamma ray cloud with a radius of a foot or so surrounding the source. The size of the cloud depends primarily on the formation scattering properties, therefore on electron density. It shrinks and expands as the density varies. The greater the density, the smaller the cloud and vice versa. The population of the cloud, however, which consists mainly of very low-energy gamma rays, depends on the absorption properties of the formation. The greater the absorption coefficient, the smaller the population and vice versa.

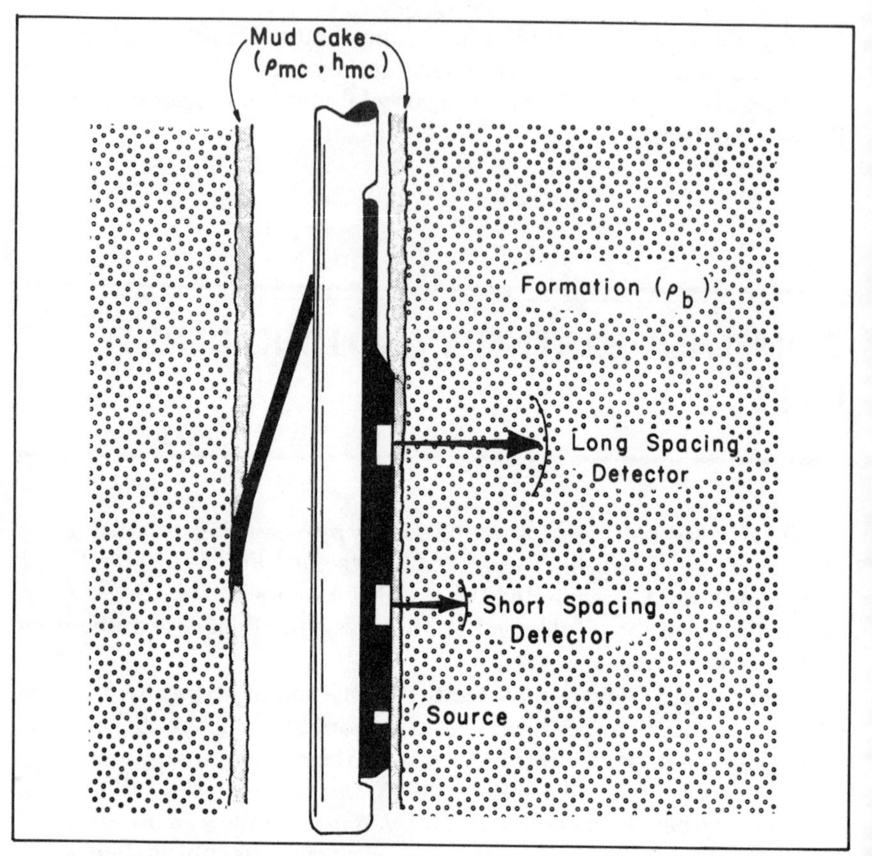

Fig. 5–1 Source-detector configuration of the Compensated Density tool (courtesy Schlumberger, © SPE-AIME)

THE COMPENSATED DENSITY TOOL*

The long-spacing detector, the primary one, generates a discrete electrical pulse for each gamma ray that happens to strike it. This detector is situated near the edge of the cloud, so it registers more pulses as the cloud expands and fewer as it contracts. It is shielded (as is the short-spacing detector) from the low-energy gamma rays and is insensitive to absorption properties of the formation. The net result is that the pulse rate depends only on electron density. Typically, it will decrease exponentially by a factor of

*Designated FDC by Schlumberger and CDL by Dresser Atlas, Welex, and Gearhart.

5–10 (depending on instrumentation details) as bulk density increases from 2.0 to 2.7 g/cc. This provides a sensitive measurement of density.[2]

The function of the short-spacing detector is to compensate for the effects of residual mud cake (not plowed away by the pad) and mud-filled hole rugosity interposed between the pad and the formation.[3] With normal mud these provide an easy path for gamma rays to channel upward from source to detector, leading to erroneously low density values if not corrected. First-generation Density tools with only single detectors suffered from this problem. The short-spacing detector provides a pulse rate that is also inversely proportional to density but which has a shallower depth of investigation than the long-spacing detector. It therefore gives greater weight to the mud cake and hole rugosity.

Pulse rates from the two detectors are sent to the surface and are combined in a computer using laboratory-derived "spine-and-ribs" response data, shown in Fig. 5–2, to provide two signals for log presentation. One is the corrected bulk density, ρ_b, obtained by extrapolating the response at a point such as O back to the spine, point P, following the direction of the ribs. The other is the correction, $\Delta\rho$, shown by the distance P-Q along the spine, which has been added to the basic long-spacing density, point Q, to eliminate the mud cake and rugosity effect.

Examination of Fig. 5–2 reveals two important points. First, $\Delta\rho$ corrections are positive for normal nonbarite mud cakes but can be negative for barite-loaded mud cakes. Such mud cakes can appear more dense than adjacent formations, in which case the density derived from the long-spacing rate is too high and must be corrected downward. The second point is that $\Delta\rho$ corrections are adequate up to about 0.15 g/cc but not beyond that point.

Log Presentation

Fig. 5–3 shows a typical Compensated Density presentation. The bulk density curve, ρ_b, is recorded over Tracks 2 and 3 on a linear scale from 2.0–3.0 g/cc. Densities typically vary from about 2.7–2.0 g/cc as porosity varies from 0–40%. The $\Delta\rho$ correction curve is in Track 3, with zero at the center and ± 0.25 g/cc at the extremes. The correction indicated has already been applied to the ρ_b curve; it is not necessary to add or subtract it again. The hole diameter, as measured by the backup arm, is presented in Track 1. If a Gamma Ray is run simultaneously, which is common, it is also recorded in Track 1.

In analyzing a Density log, it is best to first observe the $\Delta\rho$ recording. It is a quality-control curve. In smooth hole it should be close to the zero line, a

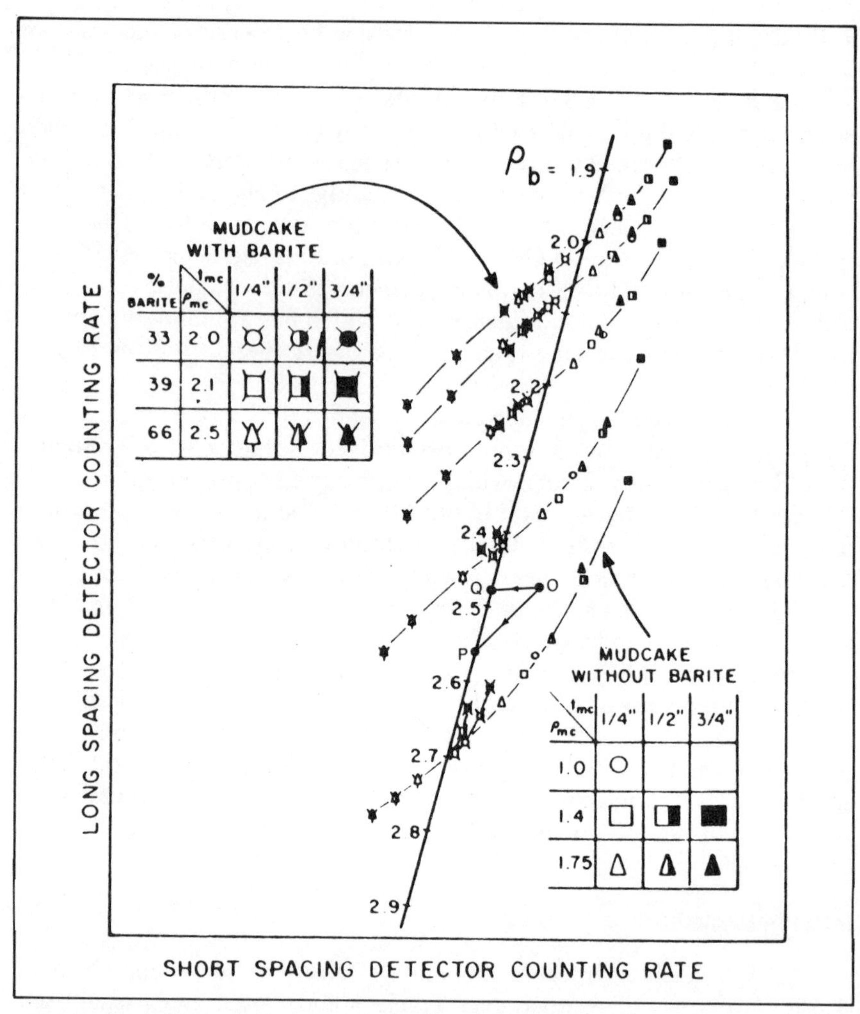

Fig. 5–2 Derivation of density and density correction from short and long-spacing detector rates (courtesy Schlumberger)

little to the right for normal (nonbarite) mud, and to the left for heavily loaded barite mud. When mud cake or hole rugosity is encountered, the $\Delta\rho$ correction will increase. As long as $\Delta\rho$ is less than 0.15 g/cc, the correction is adequate and the ρ_b curve can be trusted. Above 0.15 g/cc the correction is likely to be inadequate and the ρ_b curve in error. Corrections are barely adequate in the washed-out shales from 1,852–1,878 ft and

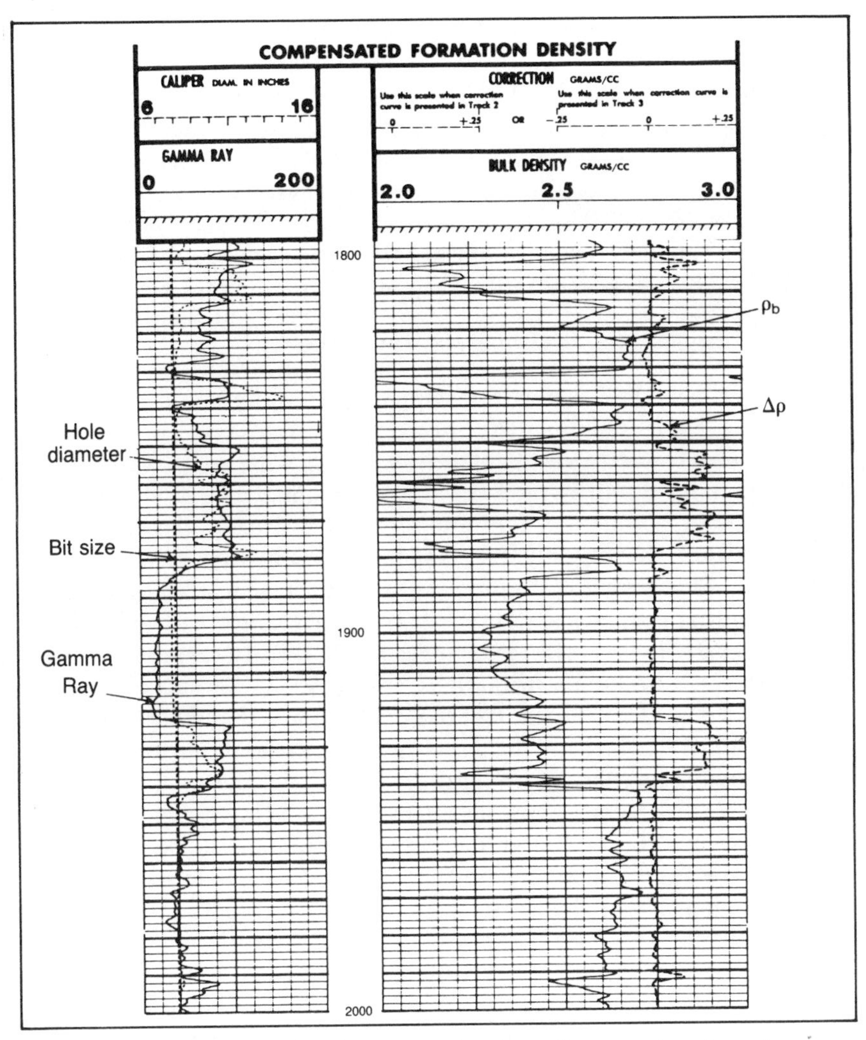

Fig. 5–3 Example of a Compensated Density recording (courtesy Schlumberger)

from 1,924–1,940 ft. The tool cannot compensate for short caves that the pad bridges if they are greater than about ¾ in. in depth. Correction will be inadequate and the ρ_b curve will read too low in normal mud. It also will not read correctly when the pad tilts on entering or leaving a sharp cave. A good example of these effects is at 1,832–1,834 ft where the low-density reading is clearly due to a short sharp hole washout but the $\Delta\rho$ curve shows almost no correction.

Because of the statistical fluctuations described in Chapter 3, the Density curve will not repeat exactly. Averages should agree, but the standard deviation between repeat runs will be about 0.04 g/cc at high density and about 0.02 g/cc at low density, with the normal averaging time of 2 sec and logging speed of 1,800 ft/hr. Nonrepeats may be aggravated by the fact that the tool can ride different sides of the hole on repeat runs. Where vugular porosity exists, formations may not be uniform around the hole.

Electron Density vs Bulk Density; Log Correction

As described, the Density tool responds to the electron density of the formation. The desired quantity, however, is bulk density. The two densities are related by the Z/A ratios of the elements making up the formation, Z being the atomic number and A the atomic weight of a given element. For all elements in sedimentary formations except hydrogen, the Z/A ratio is almost constant, varying only from 0.48 to 0.5. For hydrogen, however, the value is 1.0. Consequently, the presence of water or oil in formations significantly disturbs the usual proportionality between electron and bulk density.

The Density tool is calibrated to read bulk density correctly in fresh-water-saturated limestone formations, using test formations of precisely known bulk densities. The result of this calibration is that other formations will read a little incorrectly if their electron densities differ from those of limestone-water mixtures of like bulk densities.

The difference between true bulk density, ρ_b, and log-indicated density, ρ_{log}, is shown in Table 5–1 for various substances of interest. Differences are negligible for quartz, dolomite, and calcite. However, they are significant for beds such as sylvite, halite, gypsum, and anhydrite. The first two will record bulk densities about 0.12 g/cc lower than their true densities, and the last two will record about 0.02 g/cc higher than true densities. When these discrepancies are seen on a log, they should not be judged faulty tool operation; they are the result of the particular calibration choice.

Fig. 5–4 gives the corrections to be applied to log readings to obtain correct bulk densities. Corrections are zero or negligible in liquid-saturated limestone, sandstone and dolomite formations. For gas-saturated formations, however, log-indicated densities will be slightly low (such formations having lower electron densities than water-saturated formations of the same bulk densities). The worst case is that of very low-pressure gas or air in the pores at a porosity of 40%; the required correction is about 0.075 g/cc at 1.6 g/cc, or almost 5%. The correction becomes proportionately smaller as gas

TABLE 5-1 COMPARISON OF ACTUAL AND LOG-INDICATED DENSITIES

Compound	Formula	Actual Density, ρ_b, g/cc	Log-Indicated Density, ρ_{log}, g/cc
Quartz	SiO_2	2.654	2.648
Calcite	$CaCO_3$	2.710	2.710
Dolomite	$CaCO_3MgCO_3$	2.870	2.876
Anhydrite	$CaSO_4$	2.960	2.977
Sylvite	KCl	1.984	1.863
Halite	NaCl	2.165	2.032
Gypsum	$CaSO_42H_2O$	2.320	2.351
Anthracite coal		1.400 1.800	1.355 1.796
Bituminous coal		1.200 1.500	1.173 1.514
Fresh water	H_2O	1.000	1.00
Salt water	200,000 ppm	1.146	1.135
Oil	$n(CH_2)$	0.850	0.850
Methane	CH_4	ρ_{meth}	1.335 ρ_{meth}—0.188
Gas	$C_{1.1}H_{4.2}$	ρ_g	1.325 ρ_g—0.188

Source: Schlumberger *Log Interpretation/Principles*

density increases and gas saturation decreases in the formation pores. In normal logging no corrections are needed; the tool reads bulk density directly.

Fig. 5–4 also shows the corrections required for magnesium and aluminum. These points are significant because blocks of these materials are used as secondary calibration standards in field locations.

Depth of Penetration and Vertical Resolution

The 90% depth of investigation of the Compensated Density log is approximately 4 in. from the borehole wall at mid densities, slightly greater at lower densities, and slightly less at high densities. This means the log senses the flushed zone, which contains mud filtrate and possibly residual hydrocarbon in the pores. There is usually insufficient difference in density between water and oil for the Density to sense residual oil in the flushed zone. On the other hand, it can readily sense residual gas, especially if porosity is high and gas pressure is low.

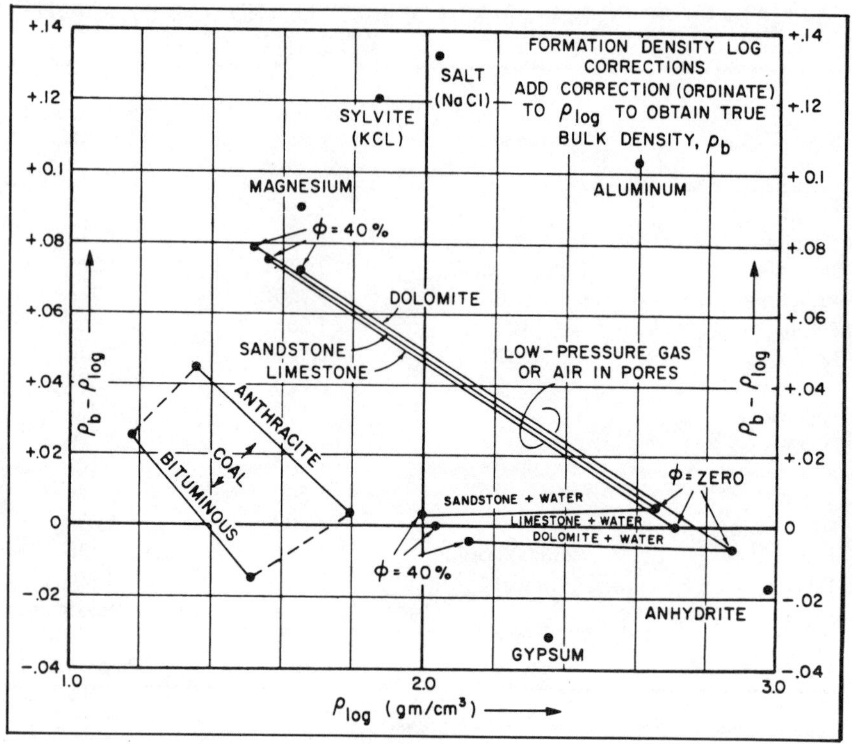

Fig. 5–4 Corrections to obtain true bulk density from log density (courtesy *Geophysics*, reprinted by Schlumberger)

The vertical resolution of the tool, if run very slowly, is approximately 1.5 ft. Formation density is averaged over that interval. However, with the usual averaging time and logging speed, bed resolution is about 3 ft (see Chapter 3).

Borehole Effects

In fluid-filled holes the Density tool response is independent of borehole size in the range of 6–9 in. This is a result of good shielding on the back side of the pad. In larger holes 0.005 g/cc should be added to the log reading for each inch of hole diameter above 9 in. for best accuracy.

The Density tool works quite well in empty holes. The response is not the same as that shown in Fig. 5–2, but the appropriate data is utilized in the surface computation of density. Tool response is independent of hole diameter in the range of 6–9 in.; above that point 0.01 g/cc should be added to the log reading for each additional inch of hole diameter. Mud cake is not generally a problem in empty holes, which are usually air drilled. However, hole rugosity may be a problem since extremely low-density material (air) is interposed between the pad and the formation. Much less rugosity can be tolerated in air-filled holes than in liquid-filled holes.

POROSITY DERIVATION FROM THE DENSITY LOG

Porosity is derived from bulk density in a very straightforward manner. For a clean formation with matrix (or grain) density ρ_{ma}, fluid density ρ_f, and porosity ϕ, the bulk density, ρ_b, is given by the summation of fluid and matrix components

$$\rho_b = \phi \cdot \rho_f + (1 - \phi)\rho_{ma} \tag{5.2}$$

from which porosity is given by

$$\phi = (\rho_{ma} - \rho_b)/(\rho_{ma} - \rho_f) \tag{5.3}$$

Matrix densities in g/cc typically are

$$\begin{aligned} \rho_{ma} &= 2.65 \text{ for sands, sandstones, and quartzites} \\ &= 2.68 \text{ for limey sands or sandy limes} \\ &= 2.71 \text{ for limestones} \\ &= 2.87 \text{ for dolomites} \end{aligned}$$

In liquid-bearing formations fluid density is typically that of the mud filtrate

$$\begin{aligned} \rho_f &= 1.0 \text{ for fresh mud} \\ &= 1.0 + 0.73\text{ N for salt mud} \end{aligned}$$

where N is the sodium chloride concentration in ppm $\times 10^{-6}$.

Porosity may be derived from Fig. 5–5, which provides a graphical solution to Eq. 5.3. Bulk density is entered on the bottom scale and porosity is read on the vertical scale for appropriate values of ρ_{ma} and ρ_f. As an example, consider the interval 1,899–1,905 ft in Fig. 5–3 where the log density averages 2.29 g/cc. Assuming the formation is limestone and the fluid density is 1.0 (fresh-mud filtrate), the derived porosity from Eq. 5.3 or Fig. 5–4 is 24.5%.

It is more important to know the precise matrix density at low porosity than at high porosity. For example, at $\rho_b = 2.6$ g/cc, derived porosities would be 3% for sand and 6% for limestone. These differ by a factor of 2 and could mean the difference between expecting commercial and noncommercial production since a cutoff is often set around 5%. On the other hand,

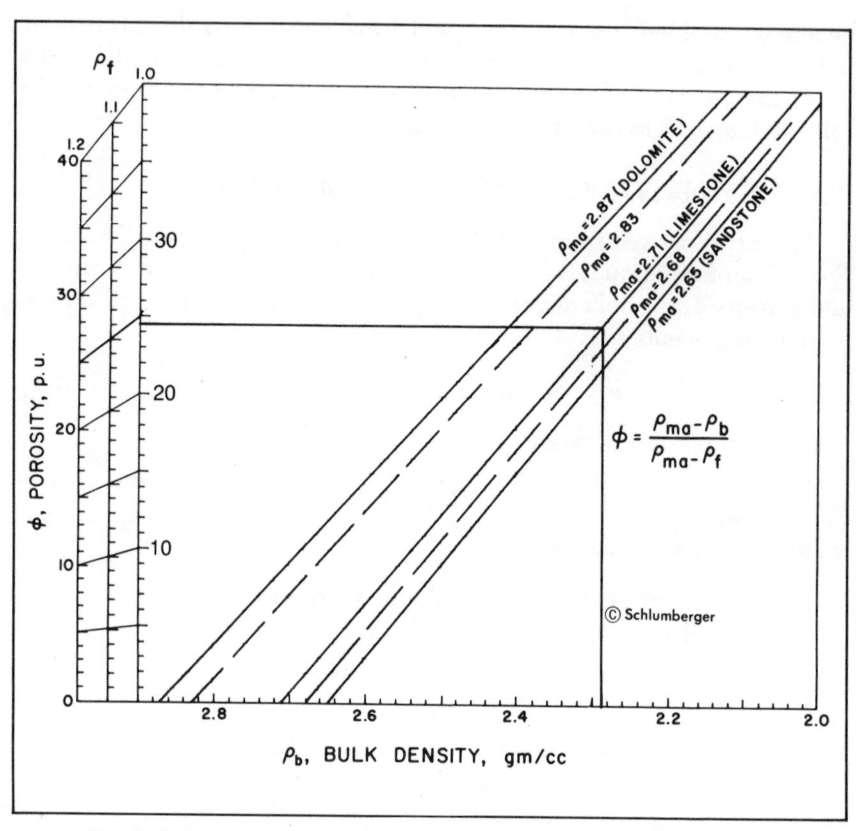

Fig. 5–5 Determination of porosity from bulk density (courtesy Schlumberger)

at $\rho_b = 2.2$ g/cc, derived porosities would be 27% and 30%, which differ only by 10%.

Many logs today have density-derived porosity curves recorded as the log is run. To effect this the logging engineer must insert values of matrix and fluid densities into the surface computer, which is continuously solving Eq. 5.3. Normal choices for matrix density are 2.65 (SS), 2.68, or 2.71 (LS); those for fluid density are 1.0 and 1.1. The logging engineer chooses values generally applicable to the area. It is important that these values be shown on the log heading since the reading must often be corrected to a different matrix value more appropriate for the particular formation being analyzed. Examples of Density porosity curves, overlain with Neutron porosity, are given later.

Heavy minerals in the formation such as pyrite (FeS_2) increase the effective matrix density and cause derived porosity to be too low if not taken into account. Occurrence is not frequent but is important in a few areas, particularly Alaska and the North Sea.[4]

Effect Of Gas

As described in chapter 1, considerable gas can be left in the flushed zone of a gas-bearing formation, bypassed by the invading filtrate. The density of the pore fluid can then be considerably less than one. Consequently, in gas-bearing formations there is a dual dilemma if the Density log is the only porosity curve run. First is recognizing that there is gas present, since the curve simply shows a decrease in bulk density that would normally be interpreted as an increase in fluid-filled porosity. Second is determining the correct porosity. It is not a straightforward matter. To apply Eq. 5.3, the fluid density, ρ_f, in the zone of investigation must be known. This depends on the water saturation in the invaded zone, S_{xo}, the mud filtrate density, ρ_{mf}, and the density, p_h, of the gas in the pores. That is

$$\rho_f = \rho_{mf} \cdot S_{xo} + \rho_h (1 - S_{xo}) \tag{5.4}$$

Gas density can be estimated from Fig. 5–6. However, S_{xo} is not known beforehand. If an R_{xo} curve is available, then

$$S_{xo} = c\sqrt{R_{mf}/R_{xo}}/\phi \tag{5.5}$$

If an R_{xo} curve is not available (usually the case with fresh mud), one can make an assumption such as

$$S_{xo} = S_w^{1/2} = [c\sqrt{R_w/R_t}/\phi]^{1/2} \tag{5.6}$$

Eqs. 5.3, 5.4, and 5.5 or 5.6 can be solved simultaneously or iteratively to give an apparent porosity ϕ_a. Allowing for the electron density effect gives a final porosity

$$\phi = \phi_a(0.93 + 0.07\rho_f) \tag{5.7}$$

As an example, consider the same interval (1,899–1,905 ft) in Fig. 5–3. Assume it is known to contain gas and that the electric logs give $R_w = 0.05$ and $R_t = 40$ ohm-m for this interval. From Fig. 5–6, $\rho_h = 0.07$ g/cc. Following the procedure outlined using Eq. 5.6, $\phi = 17.5\%$. This porosity value differs significantly from the 24.5% found on the assumption of 100% liquid saturation.

This procedure is cumbersome and inaccurate, and it is rarely used. The Density log really needs outside help to establish matrix type, identify gas,

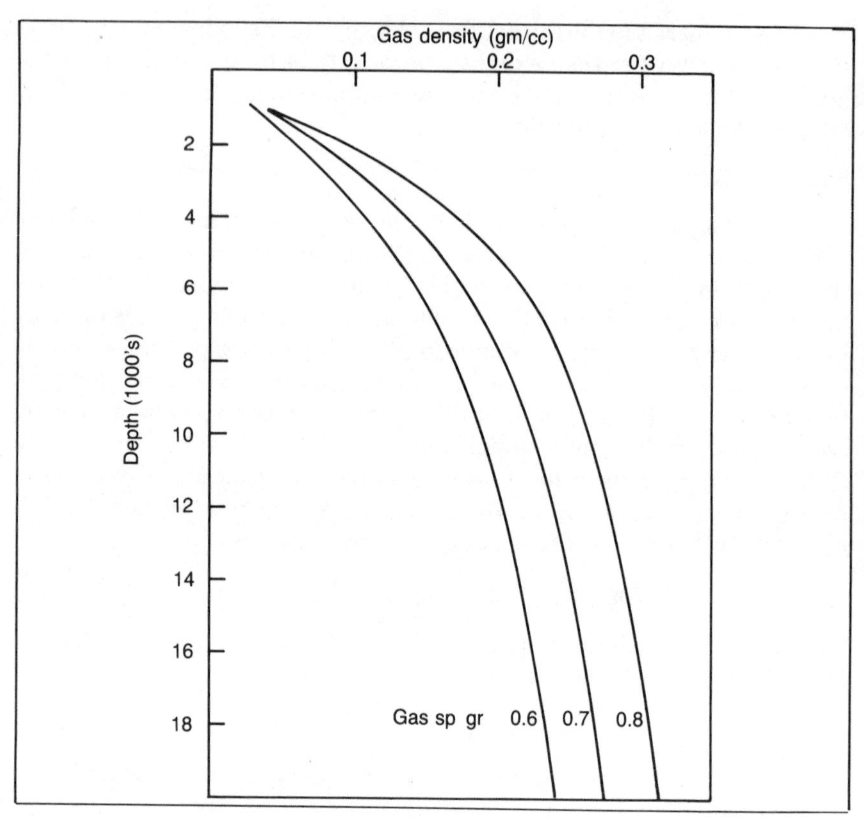

Fig. 5–6 Estimation of gas density (from *Applied Openhole Log Interpretation,* courtesy D.W. Hilchie)

and simplify porosity determination when gas is present. As we shall see, the Neutron log fulfills this requirement admirably.

THE LITHO-DENSITY LOG

The Litho-Density tool (LDT)* is a third-generation density instrument that provides in addition to the bulk density log, ρ_b, a photoelectric absorption curve, P_e.[5] This curve reflects the average atomic number of the formation and is therefore a good indicator of the type of rock matrix. It is helpful in complex lithology interpretation.

*Offered only by Schlumberger at present and designated LDT.

Measurement of P_e

The source-detector arrangement of the LDT tool is basically the same as that of its predecessor, the FDC (Fig. 5–1). The operation, however, is different. With the LDT, ρ_b and P_e measurements are made by energy selection of the gamma rays that reach the long-spacing detector. This is shown in Fig. 5–7, which is a plot of the number of gamma rays reaching the detector, as a function of their energy, for three formations having the same bulk density but different volumetric absorption indices, U, designated low, medium, and high.

The basic density measurement is made by registering only those gamma rays that fall in the high-energy region, designated H. In this range only scattering of the gamma rays is taking place and the number of gamma rays, represented by the area under the curve, depends on electron density only. Conversion of pulse rate to bulk density and correction for mud cake and rugosity is carried out in the same manner as for the Compensated Density tool. Statistical fluctuations in computed density, however, are reduced by a factor of about 2, to the range 0.01 to 0.02 g/cc, by utilization of more efficient detectors.

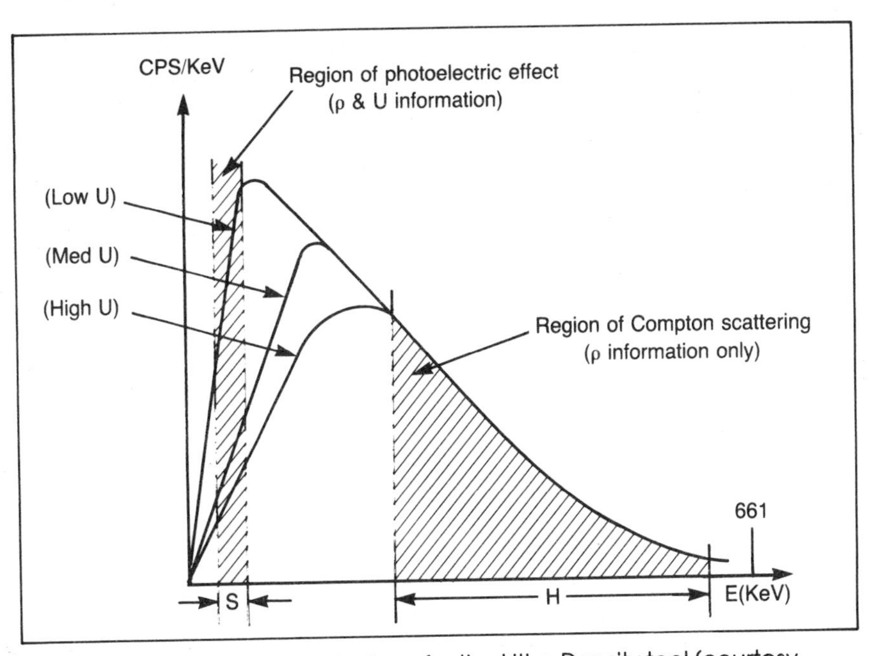

Fig. 5–7 Detection windows for the Litho-Density tool (courtesy Schlumberger and SPWLA)

The photoelectric measurement, P_e, is made by registering those gamma rays that fall in the energy window, S, positioned at very low energy. In this region gamma rays undergo photoelectric absorption as they interact with the electrons present. The absorption rate depends on the product of the absorption coefficient per electron, P_e, and the electron density, ρ_e. The pulse rate in the counting window therefore responds to a photoelectric absorption index given by

$$U = P_e \cdot \rho_e \tag{5.8}$$

The larger the value of U, the smaller the pulse rate and vice versa. With suitable calibration, the value of U for any given formation can be determined.

The electron density ρ_e* is related to bulk density (as a result of calibrating the latter in water-saturated limestone) by the relation

$$\rho_e = (\rho_b + 0.1883)/1.0704 \tag{5.9}$$

Consequently, from Eqs. 5.8 and 5.9

$$P_e = \frac{1.0704\ U}{\rho_b + 0.1883} \tag{5.10}$$

From the two independent measurements, U and ρ_b, the value of P_e is determined.

Dependence of P_e on Lithology

The parameter P_e reflects formation lithology because it is strongly dependent on the effective atomic number of the medium absorbing the gamma rays. For a single element of atomic number Z, P_e is given, in logging units (barns/electron, with 1 barn = 10^{-24} cm^2), by

$$P_e = (Z/10)^{3.6} \tag{5.11}$$

For a formation containing a number of elements, the effective P_e value is obtaining by summing the $(Z/10)^{3.6}$ values, after weighting each by its relative electron density in the mixture. Table 5–2, column 2, gives effective P_e values for common sedimentary materials. P_e values for quartz, calcite, and dolomite are quite distinct. Anhydrite and calcite have similar P_e values but are well separated in density (column 3). Minerals such as siderite

*ρ_e has units of 3.0×10^{23} electrons/cc.

($FeCO_3$) and pyrite (FeS_2) have considerably higher P_e values by virtue of the higher atomic number of iron ($Z = 26$). Barite has an extremely high P_e value; $Z = 56$ for barium. The consequences are discussed below.

Fig. 5–8 shows P_e values for limestone, dolomite, and sandstone formations of 0–35% porosity with pores containing either fresh water or methane of density 0.1 g/cc. Note that regardless of porosity or type of fluid, the P_e values for the three types of rock are well separated. Consequently, when only one matrix type is present in a formation, the P_e curve will unambiguously distinguish it.

Depth of Penetration and Vertical Resolution

Depth of penetration and vertical resolution for the ρ_b measurement should be essentially the same as for the Compensated Density. Similar values would be expected for the P_e measurment, though no verification has been published yet.

Borehole Effects

For the ρ_b curve, borehole effects are much the same as for the Compensated Density tool. Mud cake and rugosity corrections are made via the

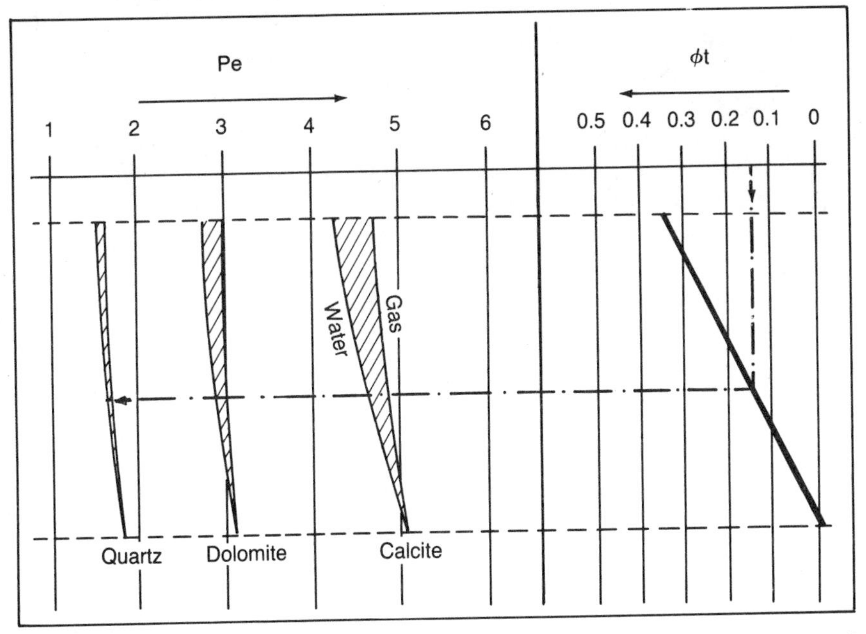

Fig. 5–8 Photoelectric absorption coefficient as a function of porosity and fluid type (courtesy Schlumberger and SPWLA)

TABLE 5-2 VALUES OF PHOTOELECTRIC ABSORPTION COEFFICIENT PER ELECTRON, P_e, AND PER CC, U, FOR VARIOUS SUBSTANCES

	P_e	Sp. gr.	$\rho_{b(log)}$	U
Quartz	1.81	2.65	2.64	4.78
Calcite	5.08	2.71	2.71	13.8
Dolomite	3.14	2.87	2.88	9.00
Anhydrite	5.05	2.96	2.98	14.9
Halite	4.65	2.17	2.04	9.68
Siderite	14.7	3.94	3.89	55.9
Pyrite	17.0	5.00	4.99	82.1
Barite	267	4.48	4.09	1065
Water (fresh)	0.358	1.00	1.00	0.398
Water (100K ppm NaCl)	0.734	1.06	1.05	0.850
Water (200K ppm NaCl)	1.12	1.12	1.11	1.36
Oil ($n(CH_2)$)	0.119	ρ_{oil}	$1.22\rho_{oil}-0.188$	$0.136\rho_{oil}$
Gas (CH_4)	0.095	ρ_{gas}	$1.33\rho_{gas}-0.188$	$0.119\rho_{gas}$

spine-and-ribs method. Corrections are adequate as long as $\Delta\rho$ is less than 0.15 g/cc. For the P_e curve the situation is less clear. No information has been published on whether the U measurement is compensated for mud cake and rugosity. These data are needed. Barite-loaded drilling mud presents a difficult problem for the Litho-Density log. If barite-loaded mud or mud cake intrudes between the pad and the formation, as is almost inevitable, the very high P_e value of barite swamps the formation value, rendering the P_e curve useless. This severely limits the utility of the LDT in areas where barite mud is used. Weighting the mud with iron compounds would cause much less effect on the P_e curve.

Litho-Density Example

An example of a Litho-Density log run simultaneously with Gamma Ray and Compensated Neutron is shown in Fig. 5–9. The ρ_b curve, converted to porosity using a limestone matrix density (2.71 g/cc), is recorded in Track 3 (solid). Overlain (dashed) is the Neutron porosity curve. The P_e curve is recorded in Track 2. Gamma Ray and Caliper logs are in Track 1. On the bottom of the example, the P_e ranges of sandstone, dolomite, and limestone are marked for the 0–12% porosity interval of interest. Clearly, the formation at level A is limestone and that at level C is almost pure sandstone. At

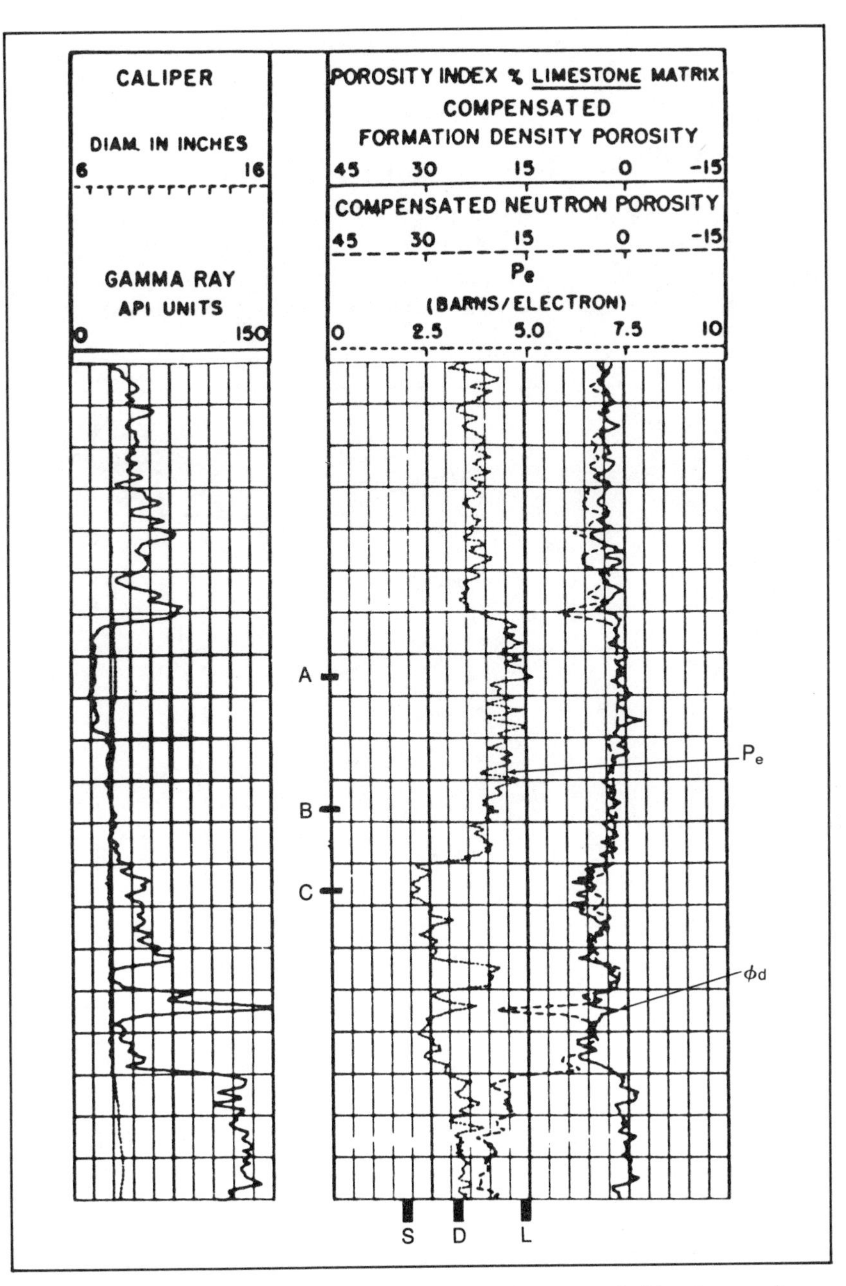

Fig. 5–9 Example of a Litho-Density log (courtesy Schlumberger)

level B, however, the matrix cannot be resolved from the P_e curve. It could be either a mixture of about 50% limestone and 50% dolomite or about 70% limestone and 30% sandstone. It could even be a combination of all three.

LITHOLOGY INTERPRETATION WITH ρ_b-P_e CURVES

If only two matrix minerals are present, the volumetric fractions of each, along with the porosity, can be derived from the combination of ρ_b and P_e values.[6,7] Fig. 5–10 shows the applicable chart. The log values of ρ_b and P_e are entered and a point of intersection is found. Selecting the almost-vertical

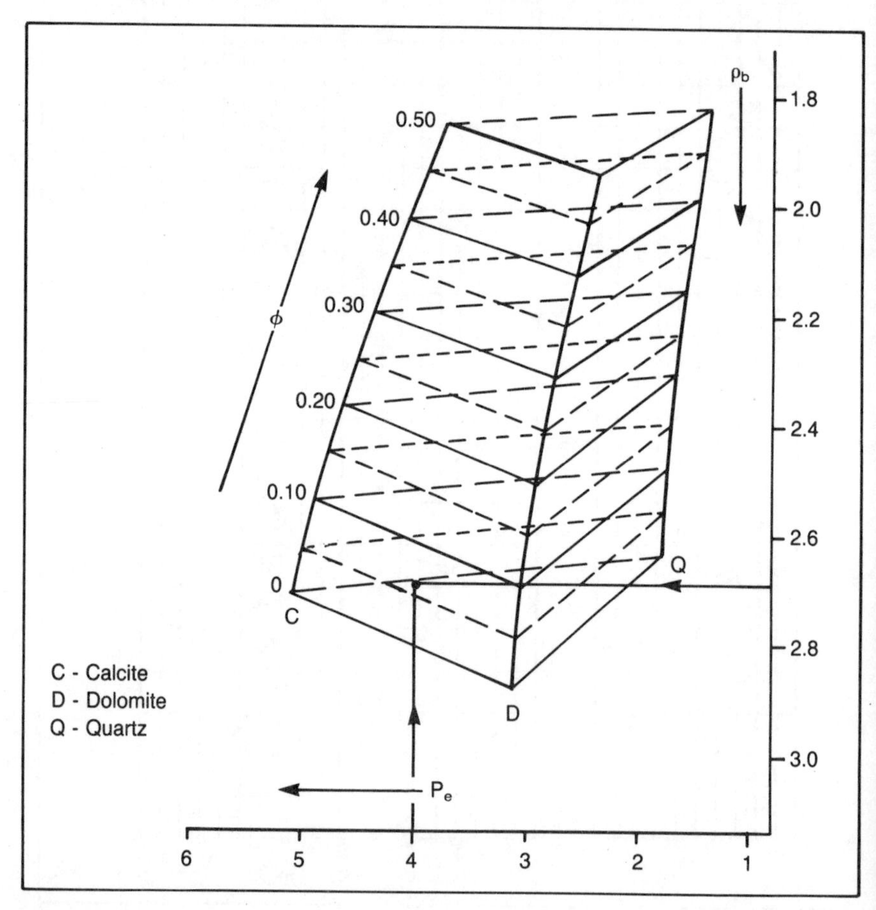

Fig. 5–10 Derivation of porosity and lithology for a two-component matrix (after Gardner and Dumanoir, Schlumberger and SPWLA)

stems representing the two minerals known to be present, porosity is deduced by interpolating between the equiporosity lines (at 5 porosity-unit intervals) joining the two stems. Matrix fraction is represented by the displacement of the point along the equiporosity line joining the stems.

As an example, consider point B of Fig. 5–9 and assume the matrix is a calcite-dolomite mixture. The log value of ρ_b is 2.68, corresponding to limestone-derived porosity of 2% (Fig. 5–5), and P_e is 4.0. Using the calcite-dolomite lines, porosity is found to be 6% and the matrix is a 50–50 mixture. If, however, a mixture of calcite and sandstone is assumed, porosity is 0% and the matrix is ⅔ calcite and ⅓ sandstone. Additional information is needed to resolve the question. Utilizing the Neutron curve shows the actual porosity is 2% and the matrix is about 60% calcite, 20% sandstone, and 20% dolomite. For this purpose the volumetric absorption coefficient, U, is used. Values of U are listed in Table 5–2. The procedure is described in chapter 6.

COMPENSATED NEUTRON AND DUAL POROSITY NEUTRON LOGS

In its simplest form a Neutron tool is illustrated in Fig. 5–11. Fast neutrons (~5 mev) are continuously emitted by the neutron source and travel out in all directions into the formation. As they progress, they are slowed or moderated by collisions with nuclei in their path. When they reach very low or thermal energies (~0.025 ev), they zigzag or diffuse aimlessly until they are absorbed or captured by the nuclei present.[8]

The element most effective in slowing the neutrons is hydrogen because a neutron and hydrogen nucleus have the same mass. In a direct collision the neutron will transfer all of its energy to the hydrogen nucleus and stop dead, as in a head-on billiard ball collision. On the other hand, other nuclei common to elements in sedimentary formations, such as those of silicon, calcium, carbon, and oxygen, are much more massive than neutrons. They are effective in scattering neutrons into different directions but absorb only a small fraction of the neutron energy even in a direct collision. Their effect on the neutron slowing-down process is much smaller than that of hydrogen, though not negligible.

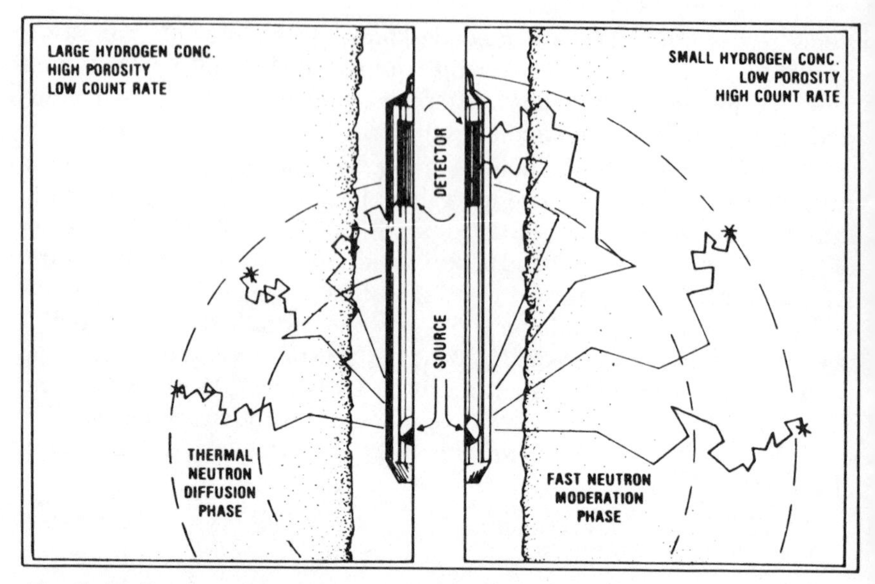

Fig. 5–11 Single detector Neutron tool in borehole environment (courtesy Welex)

The net result is that one can visualize at any instant a neutron cloud surrounding the source, extending a maximum of about 2 ft. As the hydrogen content of the formation varies, the size of the cloud expands and shrinks. The greater the hydrogen content, the smaller the cloud and vice versa. The population of the cloud, which consists mainly of thermal neutrons, is dependent on the absorbing qualities of the formation for such neutrons. The population at any instant is such that the number of thermal neutrons being absorbed each second equals the rate of fast neutron emission from the source. Consequently, the greater the absorption coefficient of the formation, the smaller the population and vice versa.

Situated toward the edge of the cloud is a detector that may be one of three types: a thermal neutron detector sensing the density of the lowest–energy neutrons in its vicinity, an epithermal detector sensing the density of neutrons just above thermal energy, or a capture gamma ray detector sensitive to the gamma rays produced by absorption of thermal neutrons in its vicinity. Regardless of the type, the pulse rate registered by the detector increases when the cloud expands (less hydrogen) and decreases when it contracts (more hydrogen). Pulse rate therefore varies inversely with porosity, since all of the hydrogen (in clean formations) is contained in the pore fluid.

NEUTRON TOOL EVOLUTION

First-generation Neutron tools, used extensively in the 1950s, were of the symmetric single-detector type shown in Fig. 5–11. They utilized either thermal neutron detectors, capture gamma ray detectors, or a combination of the two. These tools were very sensitive to borehole parameters, and conversion of log response to porosity was subject to considerable error. They have long since been superseded for open-hole logging, but a limited number still are used for correlation logging in cased hole.

The second-generation tool popular in the 1960s was the Sidewall Neutron,[9] a pad device very similar in configuration to the Density sonde.* This tool was much less sensitive to borehole parameters than its predecessor. Also it was insensitive to thermal neutron absorbers by virtue of using epithermal detection. However, depth of investigation was reduced, which resulted in more severe mud cake and hole rugosity effects and prohibited use of the tool in cased holes. For these reasons the Sidewall Neutron is now rarely used. Nevertheless, it remains the best Neutron device to run in air-filled holes, where it has better porosity response than other types of Neutron devices.

The third-generation tool, the Compensated Neutron introduced about 1970, is the current standard.[10] It utilizes a pair of neutron detectors (thermal) instead of a single one. This provides definite advantages.

A fourth-generation took, the Dual Porosity Neutron, is just being introduced. It uses a pair of thermal neutron detectors on one side of the source and a pair of epithermal detectors on the other. Each set provides a porosity curve. This tool will be discussed later.

THE COMPENSATED NEUTRON

The principle of the Compensated Neutron tool** is shown in Fig. 5–12. A fast neutron source is located near the bottom of the tool, and two thermal neutron detectors are spaced 1–2 ft above it. The ratio of the pulse rates from the near and far detectors, N_n/N_f, is measured and related to formation porosity. It has been proven, theoretically and experimentally, that ratio

*Designated SNP by Schlumberger, SWN by Dresser-Atlas and Welex, and SNL by Gearhart.

**Designated CNL by Schlumberger, CNLog by Dresser-Atlas, DSN by Welex, and CNS by Gearhart.

measurement significantly reduces borehole effects and increases the depth of penetration of the tool relative to a single detector measurment.[11] At the same time neutron absorption effects are reduced, though not eliminated.

The actual configuration of the logging instrument is shown in Fig. 5–13. The whole tool is decentralized by means of a spring, and the backside of the source-detector array facing the mud column is shielded as much as possible to minimize borehole effects.

The relationship between ratio and porosity for the Schlumberger CNL as determined by measurements in laboratory formations is shown in Fig. 5–14. The ratio increases with porosity because the thermal neutron density falls off more sharply with distance from the source as porosity increases, even though both counting rates decrease. (Ratio-porosity relationships differ among tool models, depending on detector size, placement, and shielding.)

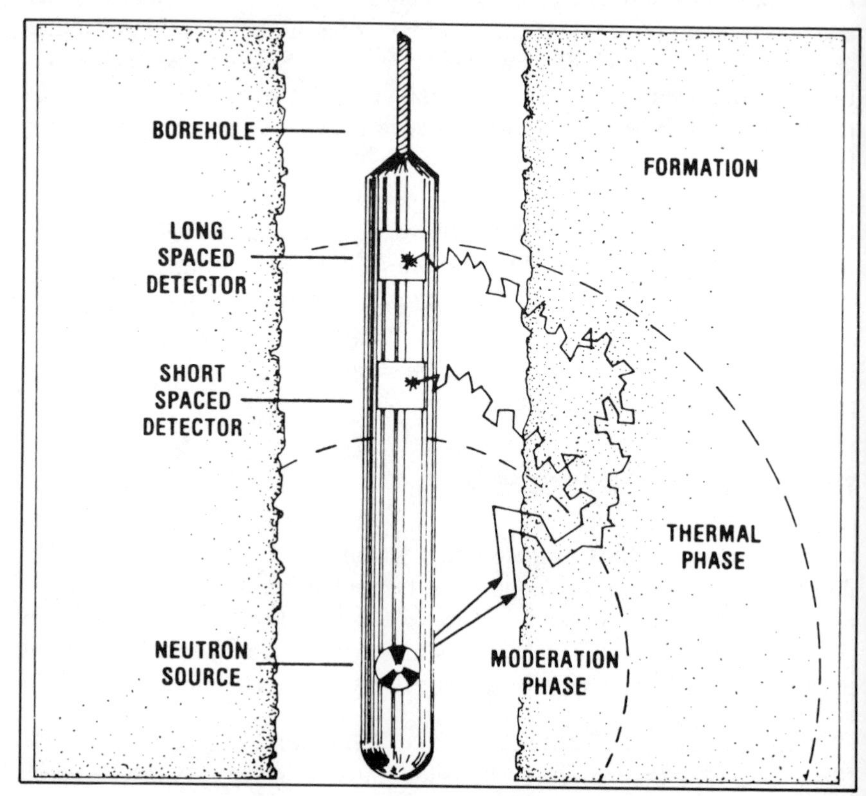

Fig. 5–12 Dual detector Neutron tool in a borehole environment (courtesy Welex)

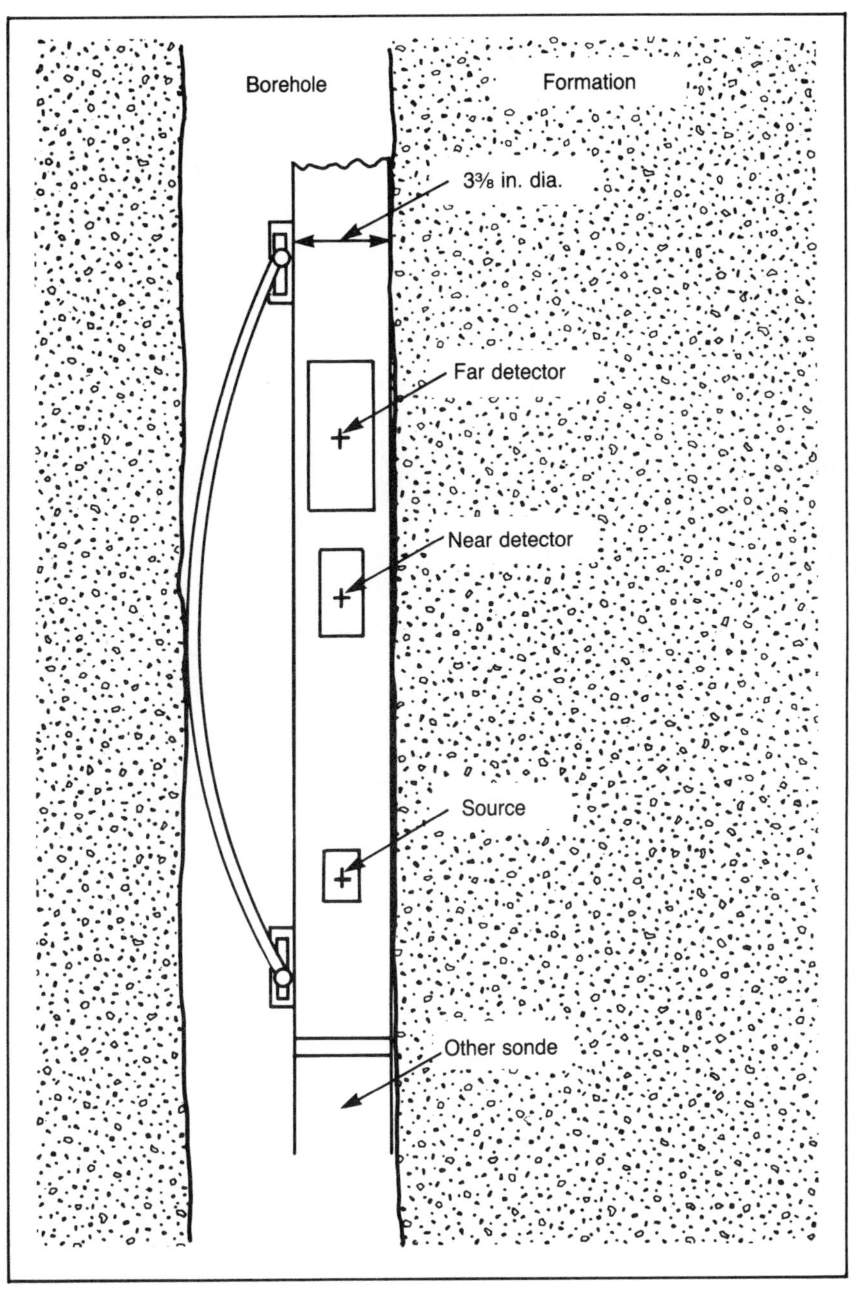

Fig. 5–13 Sketch of a CNL tool (courtesy Schlumberger, © SPE)

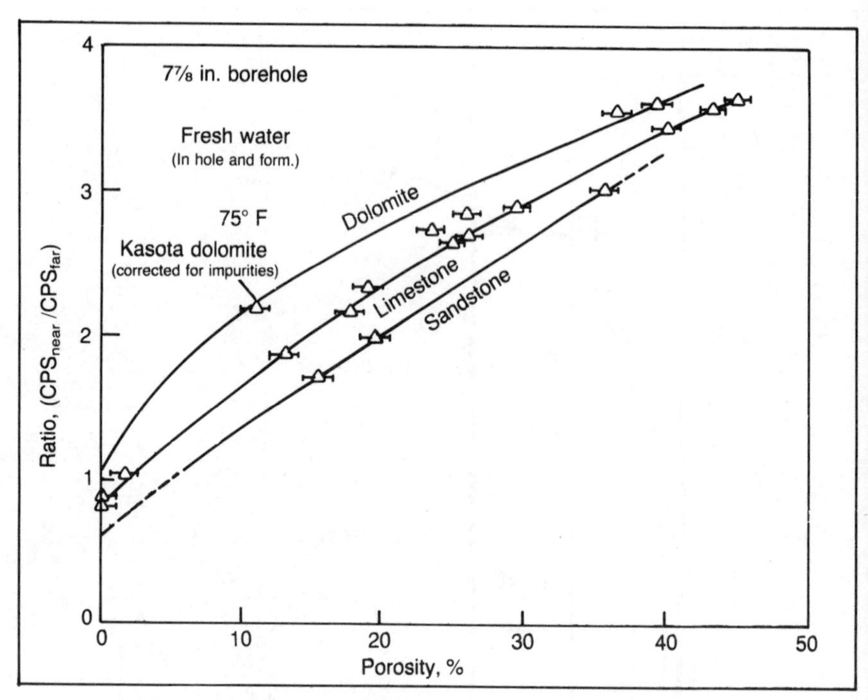

Fig. 5–14 CNL response in sandstone, limestone, and dolomite formations (courtesy Schlumberger, © SPE-AIME)

While the ratio depends primarily on porosity, there is also a signficant dependence on lithology because the matrix contributes some to the slowdown and capture of the neutrons. It is clear that to derive porosity from the ratio with any accuracy, the lithology must be known.

Porosity Equivalence

Examination of Fig. 5–14 leads to a useful concept, that of *equivalent porosities*. Equivalent porosities are obtained by reading the dolomite, limestone, and sandstone porosities corresponding to a given ratio. For example, at a ratio of 2.0 we read 8% porosity for dolomite, 15% for limestone, and 19.5% for sandstone. These are equivalent porosities. Loosely speaking, neutrons slowing down and thermalizing cannot tell whether they are in one or the other of these equivalent formations.

Plotting porosity equivalents obtained at different ratios as a function of the limestone porosity corresponding to a given ratio leads to the porosity-equivalence chart of Fig. 5–15 (dashed lines) for the CNL. Equivalent porosities are read vertically. For example a limestone porosity of 14% is equivalent to a dolomite porosity of 7% or a sandstone porosity of 18% (lines A).

Porosity equivalents for the Sidewall Neutron (SNP) are also shown on Fig. 5–15 (solid lines). Matrix effects are less than for the CNL because epithermal detection eliminates neutron absorption effects that contribute partially to the lithology differences. Because of this, some operators still prefer the Sidewall Neutron over the Compensated Neutron at low porosities. In particular there is some uncertainty in the CNL response to very low-porosity dolomites apparently because of thermal neutron absorbers that are sometimes present in these formations in trace quantities.[12,13]

When a Compensated Neutron log is run, the ratio is not recorded. Rather, the ratio is transformed to porosity, on the basis of laboratory data

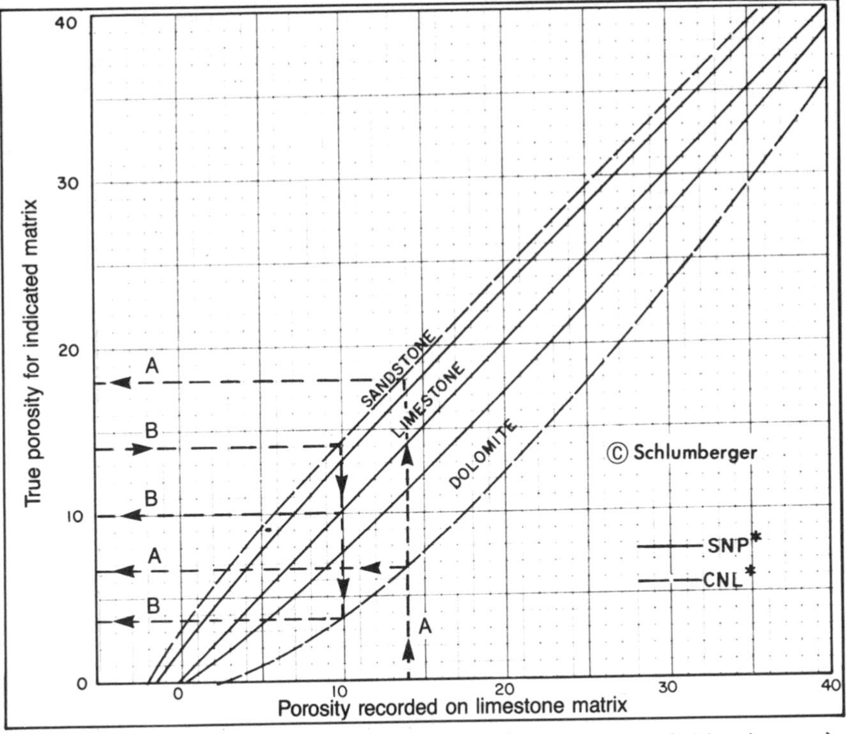

Fig. 5–15 Neutron porosity equivalence curves (courtesy Schlumberger)

such as that of Fig. 5–14, in a surface computer and a porosity curve is recorded. To effect the transformation, the logging engineer must input to the computer which matrix to use. He has a choice of limestone (LS) or sandstone (SS). The value chosen is the one most appropriate for the area and is shown on the log heading. It is usually left constant over the whole log even though the matrix may vary in intervals.

Depth of Investigation and Vertical Resolution

Fig. 5–16 shows the depth of investigation of the Compensated Neutron (CNL) tool in open hole at 22% porosity. For comparison those of the Sidewall Neutron (SNP) and Compensated Density (FDC) tools are also shown. None of the tools penetrates very deeply; but of the three the CNL has the greatest depth of investigation. It obtains 90% of its response from the first 10 in. of formation compared to 7 in. for the SNP and 4 in. for the FDC. Just as significant in terms of suppressing mud cake and rugosity effects is that the CNL receives only about 3% of its response from the first inch of formation compared to about 6% for the SNP and 17% for the FDC.

For the Neutron tools, depths of investigation will decrease slightly at higher porosites and increase somewhat at lower porosities. The reverse is true for the Density tool.[14]

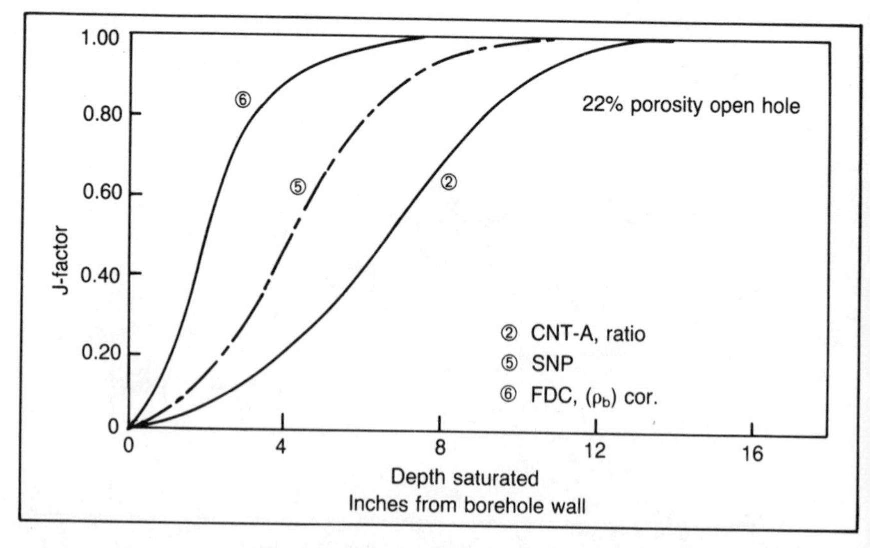

Fig. 5–16 Depth of investigation of Neutron and Density tools (courtesy Schlumberger, © SPE-AIME)

Vertical resolution of the CNL tool, if run very slowly, is approximately 15 in. However, with the usual 2-sec averaging time and 1,800-ft/hr logging speed, it is approximately 3 ft. Statistical fluctuations will average about 1 porosity unit at very low porosities and 3 pu at high porosities (opposite to the Density behavior). Consequently, sharp peak values should not be read; averages over about 3 ft should be taken.

Log Presentation

The Compensated Neutron is rarely run by itself because of substantial matrix and clay effects. It is normally run in combination with the Compensated Density and Gamma Ray in the configuration shown in fig. 5–17. The Neutron is positioned above the Density with its backup spring lined up with that of the Density so both are forcing the array against the same side of the hole.

Fig. 5–18 shows the normal presentation of curves obtained with the aforementioned combination. Gamma Ray, Caliper, and bit size are recorded in Track 1, and Neutron and Density porosities are recorded in Tracks 2 and 3 with the Neutron curve dashed and the Density solid. (In the Eastern Hemisphere the Density curve is often left on a density scale rather than transformed to porosity.) In this example the engineer chose to record porosity on the assumption of a limestone (LS) matrix.

Consider how the Neutron would have to be interpreted if run alone. The Gamma Ray shows the whole section to be clean, so clay effect is not a concern. Also the hole is smooth and close to gauge, so environmental effects are minimal.

The interval from 15,332–15,336 ft reads a porosity value of 14%. If the matrix is indeed limestone, this is the correct value. If, however, the matrix is dolomite, porosity is 7%; if sandstone, porosity is 18%, as shown by lines A of Fig. 5–15. (Note that if the log heading had shown SS, the porosities would be read as 14% if sandstone, 10% if limestone, and 3.5% if dolomite, as indicated by lines B of Fig. 5–15.) Obviously the Neutron is not much help unless lithology is known.

The same uncertainty exists if Density alone is considered. For the same interval Density porosity reads 2% if the matrix is limestone. Referring to Fig. 5–5 and assuming fluid density of 1.0 g/cc, porosity would be 10.5% if the matrix is dolomite. Sandstone is ruled out since the indicated bulk density, 2.68 g/cc, is greater than the matrix density of sand, 2.65 g/cc.

By combining Neutron and Density interpretation, the uncertainties of lithology can largely be circumvented. First, however, environmental effects and the effect of gas on the Neutron log need to be considered.

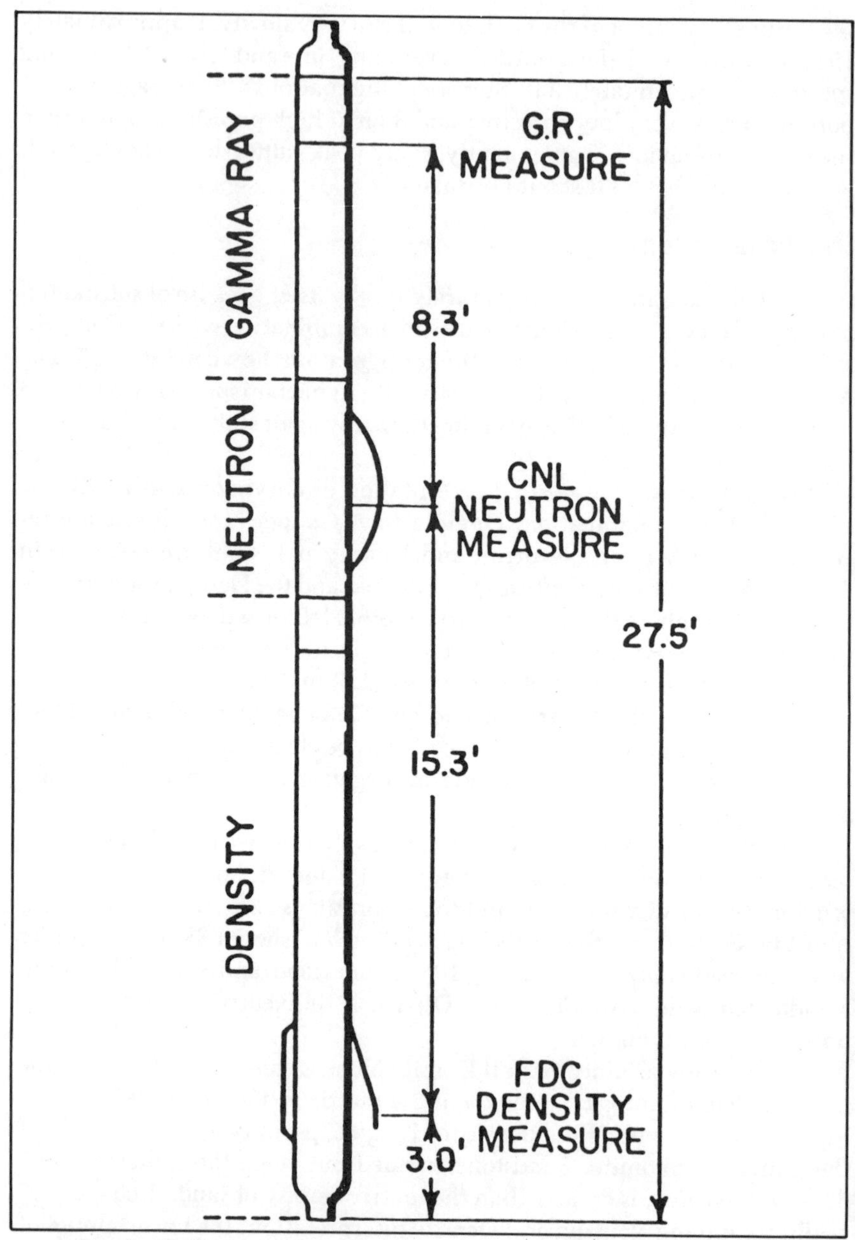

Fig. 5–17 Gamma Ray-Neutron-Density combination (courtesy Schlumberger)

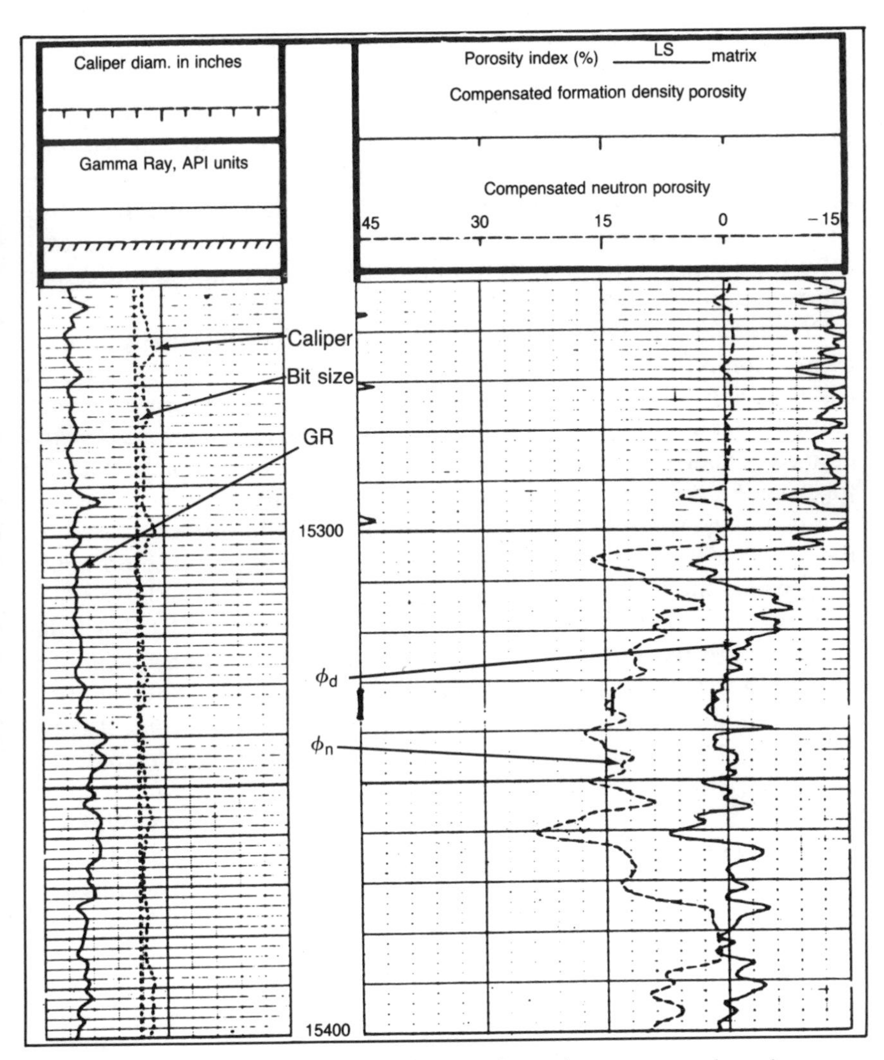

Fig. 5–18 Neutron-Density log in a carbonate sequence (courtesy Schlumberger)

Environmental Corrections

Various environmental corrections in principle should be applied to the Compensated Neutron log reading before proceeding with interpretation. These are often ignored in the first pass where a quick decision of whether to run casing in a well is required because the net result of the corrections is usually small. However, on final interpretation—particularly one made for reserve calculations—the corrections should be included.

Fig. 5–19 embodies the necessary corrections and is largely self-explanatory. Except for nomograph A, correction is made by drawing a line at the porosity of interest down through the other nomographs, B through H, and estimating each correction independently using the guidelines furnished. The corrections—some positive, some negative—are then added algebraically to obtain the net correction in porosity units, which is added to or subtracted from the log reading.

The largest correction is for borehole diameter. This is made by means of nomograph A. However, when the Neutron is run with the Density, the Caliper reading of the latter is used to make the borehole correction automatically while the log is recorded. In this case the Caliper reads borehole diameter smaller than it actually is by the amount of mud cake thickness (because it is assumed the backup shoe cuts through the mud cake but the pad does not). Consequently, the automatic borehole correction is a little too small and must be increased (reducing indicated porosity) before applying the remaining corrections, as illustrated in Example 1 of Fig. 5–19.

Example 2 of Fig. 5–19 illustrates the case when a CNL log is not Caliper-corrected during the run but a separate Caliper log is available. The difference between the actual hole size at a depth of interest and the assumed hole size set into the surface computer is entered in nomograph A. The mud cake correction is ignored and other corrections are applied as usual.

Standoff effects can be important, especially at low porosity. While the CNL has a decentralizing spring, the tool is usually run in a combination string that extends about 15 ft above the source-detector array and anywhere from 20–40 ft below it. This means that the tool will bridge many hole washouts, effectively creating local standoff.

Temperature and pressure corrections can also be significant, but fortunately they offset one another. Combined correction is significant only in high-pressure zones with low temperature or in the reverse situation.

Net corrections of 1–2 porosity units, illustrated by Example 1 of Fig. 5–19, are typical. This is the reason they can be ignored on a first pass, except in cases where porosity hovers around cutoff values of about 5%.

Gas Effect on the Neutron Log[15]

Replacement of liquid by gas in the pore space of a rock lowers the hydrogen density of the pore fluid. (Lower gas density predominates over

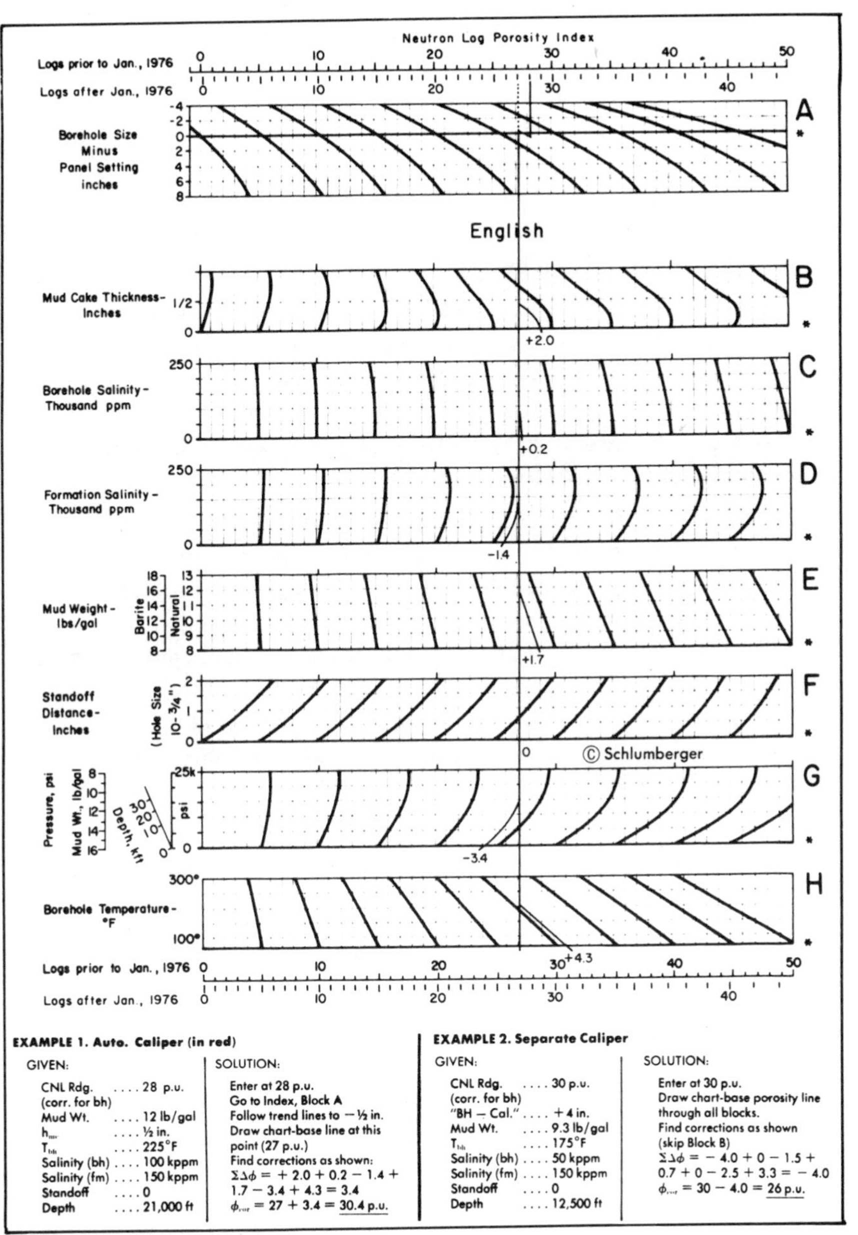

Fig. 5–19 CNL correction nomograph for open hole (courtesy Schlumberger)

greater hydrogen fraction in the fluid.) As a result the Neutron curve, calibrated for liquid-filled porosity, indicates abnormally low porosity. The effect can be large. As an example, consider the zone from 1,884 ft to 1,922 ft of Fig. 5–20. This is a gas-bearing interval of porosity close to 18 %, but the Neutron reads an average of about 5 % porosity. This implies, as a first approximation, about ⅔ of the pore space in the invaded zone is filled with gas (assumed at low pressure) and ⅓ is filled with liquid. In reality, there is somewhat more liquid and less gas than that.

If the Neutron is the only porosity log run in potentially gas-bearing zones, there is the same dual dilemma as with the Density. First is to recognize the presence of gas, since it will appear on the log simply as lower porosity; if there are other low-porosity zones, as in Fig. 5–20, the gas will not stand out. Second is to derive the correct porosity because the gas saturation is not known beforehand.

The situation is further complicated because the *excavation effect* must be taken into account.[16] This may be defined as the difference, in porosity units, between the Neutron log reading in a gas-bearing formation and that in a completely liquid-saturated formation having the same hydrogen content. The former will read lower porosity because it will contain less rock matrix, which will allow the neutrons to travel a little further. For example, a 30 %-porosity formation with 50 % water and 50 % air in the pores would not read a porosity of 15 %, as might be expected, but a porosity of 9 %. The excavation effect would be 6 pu, which is not negligible. This difference is caused by the air space not being replaced by rock matrix.

An iterative procedure can be followed to obtain porosity in suspected gas-bearing zones. It is similar to that described for the Density log, utilizing electric logs to provide values of S_{xo} (Eqs. 5.5 or 5.6) and applying the excavation correction at every iteration. However, this procedure is exceedingly cumbersome and, as with lithology determination, can be replaced with simultaneous Density-Neutron interpretation.

COMBINED DENSITY-NEUTRON INTERPRETATION

Vastly improved and simplified log analysis is achieved when Density and Neutron interpretation are combined. Fig. 5–21 shows the crossplot chart obtained when Density response is plotted against Neutron porosity. It embodies the information in both Figs. 5–5 and 5–15.

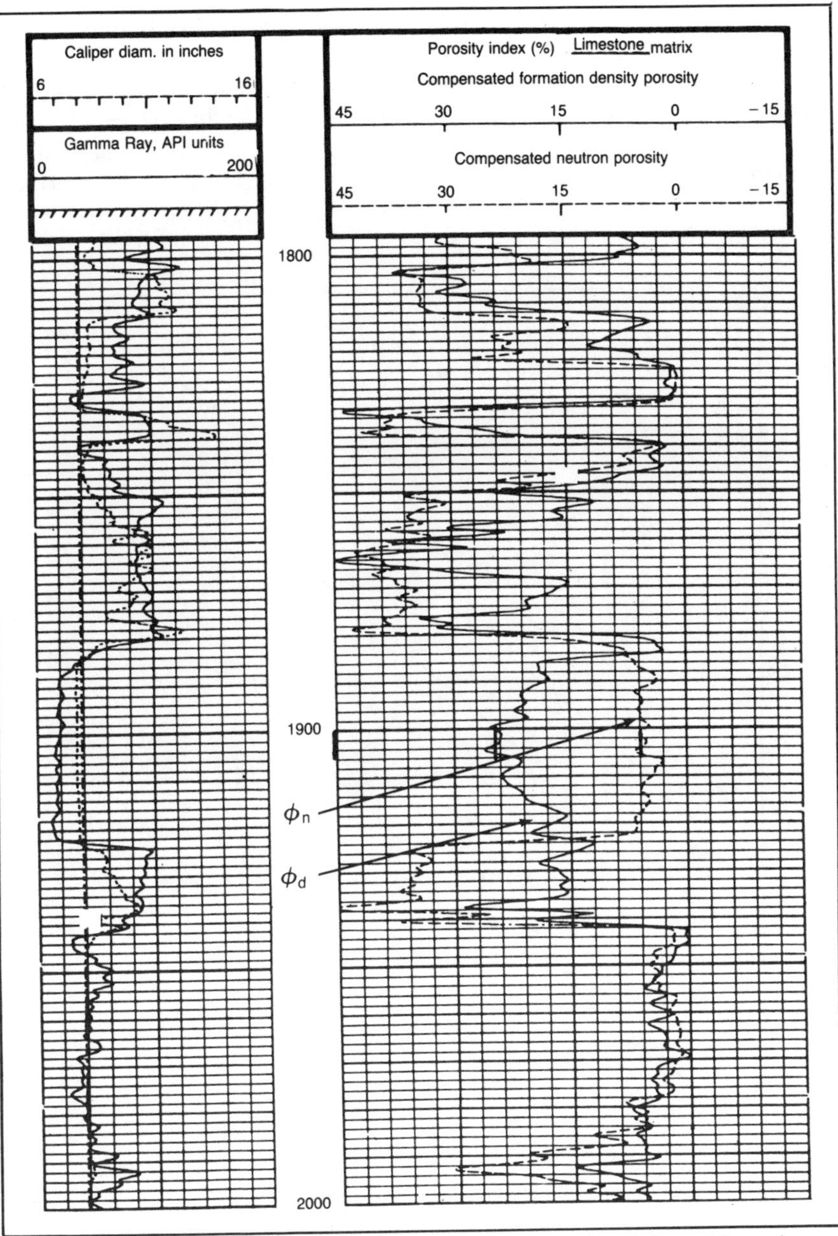

Fig. 5–20 Neutron-Density log through a gas-bearing interval (courtesy Schlumberger)

The key feature of this chart is that the equiporosity lines that join points of like porosity on the three matrix curves are virtually straight. This means that *porosities can be read without precise knowledge of lithology*. For example, the 15,332–15,336 ft interval of Fig. 5–18 where $\phi_n = 14\%$ and $\phi_d = 2\%$ gives a point of intersection A on Fig. 5–21. Interpolating between the $\phi = 5$ and 10% lines yields a porosity of 8%. Further, the matrix is

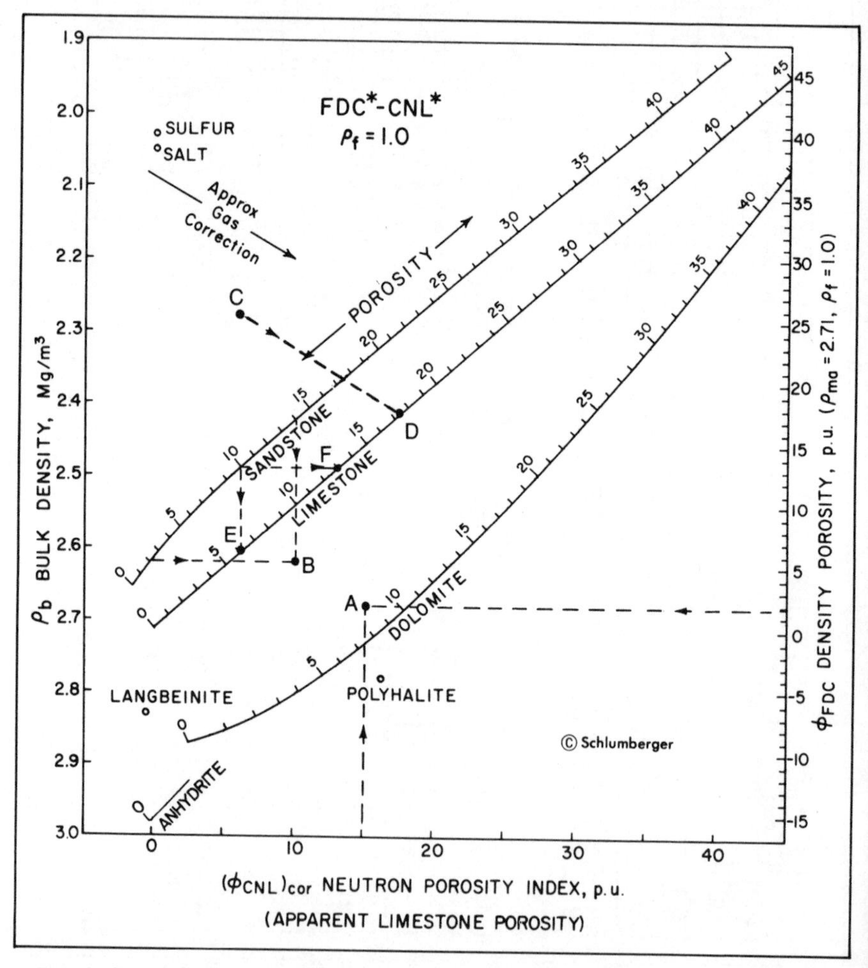

Fig. 5–21 FDC-CNL crossplot for porosity and lithology determination in fresh-mud conditions (courtesy Schlumberger)

indicated as mostly dolomite with some limestone (probable) or some sandstone (less probable) or perhaps some of both; however, it could not be a mixture of limestone and sandstone.

Note that the bottom and right-hand scales of Fig. 5–21 can only be used if the log has been recorded on a limestone matrix. Had the same log been recorded on a sandstone scale, the intersection of $\phi_n = 14\%$ and $\phi_d = 2\%$ would be found by starting at those porosities on the sandstone line and projecting vertically and horizontally, respectively, to the point of intersection B. Porosity would still be determined as 8%, but lithology would be deduced as much less dolomite and more limestone or sandstone. To avoid confusion it is good practice to disregard the bottom and right-hand scales of the chart and use only the porosities on the limestone or sandstone curves (whichever the log heading indicates) as starting points.

Liquid-Filled Formations

All liquid-filled porosity points of usual lithology will fall inside the region bounded by the sandstone and dolomite lines. For this region a good approximation to the true porosity is the average of the Density and Neutron values

$$\phi = (\phi_d + \phi_n)/2 \tag{5.12}$$

This is a most important result. Effective porosities can be eyeballed from the log as those values halfway between Density and Neutron curves for clean, liquid-filled formations. (If the Density porosity goes negative, its value should be taken as zero for this purpose.)

Looking again at Fig. 5–18 and using Fig. 5–21, we can deduce that the interval from 15,250 to 15,300 ft is primarily anhydrite of zero porosity and that at 15,381 to 15,385 ft, where the two curves agree, is limestone of 1.5% porosity. The log illustrates very nicely anhydrite, dolomite, and limestone signatures seen in tight carbonates.

Gas-Bearing Formations

As previously explained, replacement of liquid by gas in the pore space of rock causes both bulk density and hydrogen content to decrease. The Density will show higher porosity and the Neutron lower. This gives rise to the well-known *crossover effect* on Neutron-Density logs. Normally, the Neutron reads somewhat higher porosity than Density due to dolomitization and to clay effects. When the Neutron crosses over and reads lower porosity than the Density, it is an infallible indicator of gas (except for the qualifications given below). This is a very popular feature of the Density-Neutron combination.

Fig. 5–20 is an excellent example. Throughout the interval shown, the Neutron reads equal to (within statistics) or higher than the Density except in the 1,884 to 1,922 ft zone where there is a marked crossover. This interval is clearly gas bearing. The question is, what is the correct porosity?

Referring to the crossplot chart (Fig. 5–21), gas will cause intersection points to shift northwesterly and in many cases will cause them to fall above the sandstone line. For the interval 1,900 to 1,905 ft of Fig. 5–20 where $\phi_n = 6$ and $\phi_d = 24.5\%$, the point of intersection is at C. To find the porosity, the point is shifted back to the assumed lithology line (in this case limestone) in a direction parallel to the gas correction line indicated. In this case the porosity is found as 17.5% (point D).

A good approximation to the true porosity in gas-bearing zones is

$$\phi = \sqrt{(\phi_d^2 + \phi_n^2)/2} \quad (5.13)$$

In the case illustrated this formula gives $\phi = 18\%$, which agrees with the value obtained graphically. This leads to the eyeball rule that in gas-bearing zones the porosity is not midway between Neutron and Density values but is about ⅔ of the way from the Neutron to the Density reading.

The gas correction line indicated applies to the situation where both Density and Neutron are responding to the same condition, that is, when depth of invasion exceeds 12 in. (from the borehole wall) or is practically nil. Even in this case the slope of the line depends a little on the porosity and on the density and composition of the gas, which is the reason the line is labeled "approximate."

In situations where the depth of invasion is in the range 4–6 in., the Density will respond only to the invaded zone, whereas the Neutron will "see" well into the noninvaded zone. In gas-bearing intervals the invaded zone will have lower gas saturation than the noninvaded region, perhaps by a factor of two. The appropriate gas correction line then becomes significantly closer to horizontal. Use of the indicated line results in underestimating porosities. In the extreme situation of 4–6 in. of invasion and practically no gas in the invaded zone, the correction line is virtually horizontal.

Precise correction requires running an R_{xo} log and combining derived S_{xo}, d_i, and S_w values with applicable depth of investigation data for the Density and Neutron.[17] This normally is not done but could lead to improved interpretation in gas-bearing zones.

False Gas Indication

There is one circumstance where a false indication of gas can be obtained from Neutron-Density crossover. This is the situation where the porosity curves are recorded on limestone matrix but the lithology is actually sandstone. Referring to Fig. 5–21, a sandstone of porosity 10% —if recorded on limestone matrix—will show up as a Neutron porosity of 6% (point E) and a Density porosity of 13% (point F). This will appear as gas, whereas it is simply a matrix effect.

Fig. 5–22 is an example. The Cotton Valley formation illustrated is primarily sandstone, but the log has been recorded on limestone matrix. The whole interval appears gas bearing but is not. To correct the curves to sandstone matrix, 3.5 pu should be added to the Neutron and 3.5 pu subtracted from the Density (Fig. 5–21). The curves then overlay except in a few thin beds that appear to be gas bearing. However, even in these intervals much of the residual curve separation could result from hole enlargement (short cave) effects on the Density log.

The clue to false gas indication when examing logs is a 6–7 pu constant difference between the Neutron and Density. When this is seen, matrix effect should be suspected.

By the same token, gas can be missed in the case where the log is recorded on sandstone matrix and a low-porosity, gas-bearing limestone is penetrated. The gas shift might be occurring; but if it is less than 6–7 pu, it would not cause the Neutron to cross over the Density. Gas is also more difficult to see with a dolomite matrix. Regardless of limestone or sandstone recording, dolomite causes the Neutron to read much higher than Density (as shown by Fig. 5–18), which can suppress crossover in gas-bearing zones. In cases such as this, it is important to have independent determination of lithology.

The Litho-Density log can provide the necessary lithology information. Fig. 5–23 is an example from the Middle East where the Litho-Density-Neutron combination was run through a series of gas-bearing limestones and dolomites. Scales are such that the Neutron and Density curves should overlay (approximately) in liquid-filled limestones. For some intervals large crossovers occur, and in others the curves agree. Without a P_e curve these could be interpreted as gas- and liquid-bearing limestones, respectively.

The P_e curve, however, shows the true story. These zones with large crossovers are limestones ($P_e = 5$), clearly gas-bearing. Those with no crossover are dolomites ($P_e = 3$). Plotting the latter points on Fig. 5–21 shows the dolomites to be gas-bearing also. Thus, the whole interval, with the exception of a few tight streaks, is gas productive. This illustrates perhaps the prime use of the Litho-Density-Neutron combination—the separation of gas and matrix effects in tight carbonates.

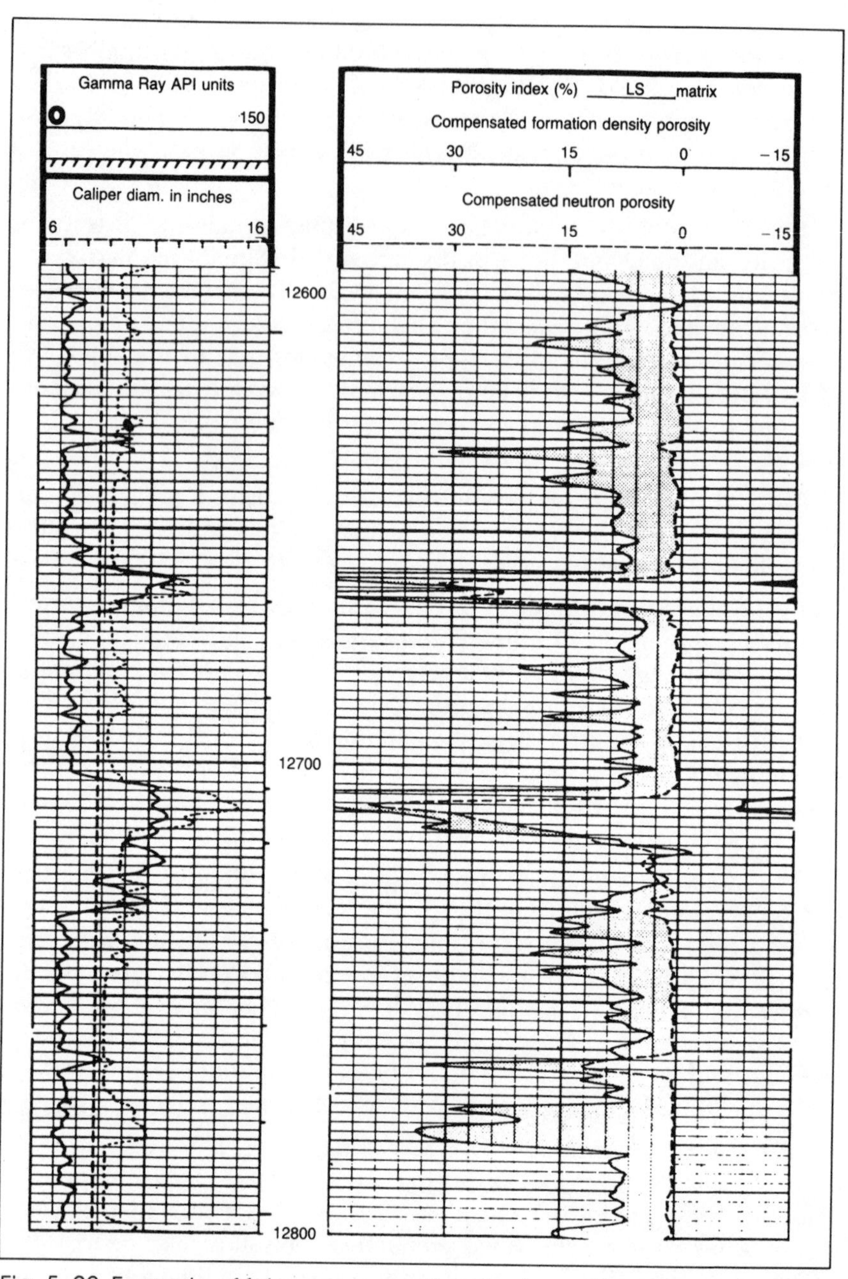

Fig. 5–22 Example of false gas indication due to matrix change (courtesy Schlumberger)

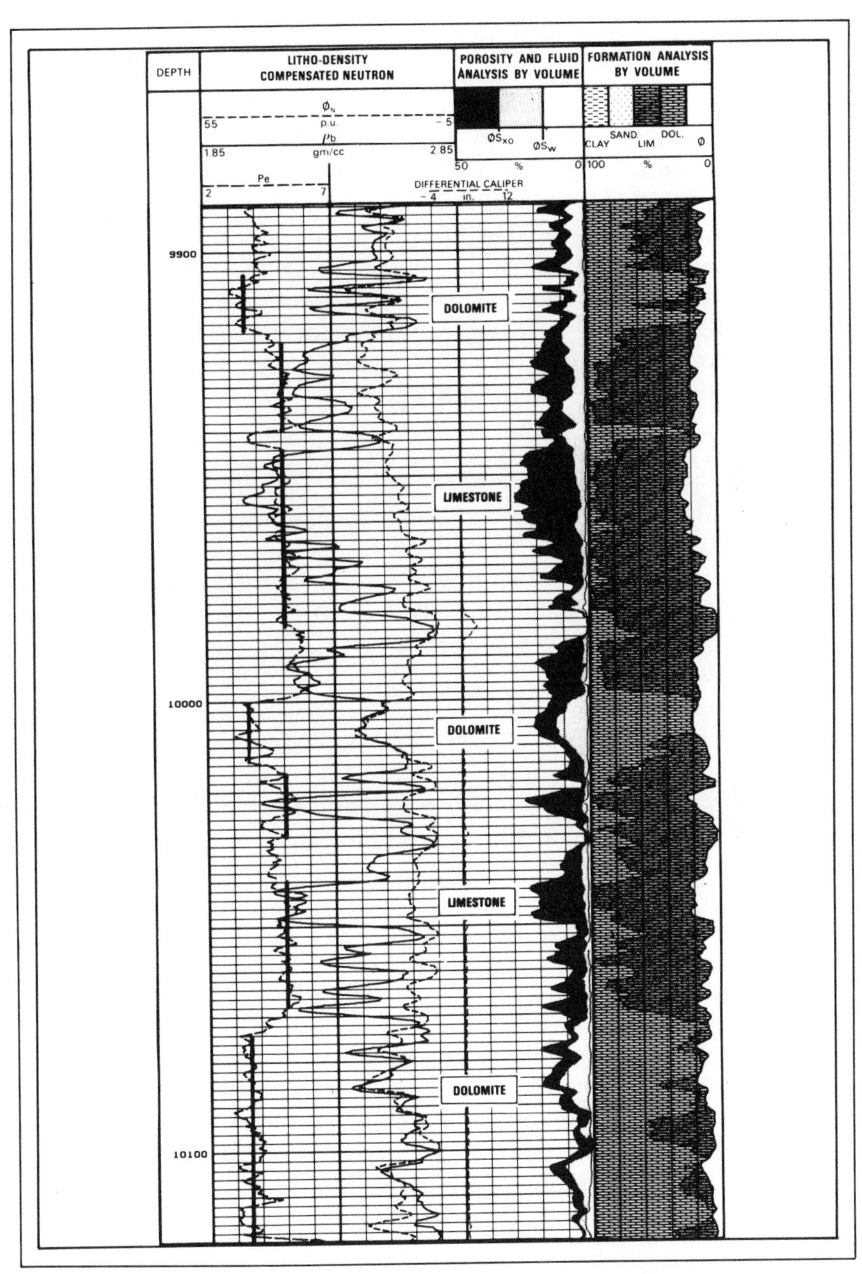

Fig. 5–23 Recognition of gas-bearing dolomites with the Litho-Density log (courtesy Schlumberger)

THE DUAL POROSITY COMPENSATED NEUTRON LOG[18]

All Neutron tools, regardless of type, respond to hydrogen bound in lattice of clays. Porosities in shales typically read 30–45 %, depending on the amount and type of clay in the shale. Shaliness of a gas-bearing sand may therefore suppress the crossover indication normally expected.

The problem is aggravated in the Compensated Neutron by virtue of the fact that thermal neutron detection makes the tool sensitive to trace amounts of strong absorbers such as boron and gadolinium in formations. Such absorbers, if present, are typically found in clays. They may contribute as much as 15 porosity units to the porosity reading in shale, though typically the effect may be more like 5 pu.

Principle of the Tool

To circumvent the absorption effect, Schlumberger is introducing a Dual Porosity Compensated Neutron designated CNT-G. Fig. 5–24 shows the configuration. Above the neutron source are two thermal neutron detectors that provide the standard CNL log. Below the source are two epithermal neutron detectors that provide a second porosity log insensitive to thermal absorbers.

Pulse rates from each pair of detectors are processed by the spine-and-ribs method (similar to that of the Compensated Density) to give porosity corrected for mud cake and rugosity, as illustrated in Fig. 5–25. This method provides better environmental correction than the ratio method and also results in a $\Delta\phi$ correction curve that can be used for quality control.[19]

Applications of the Log

The improved gas indication in shaly formations with epithermal detection is shown in Fig. 5–26. In the upper zone—marked A—which is a shaly sand, the epithermal Neutron crosses over and shows appreciably less porosity than the Density—clearly indicating gas—whereas the thermal Neutron shows no crossover. The quick-look appeal is obvious.

Another application where the Dual Porosity Neutron should prove advantageous is in very low-porosity carbonates or tight gas sands. Matrix effect with epithermal detection should be considerably less than for thermal detection, in fact similar to that shown on Fig. 5–15 for the SNP. At porosities around 5% where the effect is critical, porosity determination should be much improved with the Dual Porosity CNL.

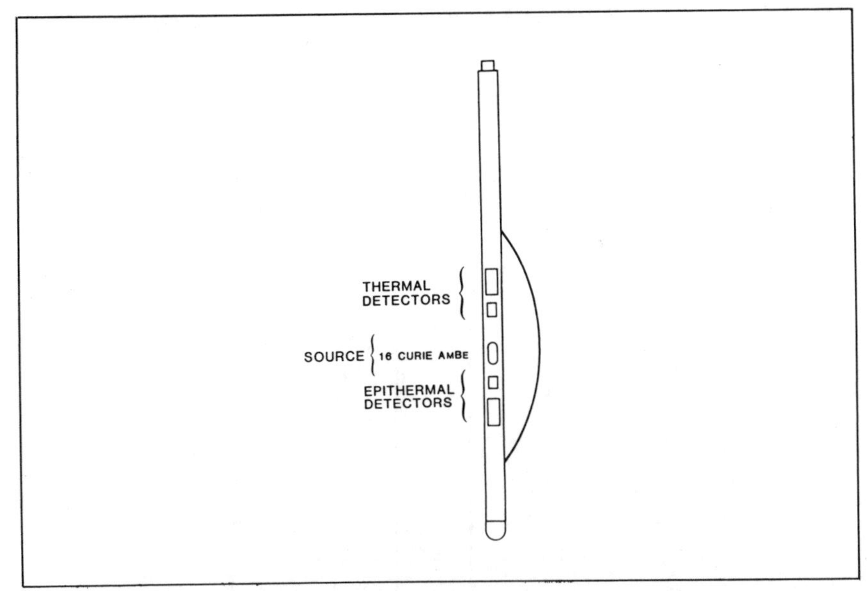

Fig. 5–24 Configuration of the Dual-Porosity Compensated-Neutron (CNT-G) (courtesy Schlumberger)

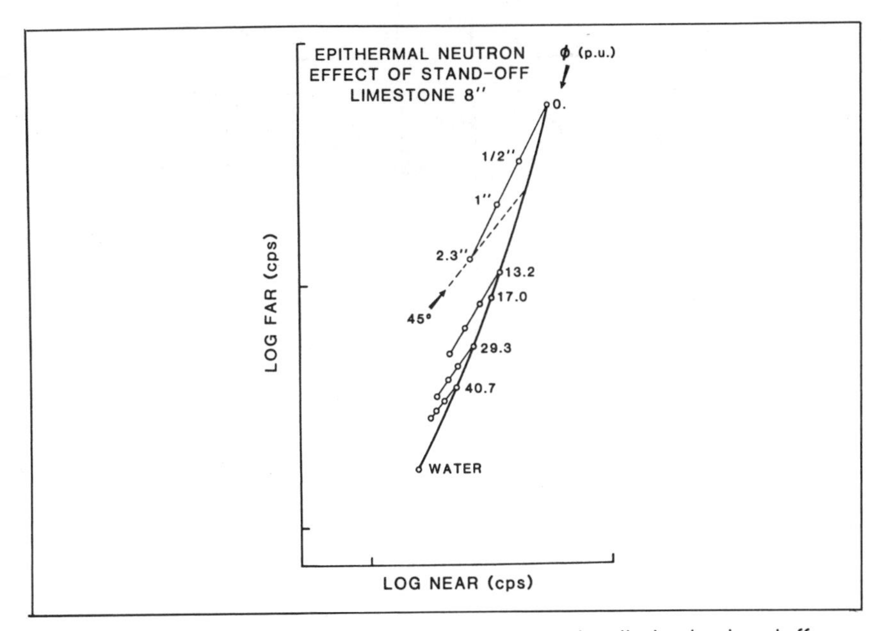

Fig. 5–25 Processing CNT-G detector response to eliminate standoff effects (courtesy Schlumberger, © SPE-AIME)

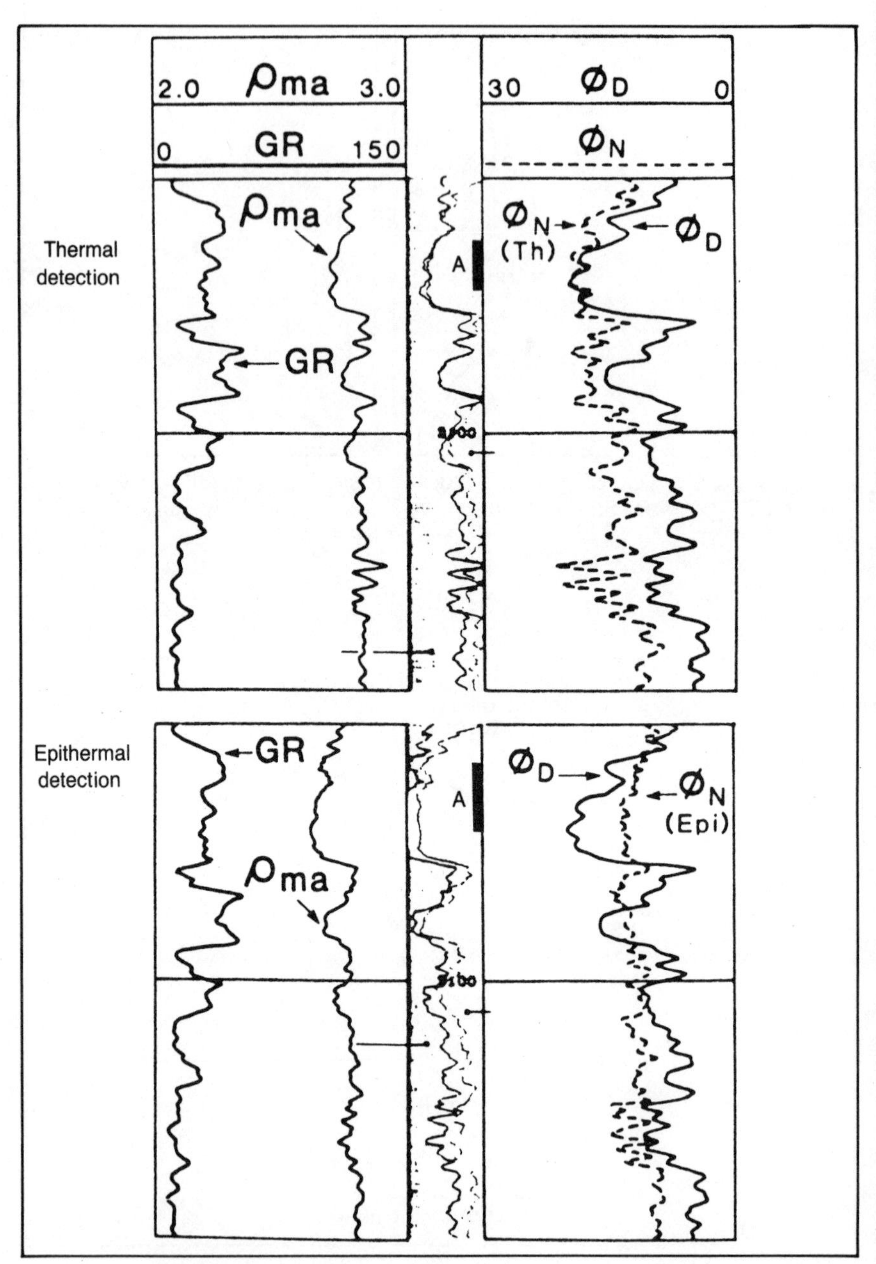

Fig. 5–26 Dual Neutron porosity comparison in a shaly gas zone (courtesy Schlumberger, © SPE-AIME)

COMPENSATED SONIC AND LONG-SPACED SONIC LOGS

A Sonic logging tool measures the velocity of sound in formations penetrated by the borehole. The principle is illustrated in Fig. 5–27. A transmitter and two receivers are positioned on a sonde with a spacing (typically) of 3 ft between transmitter and near receiver and a span of 2 ft between receivers.

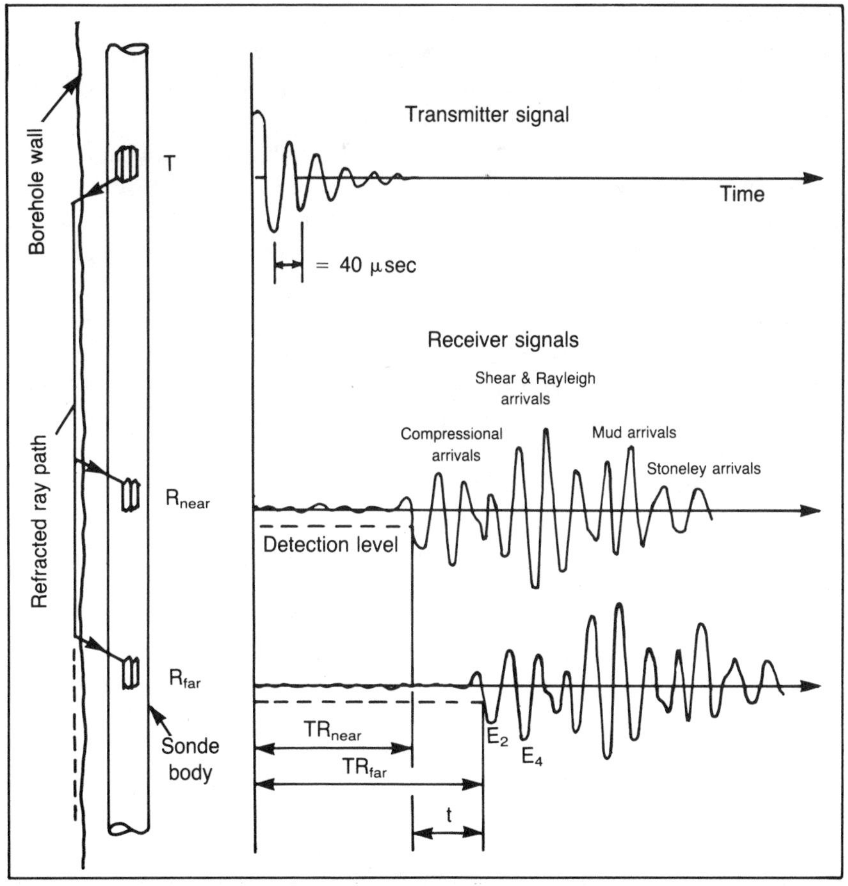

Fig. 5–27 Basic Sonic logging system (courtesy Schlumberger, © SPE)

When a pulse of current or voltage is applied to the transmitter, it generates a short oscillatory pressure pulse at about 25 kHz frequency in the mud. This initiates six different waves traveling up and down the hole: two refracted waves via the formation (compressional and shear), two direct waves (along the sonde and in the mud), and two surface waves along the borehole wall (pseudo-Rayleigh and Stoneley).[20] These waves travel at different velocities, varying from about 25,000–4,000 ft/sec.

A short time after the transmitter is pulsed, the near receiver senses the arrival of the various wave fronts. A little later they are sensed at the far receiver. The normal sequence is that shown in Fig. 5–27, with the compressional wave arriving first and the shear wave next. These actually travel as a compressional wave in the mud from the transmitter to the borehole wall, a body wave (compressional or shear) in the formation, and a compressional wave in the mud from the wall to the receiver with the initial energy following the critically refracted (minimum time) path indicated in the figure.

Closely following the shear wave is the guided pseudo-Rayleigh wave, followed by the direct mud arrival and the guided Stoneley wave. The direct sonde wave is mixed with the mud and Stoneley waves. Every effort is made in the design of the sonde to make its transmission as weak and as slow as possible.

Of primary interest are the compressional and shear waves. By definition, a compressional wave is one in which the particles in the medium are vibrating in the same direction as the energy is propagating, in this case parallel to the borehole axis. A shear wave is one in which the particles are vibrating at right angles to the propagating direction, in this case perpendicular to the borehole axis. Compressional waves travel about 1.7 times faster than shear waves.

Standard Sonic logging tools at present measure only the compressional travel time. To effect this, the transmitter is pulsed once and an electronic circuit measures the time elapsed to the first negative excursion of the compressional arrival at the near receiver (Fig. 5–27). The transmitter is pulsed again and the circuit measures elapsed time to the far receiver. The difference in arrival times is computed and divided by the span (in feet) between receivers. The result is presented on the log as formation transit time in microseconds per foot. Accuracy of the measurement is quite good, approximately $\pm$ 0.25 μsec/ft.

Compressional travel times vary from 40 μsec/ft in hard formations to 150 μsec/ft in soft ones. Corresponding velocities, which are the inverse of transit times, vary from 25,000–6,600 ft/sec. For comparison, the travel time in water (or drilling mud) is approximately 190 μsec/ft. A logging tool will occasionally read this value in an extremely large hole washout where the direct mud wave arrives first at the receivers.

It is possible, with specialized logging tools and wave form processing, to measure shear wave travel times also. This technique is evolving at present and is discussed under Long-Spacing Sonic logs.

THE BOREHOLE COMPENSATED LOG

The first-generation Sonic tool of the type illustrated in Fig. 5–27 suffered from the fact that when a hole enlargement or contraction was spanned by the receivers, the travel time across the mud was not the same at both receivers and therefore the recorded formation travel time was in error over a depth interval equal to the length of the span.[21] Very hashy logs were obtained in hard-rock holes with sharp washouts.

For this reason the tool was replaced in the 1960s by the current Borehole Compensated Sonic* shown in Fig. 5–28.[22] This tool consists of two transmitter-receiver arrays, one inverted relative to the other. A sequence of four time measurements is made as indicated. The lower transmitter is pulsed twice in succession, and the time interval ($T_2 - T_1$) between arrivals at receivers R_2 and R_1 is measured. It will be abnormally long with the hole enlargement indicated. Then the upper transmitter is pulsed twice, and the time interval ($T_4 - T_3$) between arrivals at receivers R_4 and R_3 is measured. It will be too short in the example. The correct travel time is obtained by averaging the two readings. The averaging technique also eliminates incorrect readings due to sonde tilt in the hole. With this technique the quality of Sonic logs in rough holes is vastly improved.

The Schlumberger BHC tool has a spacing of 3 ft between transmitter and near receiver and a span of 2 ft between receivers. The transmitters are pulsed a total of 20 times per second so that five complete measurements are made each second. Logging speed is 5,000 ft/hr, which means a measurement is made about every 3 in. of hole. Normally the Sonic tool is run centered so the contributions to a receiver signal from different sides of the hole will be in phase (if the hole is round) and the signal-noise ratio will be maximized. The tool can be run off center, but significant degradation in the signal-noise ratio must be tolerated.

---

*Designated BHC Sonic by Schlumberger, BHC Acoustilog by Dresser-Atlas, Acoustic Velocity log by Welex, and BCS log by Gearhart.

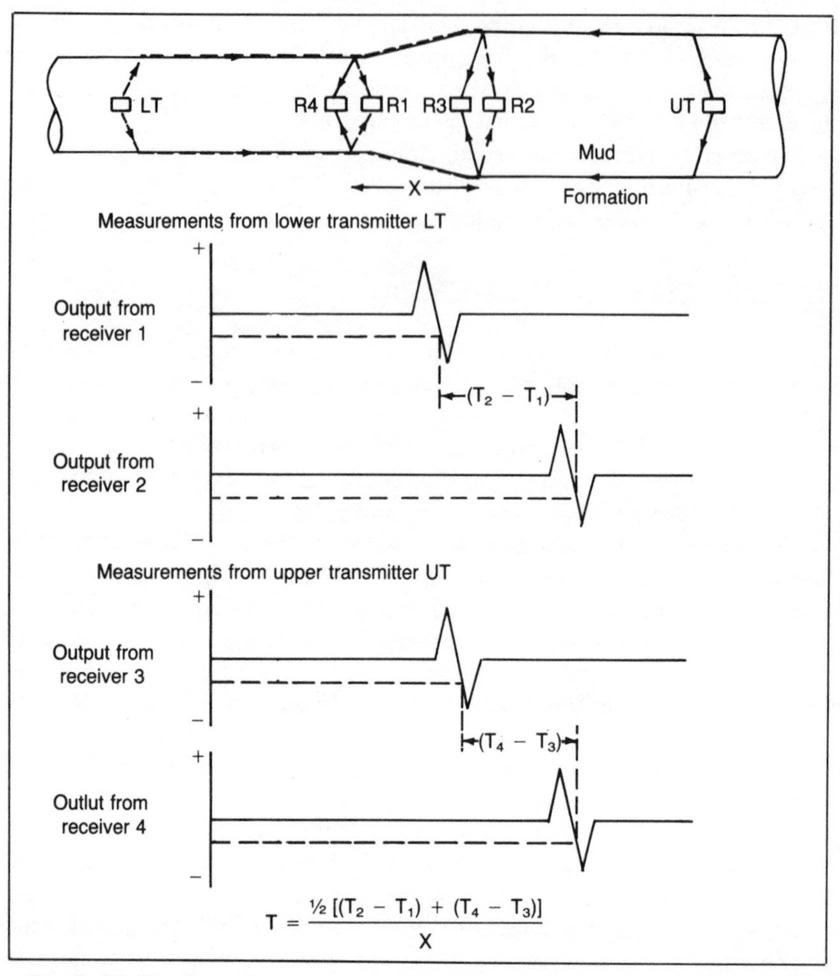

Fig. 5–28 Borehole Compensated Sonic logging system (after Thomas, courtesy SPWLA)

Vertical Resolution and Depth of Penetration

The vertical resolution of a Sonic tool is well defined. It is the span between receivers, that is, 2 ft for the standard BHC tool. The depth of penetration is not well defined but is very small; it is controlled by the basic frequency in the waveform ($\approx$25 kHz). For a homogeneous formation it is in the range of 1.0–2 in. and is independent of the spacing or span of the tools.

If, however, the formation is not homogeneous but contains an altered zone of slower velocity next to the borehole, the depth of penetration of the measurement can be increased by increasing the spacing, in the sense that the first arrival of energy at a receiver can be made to come from the deeper unaltered zone. This is discussed under Long-Spacing Sonic.

Log Presentation

Typical presentation of the Sonic log, when run by itself, is shown in Fig. 5–29. The interval transit time, in microseconds per foot, is recorded across Tracks 2 and 3. Short transit times are to the right and long transit times are to the left, such that increase in porosity deflects the curve toward the depth track consistent with Density and Neutron recording.

In the depth track are small pips, representing integrated travel time of 1 msec between each pip. Larger pips are recorded at 10-msec intervals. These are useful in comparing Sonic logs with seismic sections. Usually a caliper curve is displayed in Track 1, obtained from the combination caliper/centralizer run above the Sonic tool (another centralizer being run below). When a Gamma Ray log is run simultaneously, it is also recorded in Track 1. If a resistivity tool is run simultaneously, as is often the case, the resistivity curves are displayed in Track 2 and the Sonic travel time is restricted to Track 3.

A good check on the accuracy of a Sonic log is to observe the reading in casing. It should be 57 μsec/ft, the travel time of steel. The log may not jump immediately to this value on entering casing because there can be a drastic change in signal amplitude to which the system (or engineer) must adjust. The reading is most reliable in uncemented pipe where the casing-borne arrival will have good amplitude and will always arrive ahead of formation-borne signals, no matter how fast. The opposite can be true in cemented pipe.

Noise Spikes and Cycle Skipping

Two effects may cause erroneous spikes or shifts to occur on Sonic logs. The first is noise triggering, as illustrated by Fig. 5–30. Road noise is generated by tool centralizers rubbing against the borehole wall; a noise pulse may occasionally trigger a receiver circuit ahead of the desired arrival. This is most likely to occur at the far receiver since the trigger circuit must listen longer for a signal than at the near receiver and must also be capable of picking up a smaller signal. (Both detector channels are closed for the first

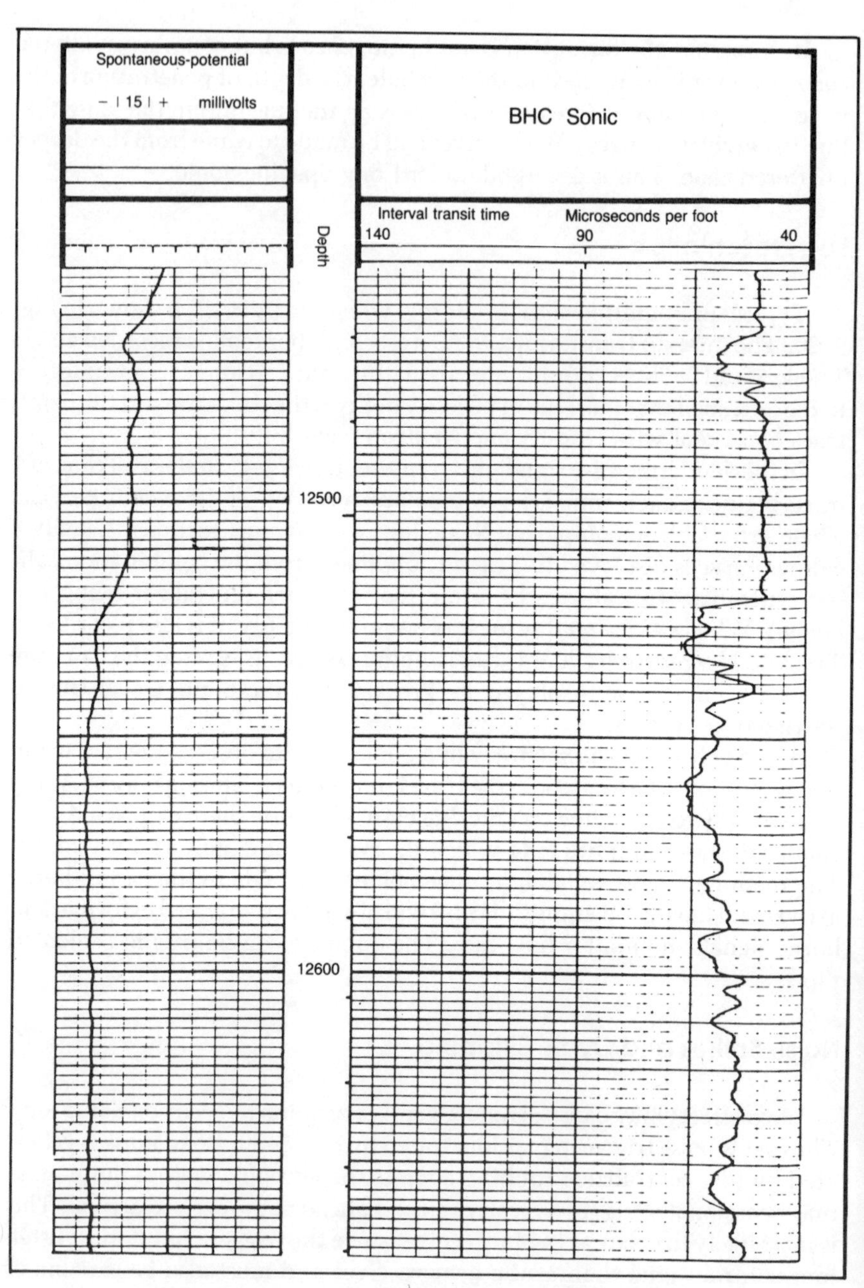

Fig. 5–29 Sonic log presentation (courtesy Schlumberger)

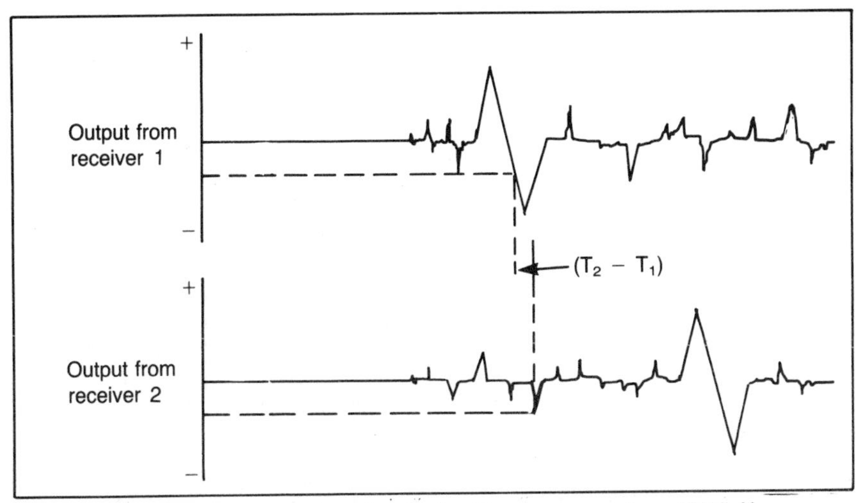

Fig. 5–30 Noise triggering (after Thomas, courtesy SPWLA)

120 μsec or so because no true signal can arrive in this interval.) In this case an abnormally short transit time is measured, with the error anywhere between 0 to about 75 μsec/ft. Abnormally long travel times can occur when noise triggers the near detector, but this happens less frequently.

The second source of error is cycle skipping, illustrated in Fig. 5–31. This occurs when the first negative arrival of the signal, usually at the far receiver, falls below the trigger level, in which case the circuit usually triggers on the next negative excursion, about 40 μsec later. If this happens on only one of the four subcycles making up a measurement, the log will momentarily read about 10 μsec/ft too long. If it happens on two of the four subcycles, the log will read 20 μsec/ft too long.

Fig. 5–32 shows a case of persistent cycle skipping around the 3,500-ft depth with one anomaly close to 3,600 ft that could be a noise spike. This behavior is most likely to occur in shallow holes with gas-saturated formations or gas-cut mud since free gas strongly attenuates Sonic energy transmission. In early days this was a problem because delicate adjustment between noise triggering on the one hand and cycle skipping on the other was required on the part of the logging engineer. In recent years, however, Sonic tools have been equipped with automatic gain control circuits that eliminate the need for the engineer to ride the gain and greatly reduce the incidence of noise spikes and cycle skips.

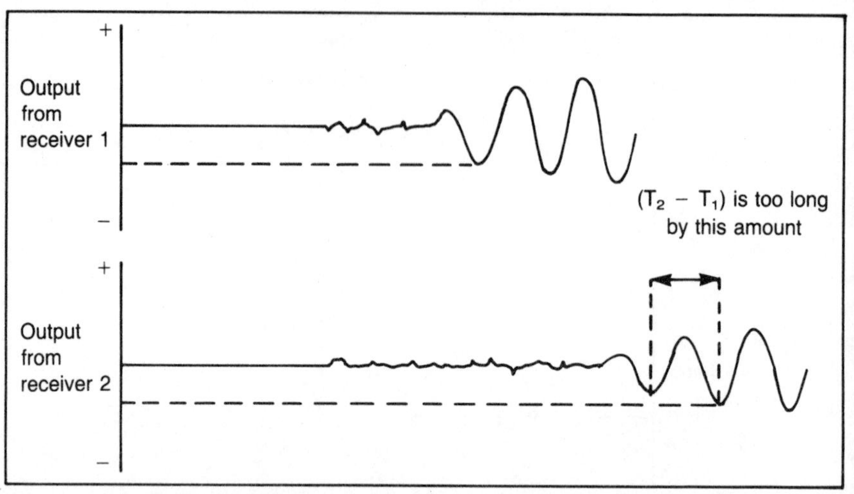

Fig. 5–31 Cycle skipping (after Thomas, courtesy SPWLA)

POROSITY DETERMINATION FROM SONIC LOGS

The velocity of compressional waves in an elastic solid is a function of the rigidity and the density of the material. The relationship is

$$V_c = (M/\rho)^{1/2} \tag{5.14}$$

where M is the rigidity (a measure of the medium's resistance to both compression and shearing) and ρ is the density.* The more rigid the medium and the lower its density, the higher the velocity.

At 25 kHz frequency, Sonic wavelengths are 4–12 in. in formations of interest. Consequently, the Sonic wave does not sense individual grains or pores. It is affected only by the overall rigidity and density of the material. The effect of porosity on velocity is then through its influence on these parameters. Intuitively, one would expect an increase in porosity to reduce rigidity faster than density and thereby reduce velocity. This is indeed the case.

Theoretical expressions exist that relate porosity to velocity in porous materials, but they are not usable in normal well-log analysis because they

---

*$M = B + 4\,\mu/3$ where B is the bulk modulus and μ is the shear modulus.

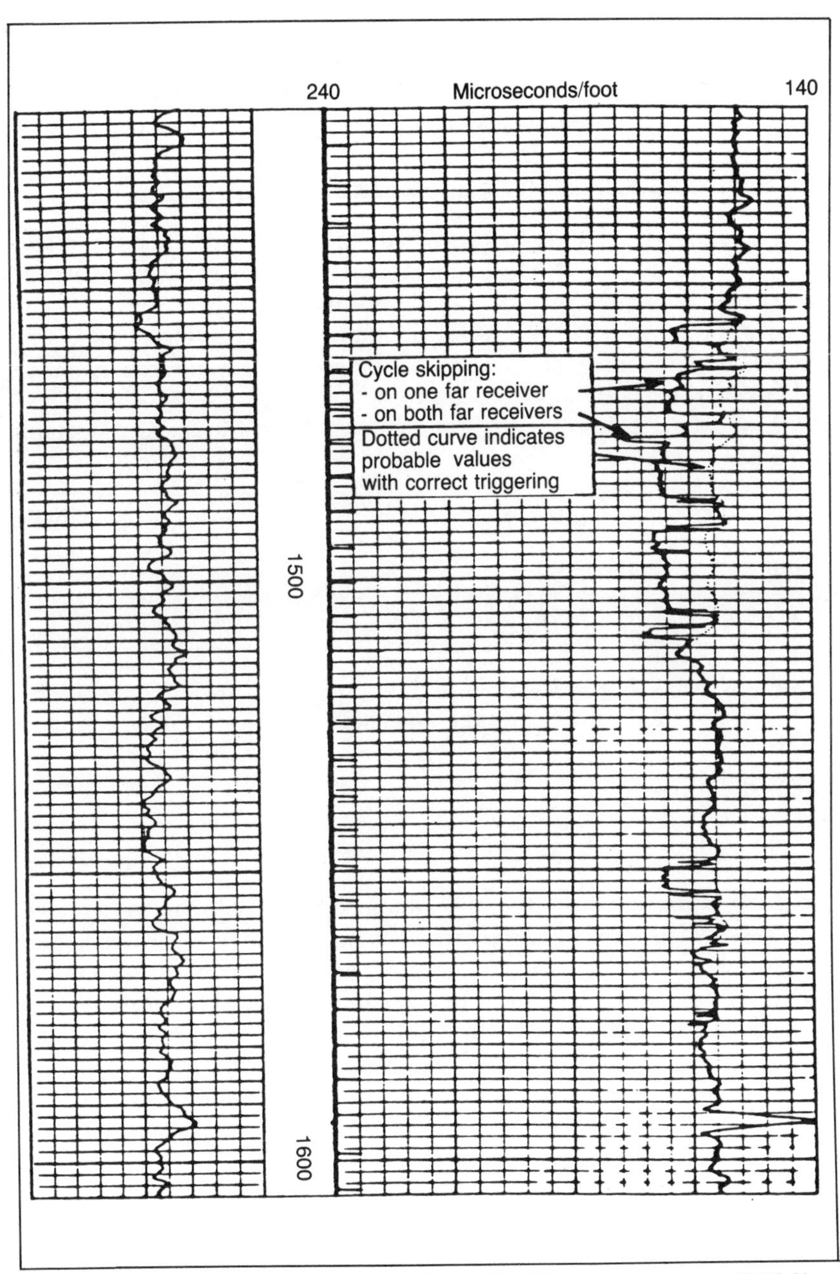

Fig. 5–32 Example of cycle skipping (after Thomas, courtesy SPWLA)

require input of elastic properties of the grain material, the pore fluid, and the empty rock frame.[23] These are not known with certainty in any particular case, especially the empty rock parameters. Therefore, it is necessary to resort to empirical relations between velocity, or transit time, and porosity.

Liquid-Bearing Formations—The Wyllie Relation

The empirical relationship universally used until recently is the Wyllie time-average formula. It is based on laboratory observations and states, in effect, that the travel time of a compressional wave through a block of porous rock is the same as if all of the matrix material in the rock were pressed into a solid piece at one end of the block and all of the pore fluid were gathered in the remainder of the space with the compressional wave traveling through the two portions in sequence.[24–26] This leads to the simple expression

$$t = \phi \cdot t_f + (1 - \phi) \cdot t_{ma} \tag{5.15}$$

where

t = travel time of the porous rock
ϕ = porosity
t_f = travel time of the fluid that occupies the pores
t_{ma} = travel time of the solid rock matrix

Solving for porosity

$$\phi = (t - t_{ma})/(t_f - t_{ma}) \tag{5.16}$$

Using this expression porosity can be obtained from the log-recorded travel time, t, provided t_f and t_{ma} are known. The fluid in the zone of investigation is typically mud filtrate. Consequently, t_f is normally taken as 189 μsec/ft in fresh mud. In salt mud a value of 185 may be used. Matrix travel times vary from 40–50 μsec/ft, depending on lithology.

The solid lines of Fig. 5–33 provide a graphical solution to Eq. 5.16. Log-derived transit time is entered on the horizontal axis, a line is projected vertically to the appropriate matrix velocity, and porosity is read on the vertical scale opposite the point of intersection.

As an example, the zone at 12,558–12,564 ft on Fig. 5–29 reads a travel time of 66 μsec/ft. Fig. 5–33 gives a porosity as low as 8% if the matrix is sandstone or as high as 18% if the matrix is dolomite. Clearly, the lithology must be known to obtain accurate porosity values. Even within a given lithology there is a range of possible matrix travel times, as indicated at the bottom of Fig. 5–33. Local knowledge dictates the value to use — although the deeper the burial of a formation, the lower t_{ma} is likely to be.

A comparison of Fig. 5–29 with Fig. 5–34, the Density-Neutron log over the same section of hole, clearly shows the advantages of the latter in defining lithology as well as porosity. Using the crossplot chart of Fig. 5–21, porosity and lithology for selected intervals of the D-N log are listed in columns 2 and 3 of Table 5–3.

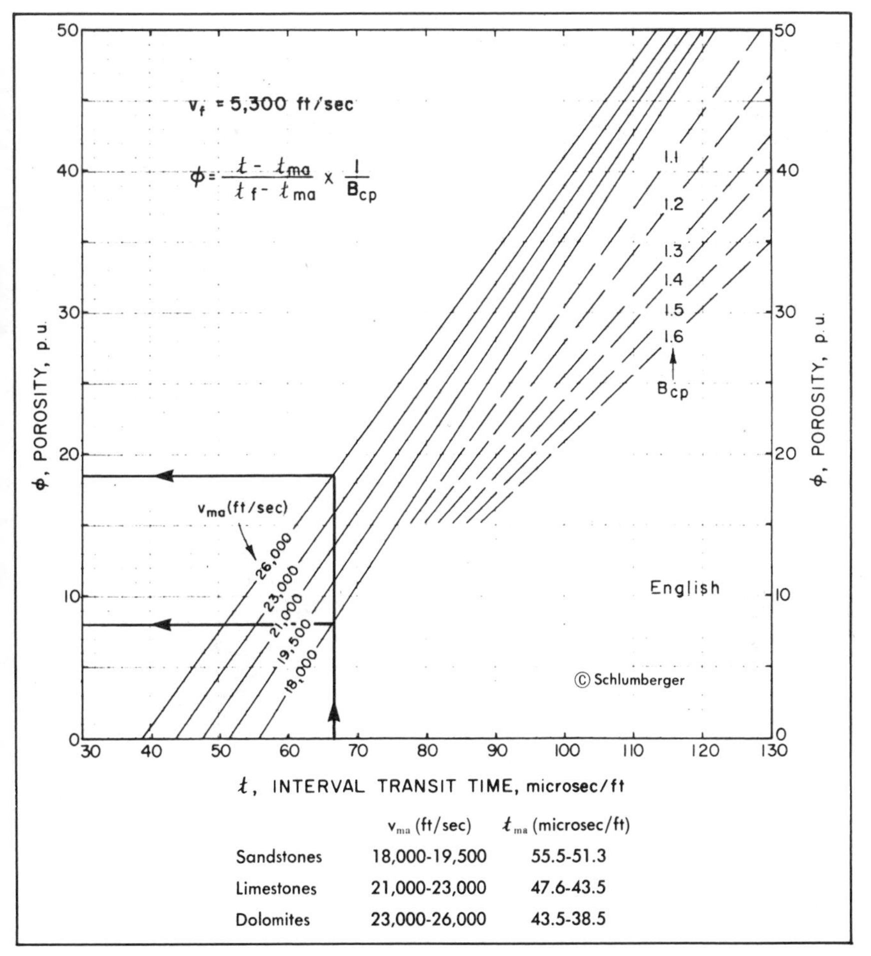

	v_{ma} (ft/sec)	t_{ma} (microsec/ft)
Sandstones	18,000-19,500	55.5-51.3
Limestones	21,000-23,000	47.6-43.5
Dolomites	23,000-26,000	43.5-38.5

Fig. 5–33 Determination of porosity from transit time by time-average method (courtesy Schlumberger)

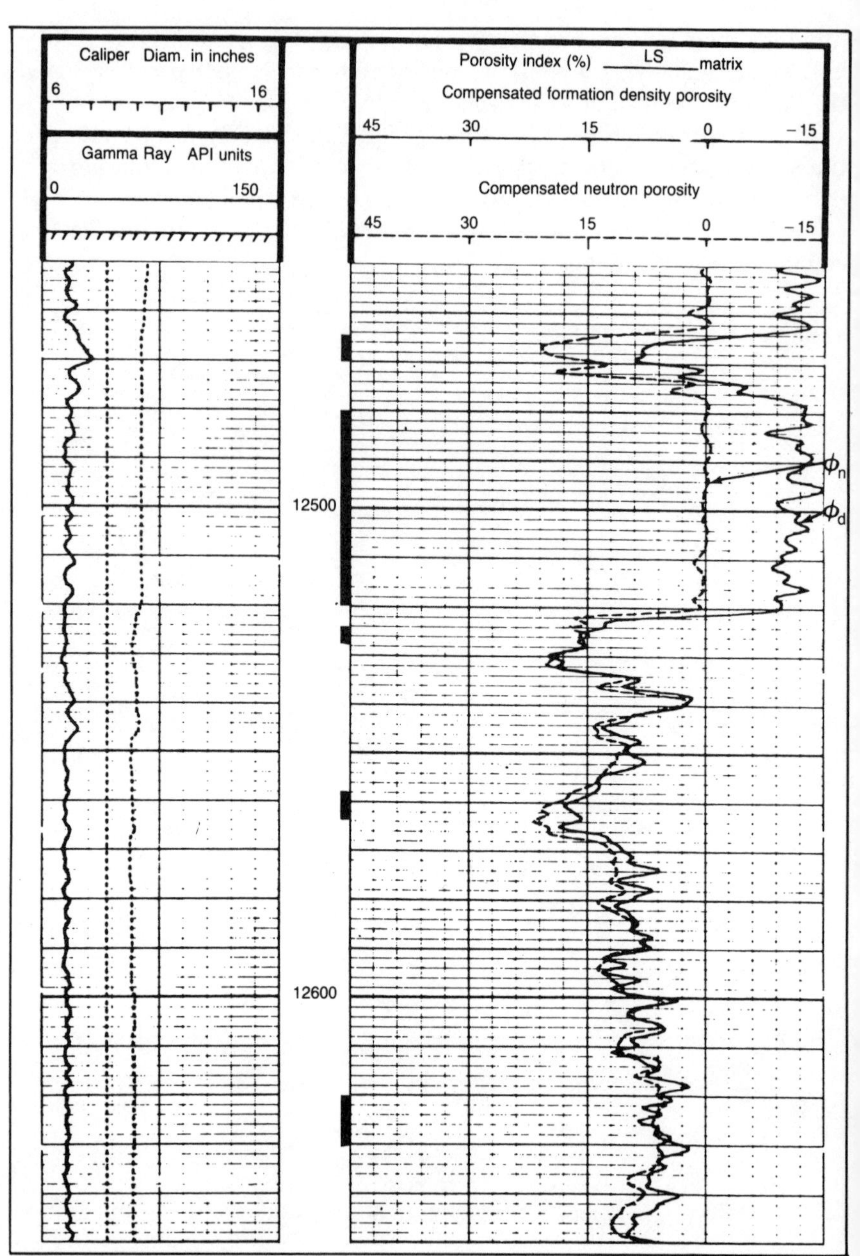

Fig. 5–34 Density-Neutron log; same interval as Fig. 5-29 (courtesy Schlumberger)

TABLE 5-3 COMPARISON OF SONIC AND DENSITY-NEUTRON POROSITIES

Interval, Ft	Density-Neutron Porosity, %	Density-Neutron Lithology	Range of Sonic Porosity, %	Sonic Porosity with D-N Lithology Fig. 5-33	Fig. 5-36
12,466–70	14	Limy dolomite	6–14	12	14
12,480–520	0	Anhydrite	0–7.5	0	0
12,525–28	16	Limestone	8–16	12	14
12,558–64	18	Limestone, slightly dolomitic	10–18	16	17
12,620–30	6	Limestone	3–11	6.5	6

In column 4 is the range of Sonic porosities possible without knowing lithology, assuming t_{ma} for sandstone is 19,500 ft/sec at the depth indicated. The range is intolerably large. However, using the D-N lithology, Sonic porosities are those given in column 5. These agree reasonably well with the D-N values of column 2.

Correction for Lack of Compaction

The time-average relation holds quite well in consolidated or well-compacted formations. Typically, these have transit times less than 100 μsec/ft. However, serious errors arise if the relation is applied without modification in shallow, unconsolidated sands.[27] If the effective pressure on the rock framework (overburden-hydrostatic) is less than about 4,000 psi, which is the case at depths less than about 7,000 ft, the sand has not reached its fully compacted rigidity and therefore its velocity. Travel times in uncompacted sands may reach as much as 150 μsec/ft, which convert to porosities far above the known maximum of 40%.

To take care of this situation, the porosity computed from Eq. 5.16 is divided by a compaction correction factor B_{cp}, as indicated in Fig. 5–33. The factor varies from 1.0 to as high as 1.8. It can be estimated by observing the transit times of shales, t_{sh}, adjacent to the zone of interest. These transit times also increase with lack of compaction. When they exceed 100 μsec/ft, B_{cp} is approximated by

$$B_{cp} = t_{sh}/100 \qquad (5.17)$$

The dashed lines of Fig. 5–33 are appropriate to use for B_{cp} values exceeding 1.0. B_{cp} is never less than unity.

Fig. 5–35, taken from a Gulf of Mexico log, exemplifies the need for a compaction correction. The water-bearing sand at 4,100 ft averages about

143 μsec/ft travel time. Eq. 5.16, with t_{ma} = 55.5 and t_f = 189, gives ϕ = 66%, which is much too high. Adjacent shales, at 4,380 ft for example, read 145 μsec/ft, which gives B_{cp} = 1.45. Corrected porosity is then 66/1.45 or 45%. This is still too high since the true value is closer to 37%. The correction factor, B_{cp}, can be determined more accurately by comparison of Sonic

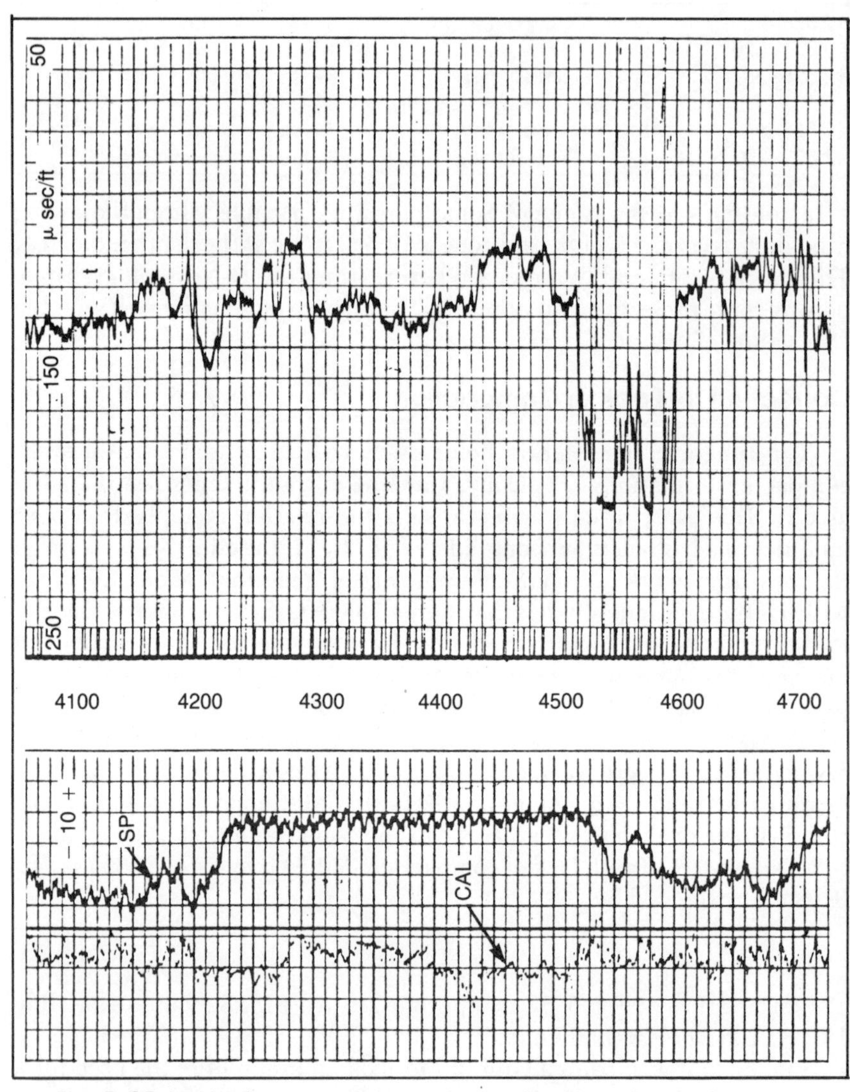

Fig. 5–35 Lack of compaction and gas effect in soft sand (courtesy Schlumberger)

porosities with Density or resistivity-computed porosities in clean water-bearing intervals. However, this negates the value of the Sonic log as a stand-alone measurement.

In effect the Sonic log is not a good porosity indicator in uncompacted formations. The Density-Neutron combination or just the Density log alone should be used for porosity evaluation when travel times exceed 110 μsec/ft.

Simpler Travel Time-Porosity Conversion

A simpler Sonic porosity transform currently coming into use is shown graphically in Fig. 5–36.[28] It is entirely empirical, based on comparison of transit times with core porosities and porosities derived from other logs. The transform can be approximated with adequate accuracy in the regions of interest by the equation

$$\phi = 0.63\,(1 - t_{ma}/t) \qquad (5.18)$$

where

t_{ma} = 54 μsec/ft for sands
= 49 μsec/ft for limestone
= 44 μsec/ft for dolomite

This is the recommended relationship. It has the dual advantage of not requiring selection of different matrix times for a given lithology and of giving reasonable porosities in uncompacted sands with transit times in the

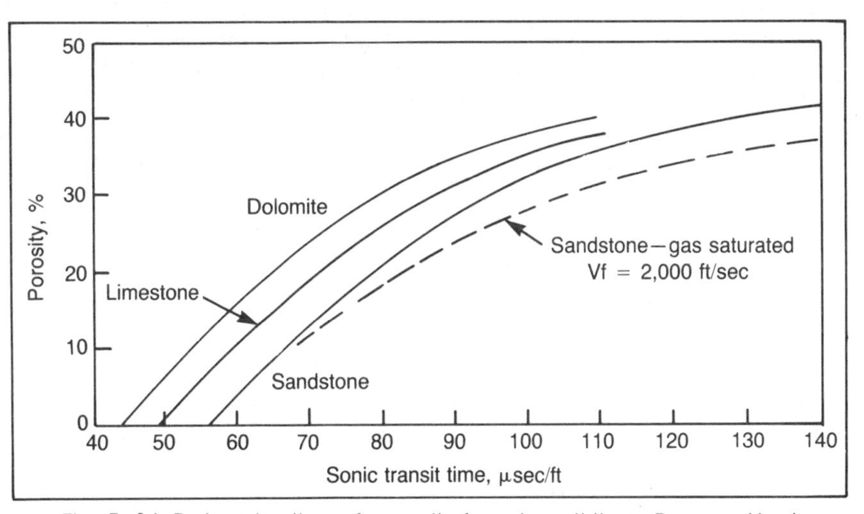

Fig. 5–36 Determination of porosity from transit time, Raymer-Hunt-Gardner method (courtesy Schlumberger and SPWLA)

range 100–150 μsec/ft. Porosity values obtained from Fig 5–36 for the selected intervals of the log of Fig. 5–29 are given in column 6 of Table 5–3. They agree very well with the Density-Neutron values in column 2.

Gas-Bearing Formations

The presence of gas in the pore space of a rock will increase the Sonic transit time over its value in the same liquid-saturated rock. Gas is very compressible; when it replaces pore liquid, it lowers the rock rigidity more than its density and decreases Sonic velocity.

The decrease in velocity is almost nil in the deeper low-porosity formations where pore volume is low and compaction pressure is high, which means that pore fluid contributes little to rock rigidity. However, it can be as high as 40% in shallow, high-porosity formations where pore volume is large and compaction pressure is minimum — in which case pore fluid has a much larger contribution to formation rigidity.

Fig. 5–37 shows calculated velocities for sands of porosities 39% to 26% at depths from 2,000–10,000 ft as hydrocarbon saturation increases from 0 to 1.0 (proceeding from right to left in the horizontal scale).[29,30] Gas saturation is represented by the solid lines, and oil saturation is represented by the broken lines. As gas saturation increases from 0 to about 15%, the velocity drops drastically, which is a result of the fluid compressibility rapidly increasing (Fig. 5–38). With further increase in gas saturation, however, velocity gradually increases again since the rapid decrease in density of the pore fluid more than compensates for the additional increase in fluid compressibility. The maximum velocity decrease predicted is about 10% for the 10,000-ft case and 35% for the 2,000-ft case.

Whether the Sonic will sense the presence of gas depends on how much residual gas is left after invasion in the 1 in. or so of formation being investigated by the tool. In medium- to high-porosity gas-bearing formations, one would expect a residual gas saturation of at least 15% in the flushed zone so that gas should be sensed by the tool. This is implied in the gas-saturated sandstone line of Fig. 5–36 where the shift in travel time from liquid to gas is consistent with the predictions of Fig. 5–37.

Fig. 5–35 shows an example of an unusually large gas response. The upper 80 ft of the sand from 4,520–4,715 ft is gas bearing with approximately 90% gas saturation in the noninvaded zone. The Sonic log shows an increase of about 50 μsec/ft in travel time, or about 35% as predicted by Fig. 5–36 (extrapolated). However, signal amplitudes in uncompacted sands can become extremely low so that increased travel times can be due to cycle

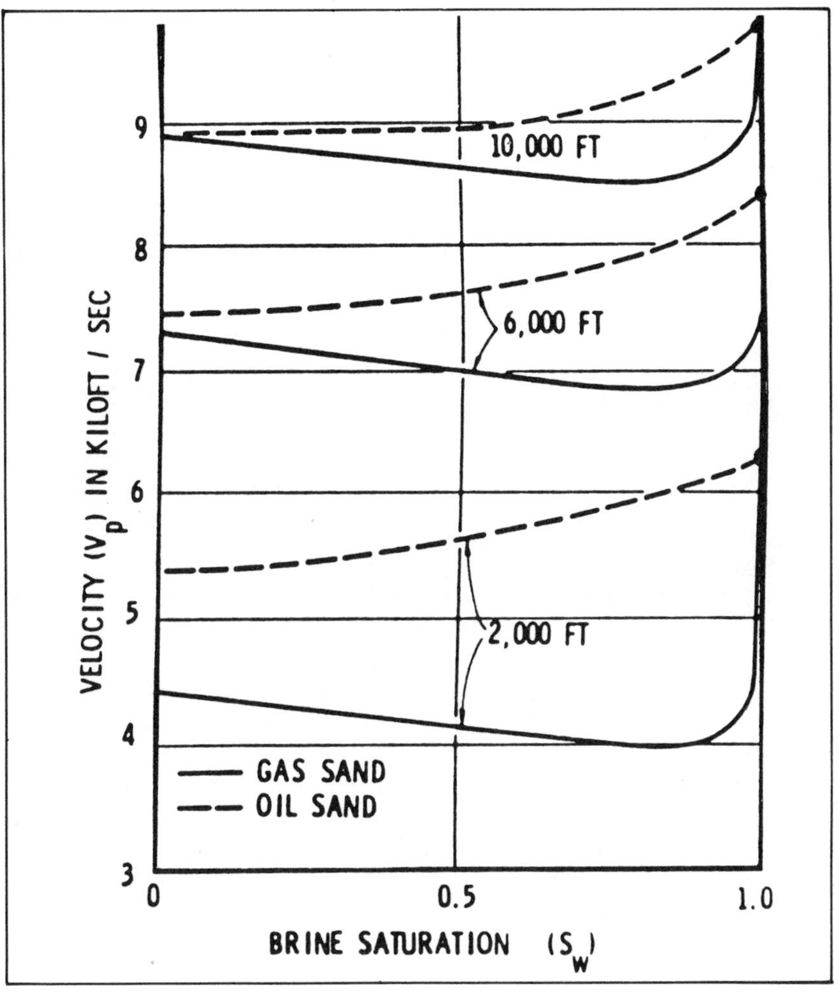

Fig. 5–37 Calculated sand velocities with varying oil and gas saturations (after Domenico, SEG)

skipping or, in the extreme, to the tool reading mud travel time. The latter appears to be the case in the 4,535– 4,550-ft and 4,572–4,580-ft intervals of the example.

Overall, the Sonic log cannot be considered a reliable gas indicator, either because the zone of investigation can be completely flushed of gas or because the more usual increase of 5–10% in travel time appears simply as

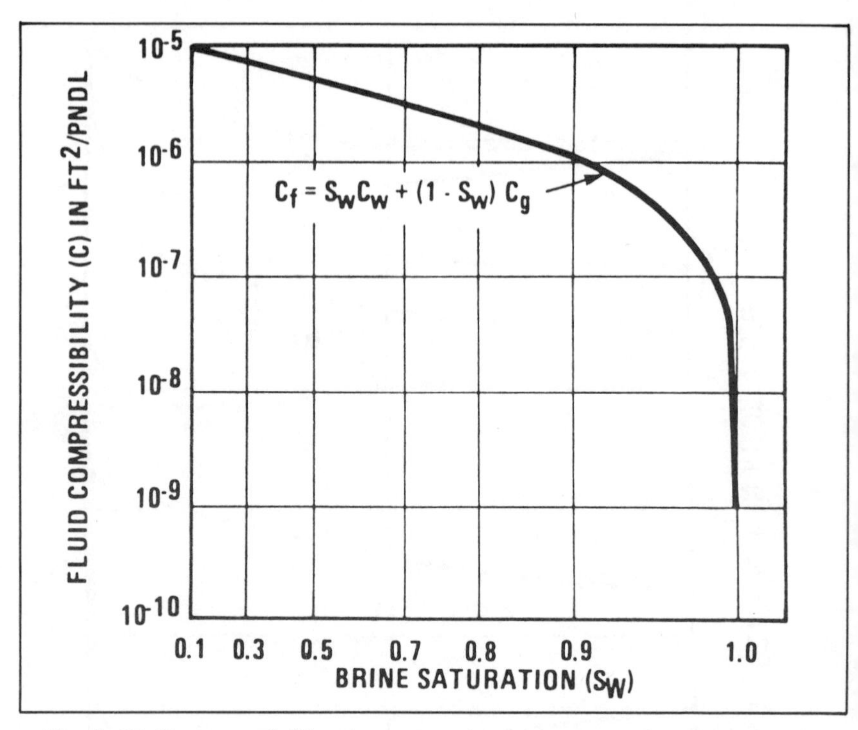

Fig. 5–38 Compressibility of a gas-water mixture as a function of water saturation (after Domenico, SEG)

an increase in porosity of similar magnitude. In the case of oil, Fig. 5–37 clearly shows that a residual oil saturation of 30% in the flushed zone will have much less effect than gas on the Sonic log and therefore will be even less distinguishable.

Secondary Porosity

The need to have a Density-Neutron log that can give both lithology and porosity in low-porosity regions does not completely negate the value of a Sonic log. It is useful as an indicator of secondary porosity, which occurs primarily in carbonates.

The Sonic log tends to ignore porosity in the form of isolated vugs or channels. Loosely speaking, the tool will not see such porosity if a vertical slice of the circumference of a hole between the two receivers is free of such vugs or channels; the first arrivals will follow that path. On the other hand, the Density and the Neutron, which respond to average porosity in their 4-in. to 10-in. depths of investigation, will average vugular porosity with intergranular porosity.

Secondary porosity may therefore be considered as the difference between the D-N and Sonic porosities ($\phi_{dn} - \phi_s$). A secondary porosity index may be defined as

$$SPI = (\phi_{dn} - \phi_s)/\phi_{dn} \quad (5.19)$$

The greater the SPI, the higher the permeability is likely to be for a given total porosity, assuming the vugs or channels are interconnected. This is valuable information for estimating formation productivity.

Porosity determinations, however, must be very accurate if the SPI is to be considered reliable. Taking the 12,525–12,528-ft interval listed in Table 5–3, SPI is calculated to be (16 − 12)/16 or 0.25 if Sonic porosity derived from Fig. 5–33 is used (column 5), but it is only 0.12 if the porosity from Fig. 5–36 is used (column 6) . The index is therefore more useful on a comparative than on an absolute basis to indicate secondary porosity trends in a given field where hole conditions, logging programs, and interpretation methods are standardized.

Hole Enlargement and Formation Alteration Effects

Sonic logs are very tolerant of borehole size variations within the normal 6–12 in. range. However, serious errors can occur in transit times recorded with a standard BHC-type Sonic tool in larger holes or those with appreciable formation alteration.

Considering Fig. 5–27, one can readily visualize that if the hole diameter is too large and the Sonic tool is centralized, the mud wave proceeding directly down the hole will arrive at the receiver ahead of the formation compressional wave. The recorded travel time will then not reflect formation travel time.

The maximum hole size tolerable depends on the shortest transmitter-receiver spacing and the formation travel time.[31] For the BHC tool with a near receiver spacing of 3 ft, the maximum tolerable hole size is 14 in. for a 150-μsec/ft formation and 22 in. for a 100-μsec/ft formation. Enlargements of this nature do occur, particularly in shallow uncompacted formations.

Formation alteration is similar in its effect. Typically, such alteration is the swelling of water-sensitive shales in contact with fresh drilling muds.[32]

The swelling begins at the borehole wall and slowly progresses outward. The swollen shale in effect has higher porosity and therefore lower velocity than the undisturbed shale behind it. If the altered zone has appreciable thickness at the time of logging, the first arrival of energy at the receivers can be through that layer and the Sonic tool will read an abnormally long transit time. A striking example is shown in Fig. 5–39 where there is a progressive increase in shale travel times (from approximately 125 to 140 μsec/ft at 9,330 ft, for example) over an interval of 3 to 35 days after drilling.[33] Little alteration occurs in the sands, such as at 9,292 ft. This is fairly typical, though shallow sands sometimes exhibit alteration.

The maximum alteration depth tolerable depends on borehole size, near receiver spacing, and both altered and unaltered zone travel times. The standard BHC tool in 10-in. hole can cope with only 2–4 in. of alteration in formations of 100–150-μsec/ft travel times.

The solution for both hole enlargement and shale swelling problems is to utilize a Sonic tool with longer transmitter-receiver spacings. A tool with 8-ft minimum spacing can tolerate almost any conceivable hole diameter and can cope with 8–12 in. alteration depths under the conditions cited previously.

In past years the main application of long-spaced tools has been for geophysical purposes, wherein Sonic logs are used to depth-calibrate seismic sections through the medium of synthetic seismograms. Travel time anomalies caused by shale alteration and hole enlargement (the two often go together) can cause false reflections on the synthetic seismograms and poor time match with the seismic section. In shallow soft rock areas such as the U.S. Gulf Coast where shale alteration is common, a long-spacing tool should always be run rather than the standard BHC.

Recently, success has been achieved in shear wave logging and in logging through casing with long-spaced sondes.[34] Increased use of such tools for these purposes in the future can be expected.

THE LONG-SPACING SONIC LOG*

Fig. 5–40 shows schematically the arrangement of the Schlumberger LSS tool. Two transmitters are spaced 2 ft apart at the bottom of the tool and two receivers are spaced 2 ft apart at the top with 8 ft of space between the nearest transmitter and receiver. With this arrangement two long-spacing

*Designated LSS by Schlumberger and LS BHC by Dresser-Atlas.

logs are recorded simultaneously, one with 8–10 ft spacings and one with 10–12 ft spacings.

Borehole compensation is accomplished by depth memorization rather than by the inverted array technique. To illustrate, the travel time at depth

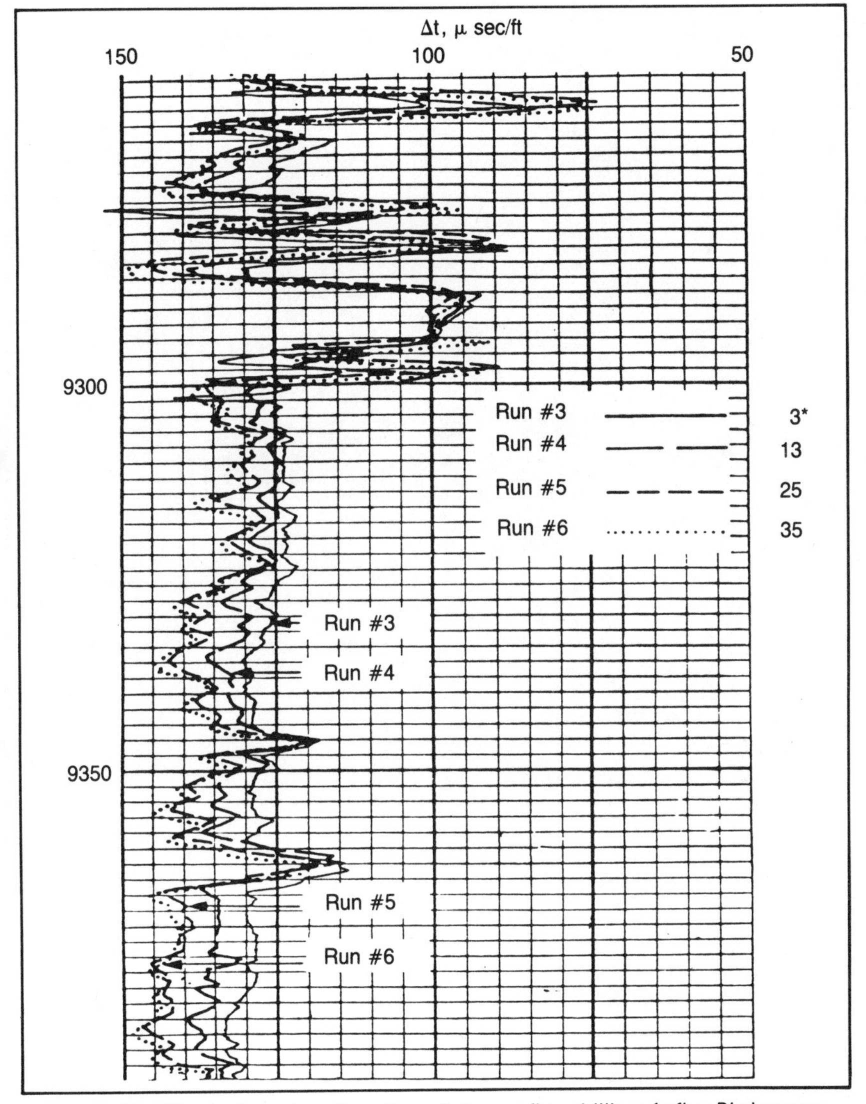

Fig. 5–39 Effect of shale alteration; *days after drilling (after Blakeman, courtesy SPWLA)

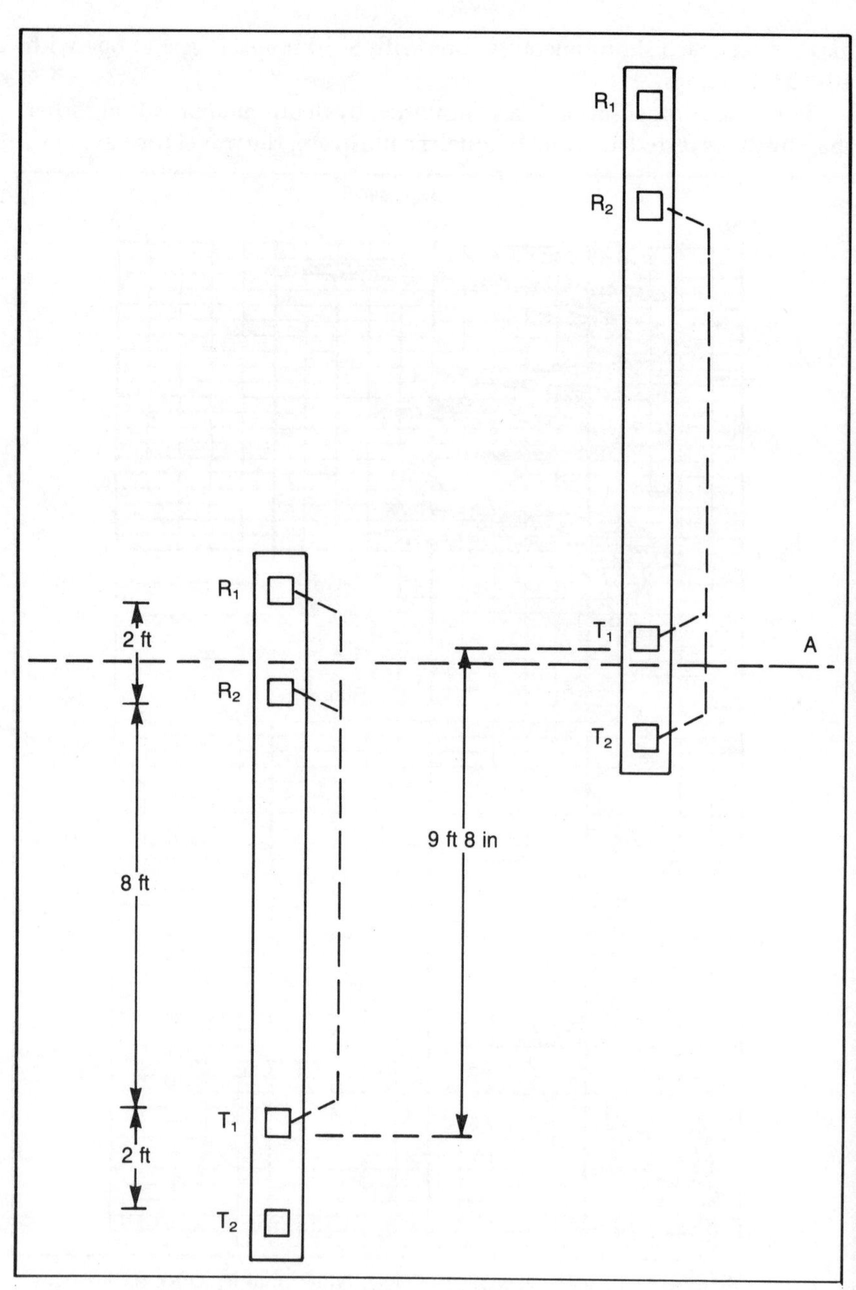

Fig. 5–40 Schlumberger Long-Spacing Sonic sonde

level A is first measured with the two receivers spanning that level. Transmitter T_1 is pulsed twice in succession and the time difference ($T_1R_1 - T_1R_2$) corresponding to compressional arrivals is placed in memory. If the hole diameter is different at the two receivers, the measured time will be in error. After the tool has moved uphole 9 ⅔ ft, the two transmitters will be spanning the same interval (between points of refraction). Each is pulsed and, using only receiver R_2, the time difference ($T_2R_2 - T_1R_2$) is measured; it will be in error by a like amount but in the opposite direction. This time difference is then averaged with the previous value, retrieved from memory, to provide an 8–10 ft spacing travel time compensated for borehole variation.

Using transmitter T_2 in the first position instead of T_1 and receiver R_1 in the second position instead of R_2 provides a 10–12 ft compensated log in like fashion.

Fig. 5–41 shows a comparison of standard BHC and long-spacing logs in a shallow section of hole through formations that are primarily shale, save for a fairly thick sand at level A. Except in the sand, where the two logs agree, the conventional log is badly in error. In the upper third of the log, the BHC curve reads a transit time too long because of shale alteration. In the sections above and below the sand, it reads much too long because of hole washout. By contrast, the LSS log reads correctly.

Anomalous Triggering Effects

Unfortunately, longer spacings lower signal amplitudes at the receivers and aggravate noise triggering and cycle skipping. When this happens, spikes occur in pairs, spaced 9⅔ ft apart, because each T-R time measurement is used twice for time computation: once at the level it is taken and again 9⅔ ft higher in the hole.

This behavior is illustrated in Fig. 5–42. The 8–10-ft log is labeled DT and the 10–12-ft log is labeled DTL. An incorrect reading recorded on the 8-ft spacing will cause two spikes of like polarity spaced 9⅔ ft apart to appear on the 8–10-ft log (pair D). Similarly, an error on the 12-ft spacing will cause a similar pair of spikes in the 10–12-ft log. An error on one of the 10-ft measurements, however, will cause a spike of one polarity on the 8–10-ft log and a displaced spike of opposite polarity in the 10–12-ft log (pairs A, B, and C). These anomalies can be smoothed out by eyeball averaging or eliminated more precisely by computer processing.[35]

We can also expect to see improper borehole compensation when sonde sticking causes erratic downhole motion. If sticking occurs, as evidenced by

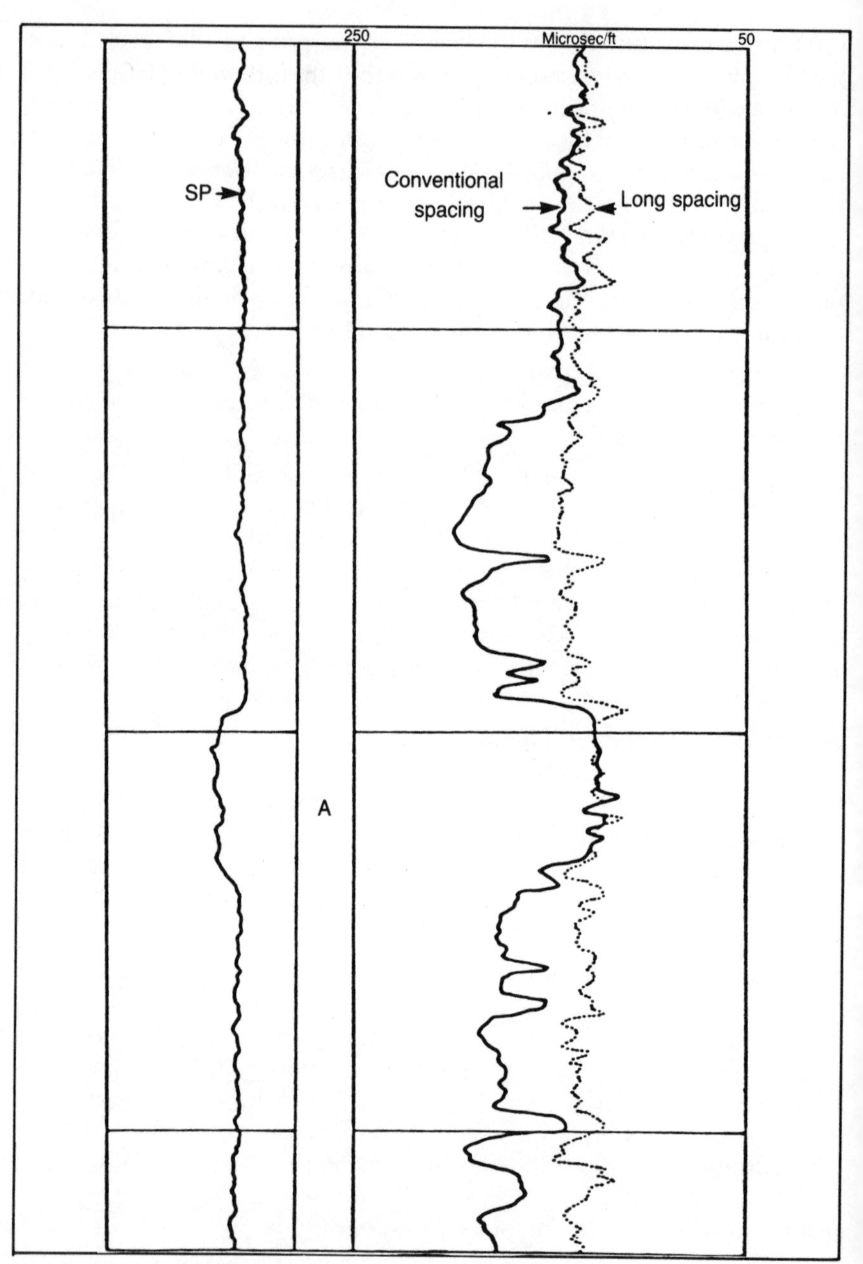

Fig. 5–41 Comparison of standard and long-spaced Sonic logs in altered and caved shale (after Thomas, courtesy SPWLA)

cable tension buildup, a given length of cable may be reeled in at the surface but the sonde will move a smaller distance downhole. In this case transit time measurements at two different depth levels are being averaged. The

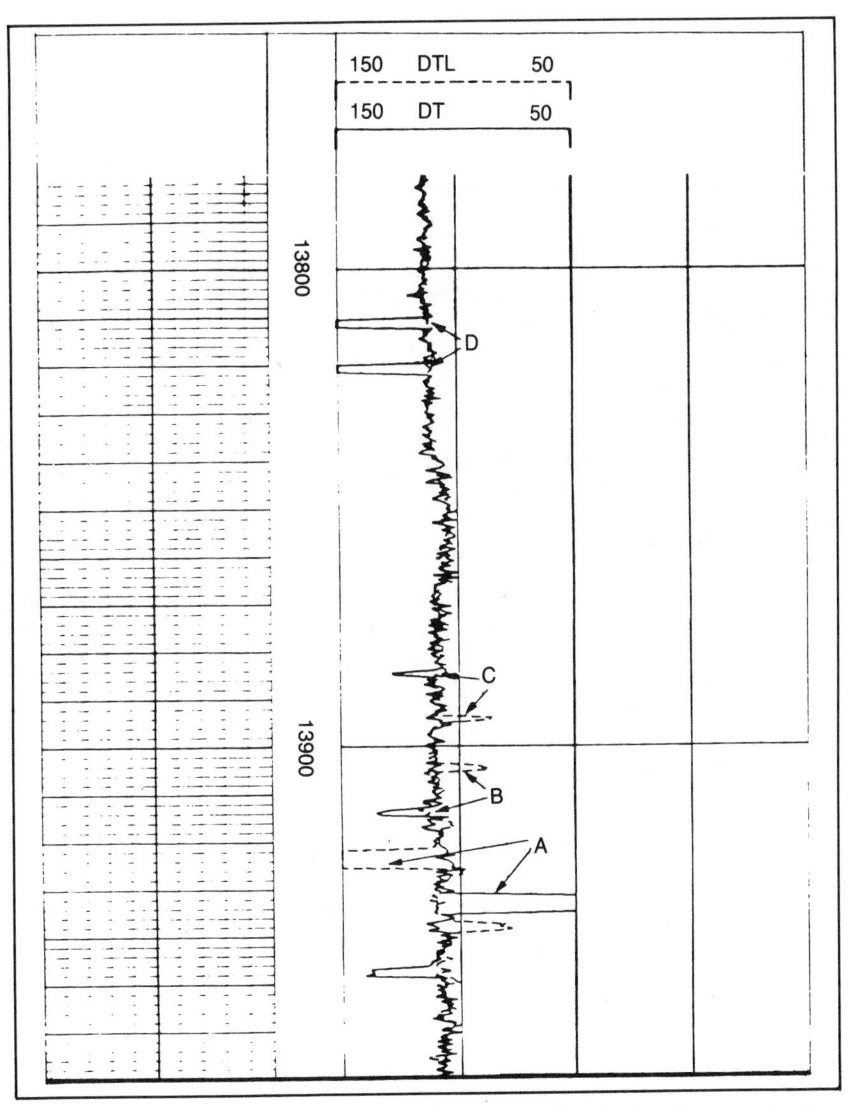

Fig. 5–42 Cycle skips and noise triggering, causing dual spikes on depth-derived Long-spacing Sonic log (after Purdy, courtesy SPWLA)

recorded travel time will be in error if the hole configuration is different at the two levels. No easy solution to this problem exists.

The accuracy of LSS measurements can be checked, as with the standard BHC, by observing that the travel time in casing reads 57 μsec/ft. It is mandatory to obtain a reading in the uncemented section of casing since signal amplitudes in cemented casing are too small at the long spacings used.

Derivation of porosity from LSS travel times is exactly the same as for the standard BHC. Figs. 5–33 and 5–36 are applicable.

Shear Travel Time Measurement

The desirability of measuring shear wave velocity in addition to compressional velocity has long been recognized. The basic reason is that, with the two velocities, the fundamental elastic constants of the medium can be calculated. With these, mechanical properties of importance such as sanding threshold and fracturing pressure can be inferred. Other applications are also coming to light for the shear wave measurement.

Shear wave measurement requires data acquisition with long spacing sondes, digitizing of sonic wave forms at the wellsite, and sophisticated wave form processing offsite.

The longer the spacing, the greater the separation in time between the compressional and shear wave packets and the easier it is to sense accurately the later shear arrivals. Fig. 5–43 illustrates the point. It is a variable-density recording of receiver wave forms at 8-ft spacing. Once per foot of hole, the receiver signal modulates the intensity of an electron beam sweeping across the face of a cathode ray tube (CRT) with the sweep starting when the transmitter is pulsed. The modulated sweeps are recorded on a moving film to provide a display in which wave amplitudes are translated into black-and-white contrasts.

The compressional arrivals are seen clearly at about 0.3 msec and the shear arrivals at approximately 0.6 msec. They are well separated, which facilitates shear wave extraction.

While shear transit time can be determined from variable-density displays, it is a tedious and imprecise process. Accurate determination requires that the wave forms be digitized at 5- to 10-μsec intervals and processed through correlation and filtering techniques to suppress the direct compressional wave and reflections associated with it and to enhance the shear arrival.[36] A tremendous amount of data must be handled, so the processing must be done in computer centers at present.

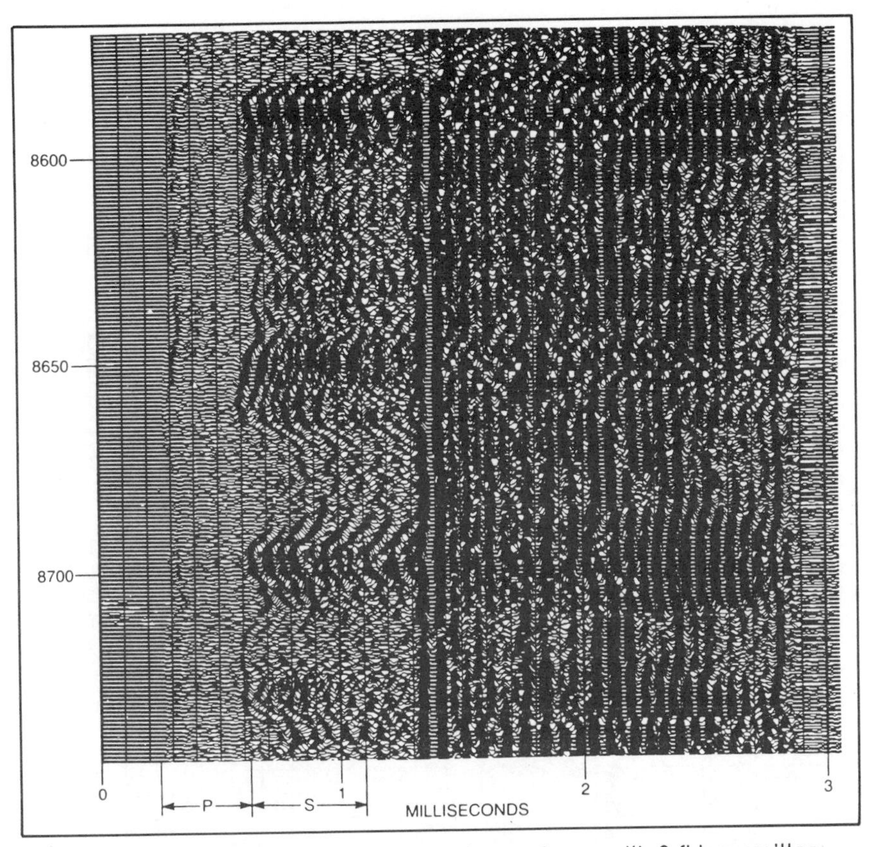

Fig. 5–43 Variable density display of waveforms with 8-ft transmitter-receiver spacing (after Siegfried and Castagna, courtesy SPWLA)

Schlumberger's computation method, *direct phase determination,* is commercially available.[37] It computes compressional and shear travel times, their ratio, and energies of the compressional and shear packets.

Fig. 5–44 is an example of compressional and shear logs recorded in a sand-shale series.[38] Travel time curves are displayed in Track 5, and the ratio of shear to compressional travel times (denoted $\Delta T_s/\Delta T_c$) is plotted as a solid line in Track 4.*In Tracks 1, 2, and 3 are the grain densities, fluid

---

*Along with the derived Poisson's ratio (dashed line).

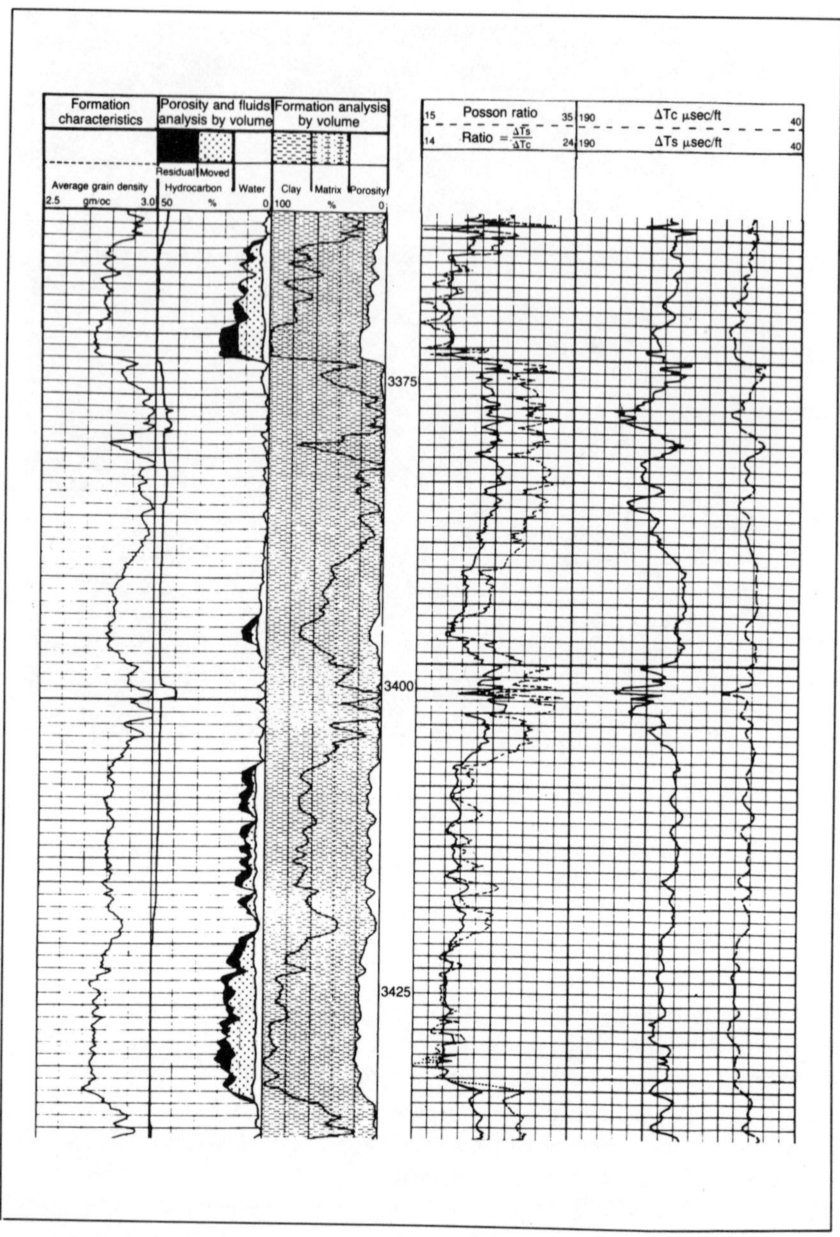

Fig. 5–44 Example of shear and compressional travel time recording in sand-shale series (courtesy Schlumberger)

analysis, and matrix analysis computed from a standard logging suite. The most obvious points are that the shear log has much the same character as the compressional log, and that Poisson's ratio correlates well with the clay concentration shown in Track 3.

The various uses of shear wave logging are presented in the following sections.

Mechanical Properties Determination

This is the most clear-cut application of shear transit time logging. The fundamental elastic constants of a medium can be expressed in terms of compressional and shear velocities and therefore calculated once these velocities are measured. The relations are

POISSON'S RATIO $$\upsilon = (0.5r^2 - 1)/(r^2 - 1) \tag{5.20}$$

where $$r = V_c/V_s$$

SHEAR MODULUS $$\mu = \rho V_s^2 \tag{5.21}$$

BULK MODULUS $$B = \rho (V_c^2 - 1.333 V_s^2) \tag{5.22}$$

YOUNG'S MODULUS $$E = 2\mu(1 + \upsilon) \tag{5.23}$$

Poisson's ratio is the ratio of lateral to associated linear strain in a medium. The moduli represent the resistance of the medium to shearing, volume compression, and linear elongation. They largely determine mechanical properties such as sanding threshold, fracture pressure, and rock drillability.

Soft unconsolidated sands will break down and flow during production if production rates or drawdown pressures are too high. When this happens the well sands up and chokes its own flow. Empirically, breakdown should not occur if the product of shear and compressional moduli exceeds a threshold value given by

$$\mu B = 0.8 \times 10^{-12}\ \text{psi}^2 \tag{5.24}$$

This product is recorded on mechanical properties logs.[39]

Prior to shear wave logging, an empirical relation between the fraction of clay in the pore space (as determined from logs) and Poisson's ratio was used to obtain the latter and thus V_s from Eq. 5.20. The product μB could then be calculated. With shear wave logging there is no need for the empirical relation (which held for the U.S. Gulf Coast but not necessarily elsewhere). V_s and V_c are directly measured, leading to a more accurate determination of μB.

Fracture pressure is that hydrostatic pressure in the borehole that will cause the adjacent formation to split, generally along a vertical plane, and permit borehole fluid to flow into the fracture. During drilling it is important not to exceed the fracture pressure or circulation will be lost. On the other hand, when fracturing tight formations to improve their productivities, fracture pressures must be exceeded. Fracture pressure may be written[40]

$$P_f = \frac{2\nu}{1-\nu} \cdot P_o + \alpha \left(\frac{1-3\nu}{1-\nu}\right) P_r \tag{5.25}$$

where

P_o = overburden pressure (0.9 to 1.0 psi/ft)
P_r = pore pressure (measurable with a formation tester)
α = constant, approximately 0.5

The fracture pressure is very dependent on Poisson's ratio. Consequently, it is important to have an accurately measured value rather than a value empirically obtained as indicated above. Fig. 5–44 shows that Poisson's ratio can vary from 0.15–0.33, a wider range than the empirical method predicts.

Improved determinations of other derived parameters, such as fracture widths and heights and rock drillability, can be expected with more precise knowledge of the elastic moduli.[41]

Lithology Identification

The ratio of shear to compressional travel time appears to reflect lithology for the three major sedimentary series. Typical values for clean formations are

Lithology	t_s/t_c
Sandstone	1.58–1.78
Dolomite	1.8
Limestone	1.9

A statistical plot from several wells is shown in Fig. 5–45. The limestone and dolomite ratios are independent of porosity over the 0–20% range covered. The sandstone points for liquid-filled pores in the 15–25% range show an increase in the t_s/t_c ratio from 1.6 to 1.8 as porosity increases. The dashed line is the sandstone line from earlier laboratory work with cores.[42] It agrees with the well log data.

The t_s/t_c ratio, however, is not as definitive an indicator of lithology as the P_e curve of the Litho-Density tool whose values increase from 1.8 to 5.0, progressing from sand to dolomite to limestone. On the other hand the t_s/t_c ratio can be useful in wells drilled with barite-loaded mud where the P_e curve is not usable.

The constancy of t_s/t_c implies that shear travel time has much the same dependency on porosity as compressional velocity. In fact, the abscissa scale of Fig. 5–36 can be multiplied by the appropriate ratio to produce an approximate porosity vs shear travel time curve for each lithology.

Clay Indication

In sand-shale sequences the presence of clay and silt increases the t_s/t_c ratio from a nominal value of 1.6 to as high as 1.9, as illustrated by Fig. 5–44. Once again the range of variation is not nearly as great as for the Gamma Ray or SP. However, there is evidence that shear attenuation may be sensitive to the disposition of clay as well as to the content. If this could distinguish between dispersed and laminated clay, it would significantly enhance formation evaluation.

Gas Effect

The presence of gas in pore space should decrease the shear wave travel time. This is because shear velocity, V_s, of an elastic solid is given by

$$V_s = (\mu/\rho)^{1/2} \tag{5.26}$$

where μ is the shear modulus and ρ is the density. Replacing liquid by gas in the pores should not affect the shear modulus, since fluids do not support shear, but it will lower the density. A gas saturation of 15% in the flushed zone should lower the travel time by about 5%. It will increase the compressional travel time by about 10%, so the t_s/t_c ratio should decrease by about 15%. Fig. 5–45 shows a 10% decrease in ratio, from 1.8 to 1.62, most of which was due to change in t_c. The conclusion is that the shear travel time log will be relatively insensitive to gas.

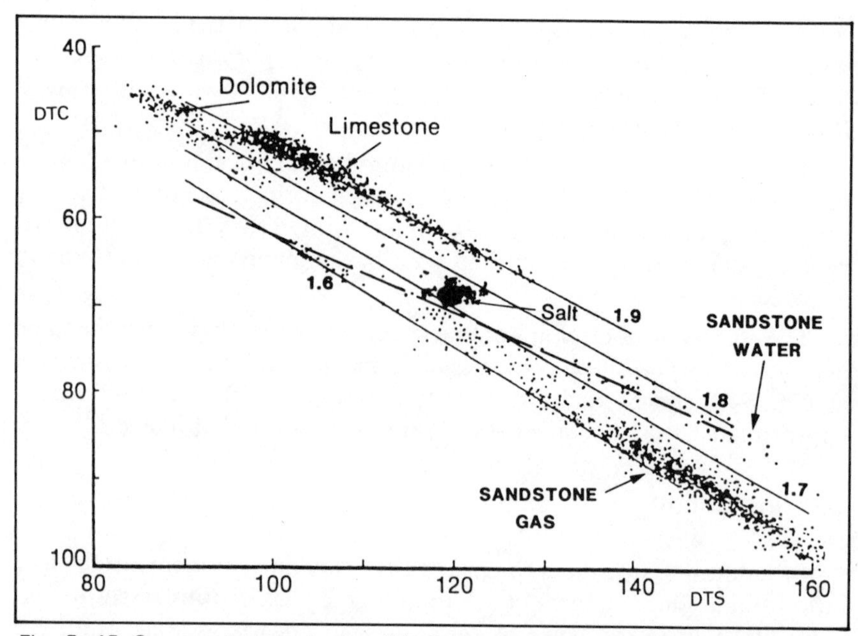

Fig. 5–45 Comparison between lithology and shear to compressional travel time ratio with data from several wells (courtesy Schlumberger and SPWLA)

ELECTROMAGNETIC PROPAGATION-MICROLOG COMBINATION

The Electromagnetic Propagation-Microlog (EPT-ML) combination is a marriage of one of the earliest and simplest electrical tools (ML) with one of the latest and most sophisticated electromagnetic tools (EPT). What they have in common is that

- both are pad devices
- both have extremely high vertical resolution and very shallow depth of investigation, a few inches in each case
- both display information concerning hydrocarbon producibility
- both are best suited to fresh mud conditions
- logging speed for both is approximately 2,000 ft/hr

Use of Micrologs

The prime use of the Microlog is to indicate permeable zones. It does this in great detail by sensing the presence or absence of mud cake. Permeable zones allow invasion of mud filtrate, which results in mud cake buildup; impermeable zones do not invade. The Microlog is still the best permeable zone or "sand count" indicator today. However, it gives no indication of the magnitude of the permeability.

Use of EPT Logs

The Electromagnetic Propagation Log essentially gives the water-filled porosity in the first few inches of formation, largely ignoring hydrocarbon-filled porosity.* Its principal use is for determining the water saturation of the flushed zone. This is done by comparing the EPT porosity, ϕ_{EP}, with the total fluid porosity, ϕ, obtained from the Density Neutron combination. Flushed-zone water saturation is approximated by

$$S_{xo} = \phi_{EP}/\phi \tag{5.27}$$

This approach is independent of the salinity of the water in the flushed zone. It does not require a knowledge of mud filtrate resistivity or of whether mud filtrate completely replaced connate water in the flushed zone.

Applications of the EPT fall into the following categories.

1. Determination of hydrocarbon movability by comparing the flushed-zone saturation, S_{xo}, with the undisturbed-zone saturation, S_w. Moved hydrocarbon, as a fraction of the pore space, is $(S_{xo} - S_w)$. The technique is most applicable to fresh-mud situations where R_{xo} logs (MLL, PL, or MSFL) are seldom run and where, if run, may not provide reliable S_{xo}values.

2. Detection of hydrocarbons in freshwater areas where electrical logs cannot distinguish water from oil. This situation is common to shallow producing areas, often where oil is fairly heavy. The heavier the oil the greater the residual oil saturation in the flushed zone, so the more readily it is detected as a difference between ϕ_{EP} and ϕ. Heavy oil regions of California and the Athabasca tar sands are examples. In such areas there is so little displacement of oil by mud filtrate that S_{xo} approximates the actual water saturation, S_w, of the reservoir.

3. Delineation of 100% water-bearing zones, where $\phi_{EP} = \phi$, which then can be used for determination of R_w. Wet zones can readily be picked

---

*Available only from Schlumberger at present.

on resistivity logs in porous Miocene sands, for example, but may not be evident in carbonate sequences where porosities are lower or in regions where uplifting or overthrusting of sediments cause R_w values to change rapidly.

THE EPT-ML SENSOR ARRAY

Fig. 5–46 shows the portion of the EPT-ML tool that contains the sensors. On the upper part is a fixed pad, about 1-ft long, with EPT transmitters and receivers. This pad is forced to ride the side of the hole by a strong backup arm on which a Microlog pad is affixed. The arm also measures gross hole diameter. At the same level but in line with the EPT sensors is a small

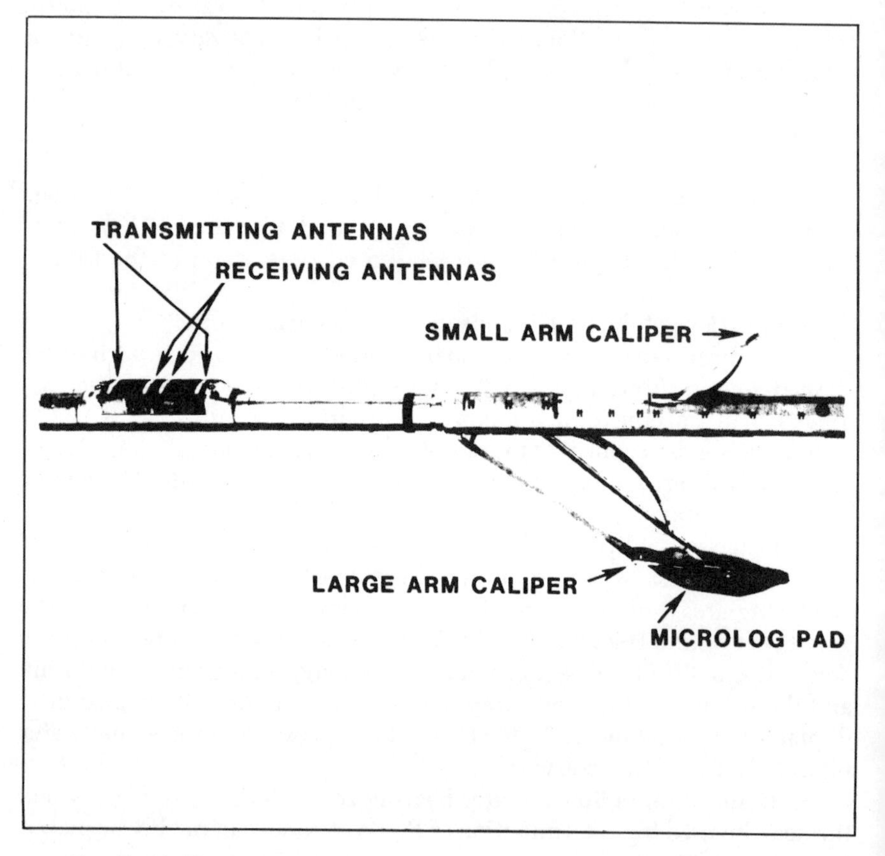

Fig. 5–46 Electromagnetic Propagation-Microlog sonde (courtesy Schlumberger, © SPE-AIME)

weak arm that measures the rugosity of the wall traversed by the EPT pad. The sum of the large and small arm displacements is recorded as hole diameter. Both arms are hydraulically retracted for descent into the hole and opened for logging on ascent. Minimum hole size that can be logged is 7½ in. with the ML pad or 6½ in. without it.

The EPT-ML tool can be run in combination with the GR-CNL-FDC (or LDT) array, as shown in Fig. 5–47. Combined tool length is approximately 80 ft; logging speed is 1,800 ft/hr. A high-speed telemetry cartridge transmits all information to the surface in digital form where it is processed in the logging unit computer. This telemetry system, becoming universal in Schlumberger, greatly extends tool combinability.

THE MICROLOG

The Microlog* tool is a simple three-electrode, nonfocused electrical logging device.[43] Three button electrodes, spaced 1 in. apart are mounted on a hydraulic pad. The pad is actually a rubber bladder filled with oil so that it can conform to hole irregularities. It is important to the quality of the log that pad contact with the wall be sufficiently good to prevent drilling mud from directly contacting the electrodes.

A constant survey current is emitted from the lower button. This current passes into the formation and returns to the sonde body. A Micronormal resistivity curve, denoted $R_{2''}$ or MNOR, is recorded by measuring the potential between the upper button and sonde mass. Similarly, a Microinverse curve, denoted $R_{1'' \times 1''}$ or MINV, is recorded by measuring the potential between the middle and upper buttons. The depth of investigation of the Micronormal is approximately 4 in. and that of the Microinverse 1.5 in.

When no mud cake is present, as in impermeable zones, both curves read the same value. Thus, the curves overlay in shales or in impermeable sands or carbonates if resistivity is not too high. When mud cake exists, however, the cake generally has lower resistivity (essentially higher porosity) than the flushed formation immediately behind it. Since the Microinverse curve has shallower penetration, it is more influenced by the mud cake and reads lower resistivity than the Micronormal. Positive separation between the curves, with MNOR > MINV, indicates permeability.

---

*Denoted Microlog (ML) by Schlumberger, Minilog by Dresser Atlas, Contact log by Welex, and Microelectric log (MEL) by Gearhart.

Example Log

Fig. 5–48 is an example of a Microlog run in combination with GR Density and Neutron logs. In Track 1 is recorded the caliper and GR. The latter shows two major sands at intervals A and E along with indications of thin sands at B, C, and D. The Microlog curves are recorded in Track 2. Positive separation, indicated by the dot coding, clearly shows the permeable intervals, namely the top 7 ft of A, about 2 ft in B, and 24 ft in E, for a total sand count of 33 ft. Zone C is indicated as tight and zone D as imperme-

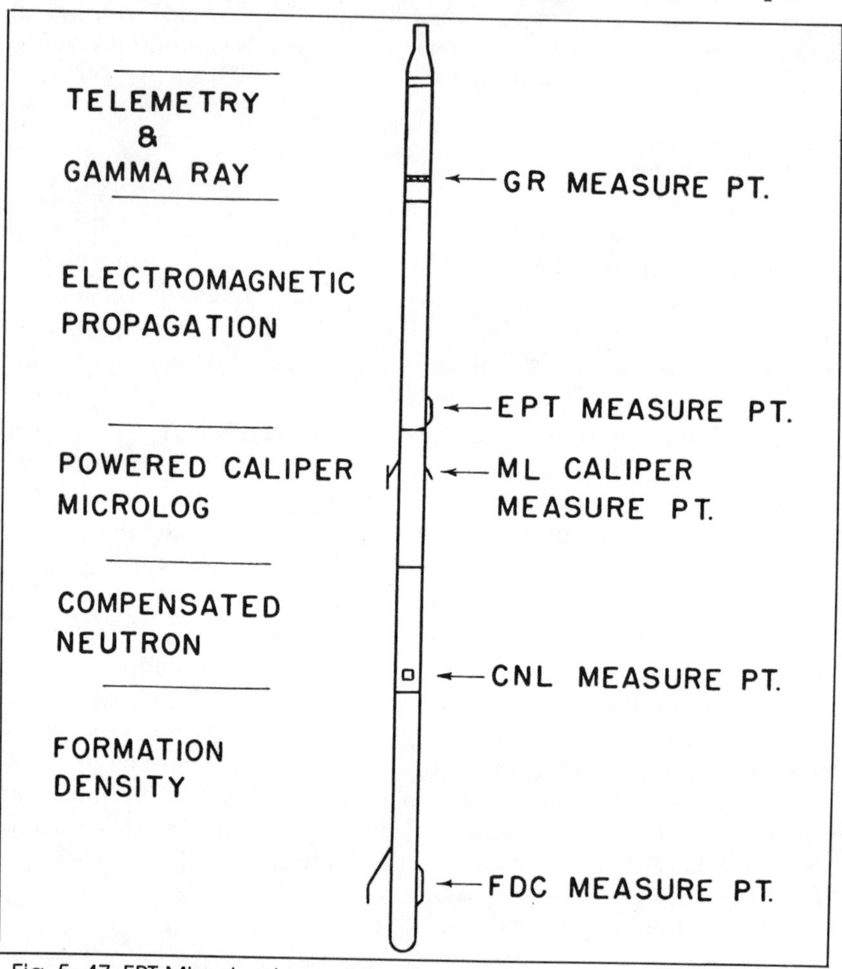

Fig. 5–47 EPT-Microlog in combination with FDC, CNL, and GR (courtesy Schlumberger)

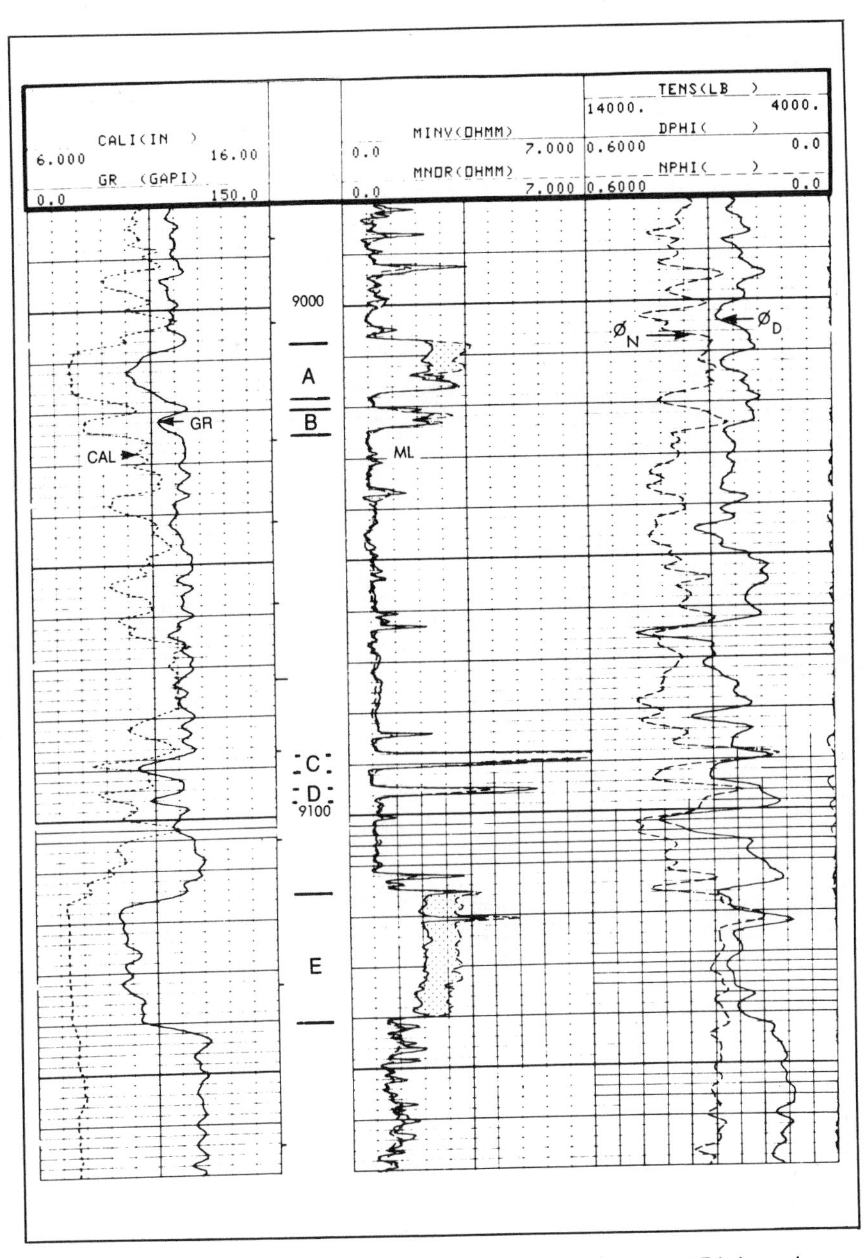

Fig. 5–48 Microlog showing preamble zones at A, B, and E intervals (courtesy Schlumberger)

able. Readings greater than 10 times the mud cake resistivity, R_{mc} (estimated as about 1 ohm-m in this case), indicate tight zones. Separation of the curves at high resistivities is meaningless due to current leakage around the pad.

Mud-Cake Thickness Estimation

An approximate value of mud-cake thickness can be obtained from the chart shown in Fig. 5–49. Mud-cake resistivity, R_{mc}, must be obtained from the log heading and converted to the temperature of interest. The ratios $R_{2''}/R_{mc}$ and $R_{1'' \times 1''}/R_{mc}$ give a point of intersection from which mud cake thickness, h_{mc}, can be estimated. For zone E of Fig. 5–48, $R_{2}'' = 2.9$ and $R_{1'' \times 1''} = 2.1$ ohm-m. Assuming $R_{mc} = 1$, h_{mc} is estimated at ⅜–½ in. as indicated on Fig. 5–49. However, if R_{mc} were 0.5 ohm-m, h_{mc} would be estimated as about ¼ in. Consequently, the procedure is not very accurate

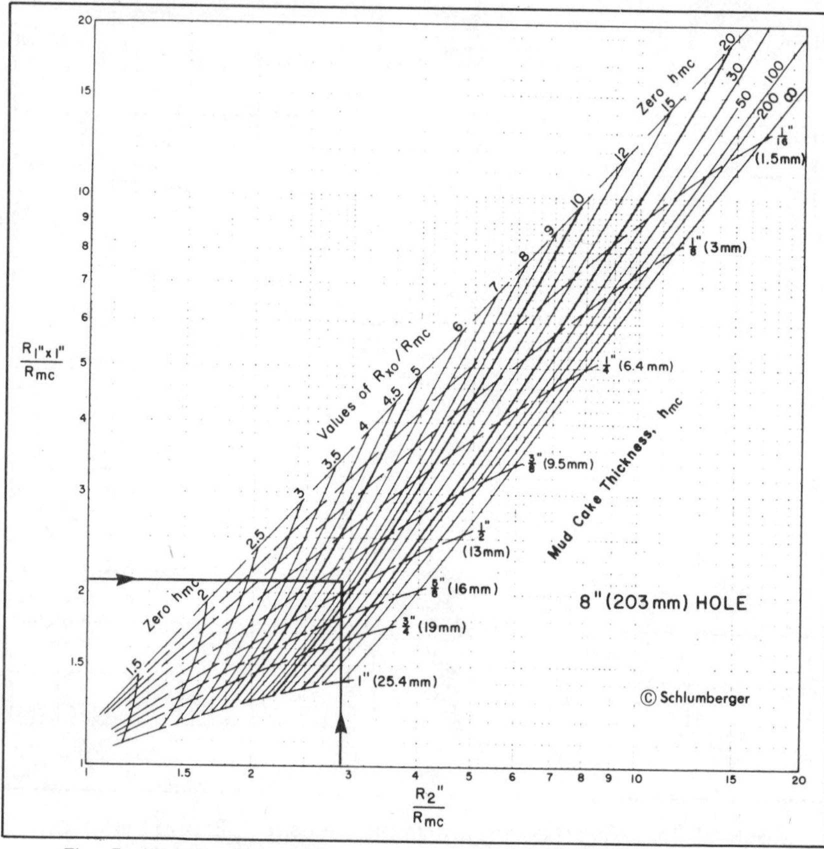

Fig. 5–49 Microlog interpretation chart (courtesy Schlumberger)

because R_{mc} values on the log heading are not measured values but are derived from the R_{mf} measurement, assuming average mud properties. The main value of the Microlog is indicating the presence of mud cake.

The Microlog works best in fresh mud. It is not suited to salt mud because mud cakes tend to be very thin in such muds and because the high conductivity of salt mud tends to short-circuit the survey current.

In early days before porosity logs were available, the Microlog was used to estimate porosity.[44] R_{xo} would be estimated from Fig. 5–49, S_{xo} would be guessed and ϕ would be derived from Eq. 2–12. However this was — and is — a very inaccurate method. The Microlog is not a porosity device nor even an R_{xo} measuring device. Microlog curves bear little resemblance to true R_{xo} curves obtained with focused MLL, PL, or MSFL electrode systems.

THE ELECTROMAGNETIC PROPAGATION LOG

The EPT tool measures the travel time and attenuation rate of microwaves propagating along the borehole in the first few inches of formation. At the very high frequency utilized, 1.1×10^9 Hz (1.1 gigaHz), the rate of travel of such waves is determined almost entirely by the dielectric properties of the formation and very little by its resistivity.[45,46] (The opposite is true for Induction and Laterolog tools, which operate at low frequencies.) In turn the dielectric permitivity is largely a function of the water content of the formation.[47,48]

Fig. 5–50 shows the EPT sensor arrangement. Two microwave transmitters (T_1, T_2) and two receivers (R_1, R_2) are mounted on a brass pad forced against the borehole wall. Spacing between transmitter and nearest reservoir is 8 cm and between the two receivers is 4 cm. The two transmitters are alternately pulsed, and upgoing and downgoing travel times measured between the two receivers are averaged. This eliminates first-order effects of uneven mud-cake thickness, pad tilt, and instrumentation imbalances. Travel time is measured by sensing the phase difference in received signals at the two receivers. A complete measurement of travel time and signal attenuation is made every 1/60 of a second and transmitted to the surface. There the measurements are averaged over 2-in. or 6-in. depth intervals.[49,50]

Vertical Resolution, Depth of Penetration, and Borehole Effects

Vertical resolution of the EPT log is extremely good. It is essentially the span between receivers, about 2 in. Depth of penetration is quite small, varying from about 1 in. in low-resistivity formations to about 6 in. in high-resistivity zones. The lower limit of formation resistivity for proper tool operation is approximately 0.3 ohm-m.

Borehole size has no effect on the EPT measurements as long as the pad makes good contact with the wall. However, thick mud cake or excessive standoff can be a problem. With fresh muds of usual resistivities, mud cake thicknesses up to ⅜ in have no effect.[51] For greater thicknesses the travel time increases until the tool reads mud cake travel time at about ¾ in. mud cake thickness. Abnormally high or low mud resistivities are undesirable. The former reduces tolerance to mud cake thickness and the latter increases signal attenuation.

Operation of the EPT is not recommended in empty hole or oil-base mud. Microwaves have high velocity in air or oil, so even very thin layers of these fluids between the pad and the formation effectively short-circuit the formation signal.

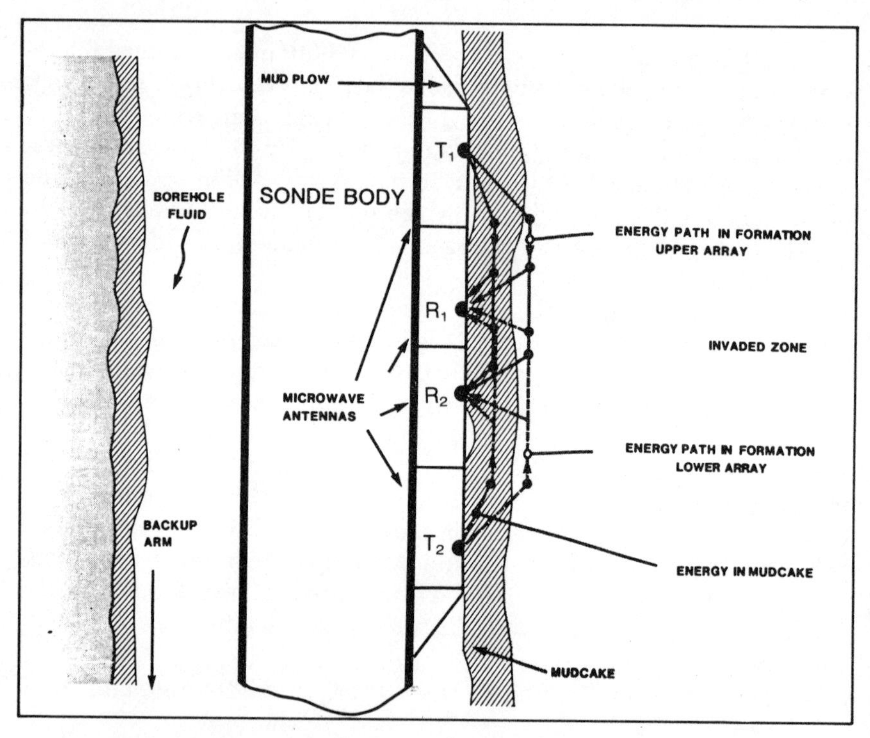

Fig. 5–50 EPT pad configuration and signal paths (courtesy Schlumberger, © SPE-AIME)

Log Presentation

Fig. 5–51 shows a recording of the basic EPT curves along with an accompanying GR log. The GR curve, the solid line in Track 1, shows two sands, denoted A and B, with the remainder of the interval appearing as shale. In Tracks 2 and 3 is recorded the measured travel time, t_{pl}, in nanosec/meter (1 ns = 10^{-9} sec). The scale is 5–25 ns/m, which encompasses all formations of interest. Sand A is a very uniform interval, reading a travel

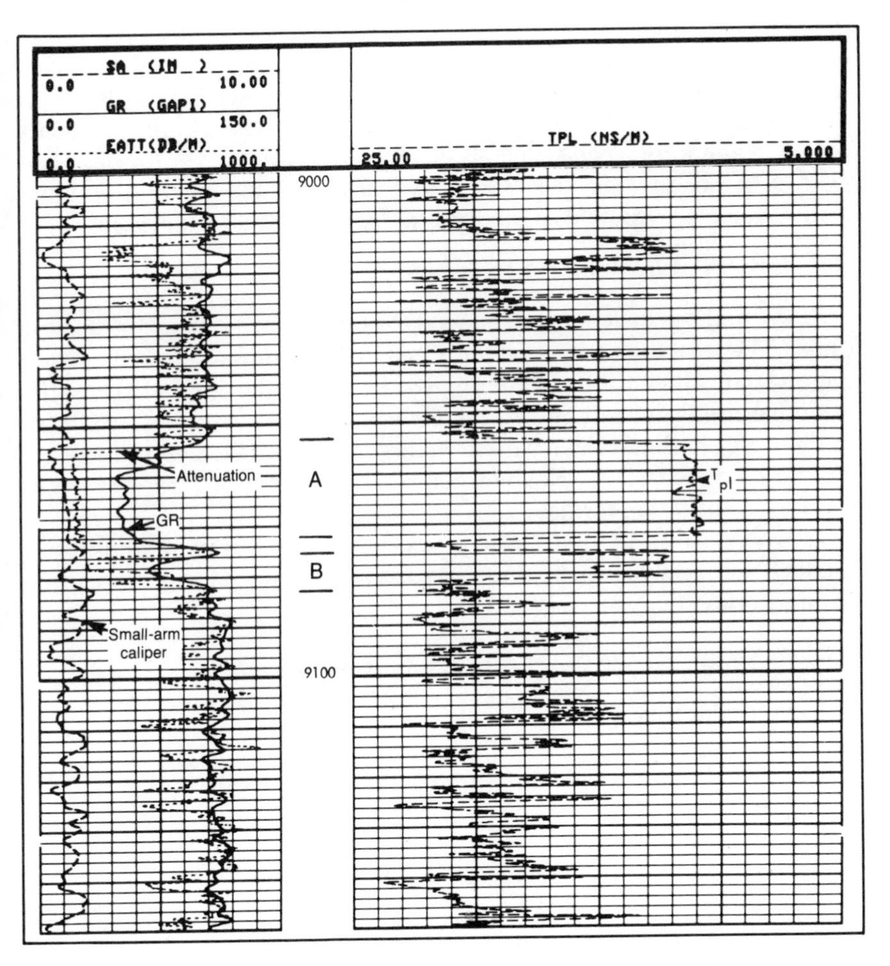

Fig. 5–51 Example of EPT travel time and attenuation recording (courtesy Schlumberger)

time of 11 ns/m. Sand B is less uniform and reads a higher travel time. The remainder of the section is shown by the EPT as a series of intermixed shale and shaly sand stringers. The shale readings, those at highest travel times, average about 22 ns/m.

Track 1 shows the attenuation curve (dotted) on a scale from 0 to 1,000 decibels/meter. Attenuation in sand A is low, about 150 db/m, while that in shales is high, approximately 750 db/m. About 60 db/m of the attenuation is due to geometrical spreading of the propagating wave from its source. The remainder is primarily a function of the clay content of the formation. The higher the clay content, the greater the signal attenuation. The range is tremendous, a factor of 4,000 in this case between sand and shale levels for the 12-cm spacing. It requires that transmitter power be continuously varied to maintain receiver signals in a workable range.

Also shown in Track 1 is the small arm caliper reading on a 0–10-in. scale. Diameter variations of about 2 in. due to rugosity are evident.

Conversion of Travel Time to Porosity

Two methods have been developed for computing porosity from travel time. They are the t_{po} method and the CRIM (complex refractive index) method. Only the former will be discussed since it is the simplest. The latter is more complicated but does in principle permit calculation of the salinity of the water in the flushed zone as well as the water-filled porosity. However, it assumes perfectly clean formations, which may be rare. The reader is referred elsewhere for details.[52]

Travel time of microwaves in clean, lossless (low attenuation) porous media is accurately given by the sum of the travel times through the constituent portions. That is

$$t_{po} = \phi \cdot t_{pf} + (1 - \phi)\, t_{pm} \qquad (5.28)$$

Solving for porosity

$$\phi = (t_{po} - t_{pm})/(t_{pf} - t_{pm}) \qquad (5.29)$$

where

t_{po} = loss-free travel time of the medium, ns/m

= $[t_{pl}^2 - (A - 60)^2/3{,}600]^{1/2}$ (5.30)

t_{pl} = measured travel time of the medium ns/m

A = measured attenuation of the medium, db/m

t_{pm} = travel time of the rock matrix, ns/m

t_{pf} = travel time of the pore fluid, ns/m

Values of t_{pm} and t_{pf} for various rock matrices and pore fluids of interest are given in Table 5–4. Note that water has a much higher value than any other constituent. This is a result of its high dielectric constant. Values of the constant, relative to air = 1, are listed in the ϵ_r column. Propagation time in nanoseconds per meter is related to dielectric constant by the simple relation

$$t_p = \sqrt{11.1\epsilon_r} \qquad (5.31)$$

TABLE 5-4 RELATIVE DIELECTRIC CONSTANTS AND PROPAGATION TIMES FOR VARIOUS MINERALS

Mineral	ϵ_r	t_{pm}, ns/m	t_{pf}, ns/m
Sandstone	4.65	7.2	—
Dolomite	6.8	8.7	—
Limestone	7.5	9.1	—
Anhydrite	6.35	8.4	—
Dry colloids	5.76	8.0	—
Halite	5.6–6.35	7.9–8.4	—
Gypsum	4.16	6.8	—
Shale	5–25	7.5–16.6	—
Oil	2.2	—	4.9
Gas	3.3	—	6.0
Fresh water, 25°C	78.3	—	29.5

Consequently, the measured travel time of a medium is a strong function of its water content.

EPT porosity, ϕ_{EP}, is the porosity derived from Eq. 5.29 on the assumption of water-filled pore space. That is

$$\phi_{EP} = (t_{po} - t_{pm})/(t_{pw} - t_{pm}) \qquad (5.32)$$

where t_{pw} is the lossless travel time of water. This value is not constant but depends on temperature and slightly on pressure. Over the range 100–300°F with usual pressure variation, it is given to sufficient accuracy by

$$t_{pw} = 31.1 - 0.029\,T \qquad (5.33)$$

where T = formation temperature, °F.

As an example we can compute ϕ_{EP} for sand A of Fig. 5–51. Average travel time t_p is 11.1 ns/m and attenuation A is 150 db/m. By Eq. 5.30

$$t_{po} = [11.1^2 - (150 - 60)^2/3{,}600]^{1/2} = 11.0 \text{ ns/m}$$

This shows the correction for attenuation is negligible. The formation is known to be sand, so t_{pm} = 7.2 ns/m. Temperature at the 9,060-ft depth is estimated to be 170°F, so that by Eq. 5.33

$$t_{pw} = 31.1 - 0.029 \times 170 = 26.1 \text{ ns/m}$$

Application of Eq. 5.32 gives

$$\phi_{EP} = (11.0 - 7.2)/(26.1 - 7.2) = 0.20$$

Consequently, the water-filled porosity of sand A is close to 20%. The actual porosity will be larger if the sand also contains hydrocarbons.

Fig. 5–52 is a plot of ϕ_{EP} vs t_{po} for sandstone and limestone, based on Eqs. 5.32 and 5.33. For clean formations of such types, the attenuation is negligible so that t_{po} is the actual t_{pl} value read from the log. It is apparent from the plot that matrix type must be known with certainty to obtain accurate porosity values and, in addition, temperature must be known to within 10°F.

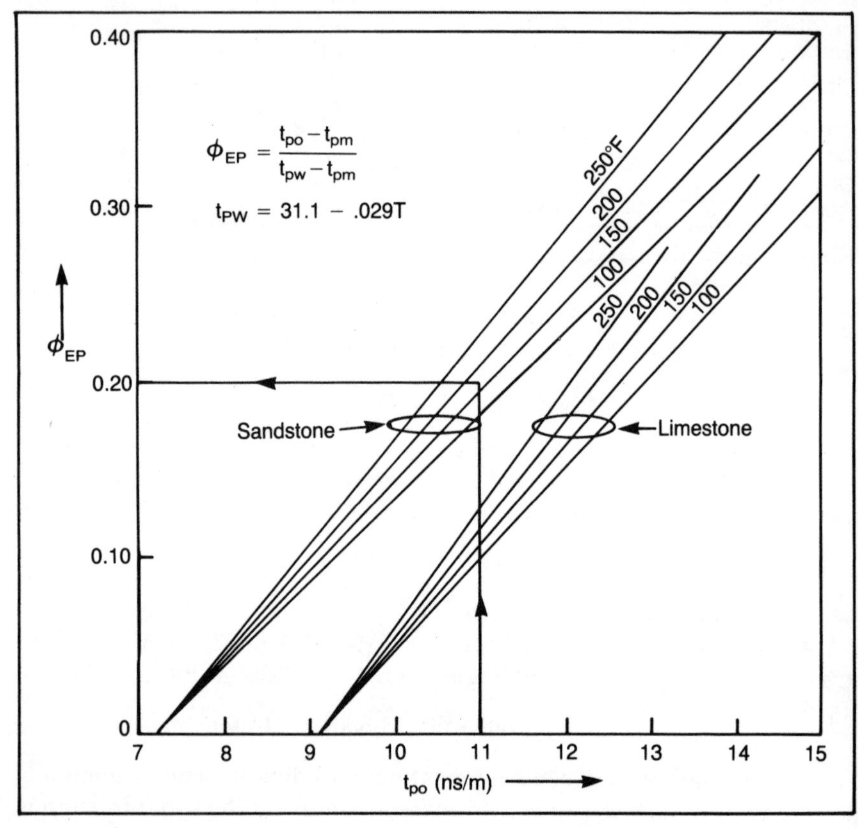

Fig. 5–52 Conversion of EPT travel times to water-filled porosity

Where the matrix is a mixture of several minerals, the matrix travel time is a linear combination of the matrix travel times of the constituents. That is

$$t_{pm} = \sum_{i=1}^{n} (V_i \cdot t_{pmi}) \tag{5.34}$$

where V_i is the fractional volume of the i_{th} component and t_{pmi} is the matrix travel time of that component. Relative volumes of usual components—silica, calcite, and dolomite—are best determined by the $(\rho_{ma})_a$ – $(U_{ma})_a$ method described in chapter 6, utilizing Litho-Density and Neutron logs. It is almost mandatory to run the LDT-CNL in conjunction with EPT for good interpretation.

EPT Flushed-Zone Water Saturation

The EPT porosity, ϕ_{EP}, is the true porosity if all pores are water filled. It will approximate the water-filled porosity in formations containing hydrocarbons because hydrocarbon appears much like rock matrix to microwaves; travel times in the two media are similar. Consequently, an apparent water saturation for the flushed zone is simply

$$(S_{xo})_a = \phi_{EP}/\phi \tag{5.35}$$

where ϕ is total fluid porosity, generally obtained from Density-Neutron logs.

The apparent saturation, $(S_{xo})_a$, is close to the true saturation, S_{xo}, at high values but not at low values. To calculate the difference, Eq. 5.28 can be expanded for the case of both hydrocarbon and water in the pores to

$$t_{po} = S_{xo} \cdot \phi \cdot t_{pw} + (1 - S_{xo}) \cdot \phi \cdot t_{ph} + (1 - \phi) \cdot t_{pm} \tag{5.36}$$

where t_{ph} is the travel time of the hydrocarbon and other terms are as already defined. This equation can be solved for S_{xo} and rearranged to give

$$S_{xo} = K + (1 - K)\,(S_{xo})_a \tag{5.37}$$

where K is a constant, depending only on matrix and fluid travel times, given by

$$K = (t_{pm} - t_{ph})/(t_{pw} - t_{ph}) \tag{5.38}$$

Fig. 5–53 is a plot of S_{xo} vs $(S_{xo})_a$ from Eqs. 5.37 and 5.38. The difference is small at high values of $(S_{xo})_a$ but appreciable at low values. In normal hydrocarbon-bearing formations $(S_{xo})_a$ is greater than 0.7, in which case it is close to being correct. In heavy-oil situations, however, $(S_{xo})_a$ could calculate less than 0.4. In this case the true S_{xo} would be significantly higher than the apparent $(S_{xo})_a$.

Example of Moved Oil Estimation

Fig. 5–54 is a recording of the basic Dual Induction-SFL, SP, GR, Density, and Neutron curves over the same section of hole as depicted in Fig. 5–51. It is fairly obvious from the R_{ild} curve that sand A is hydrocarbon bearing and sand B is water bearing.

This is verified by the computed R_{wa} curve in Track 1. From the reading of that curve in sand B, R_w is estimated to be 0.08 ohm-m. R_{wa} in zone A

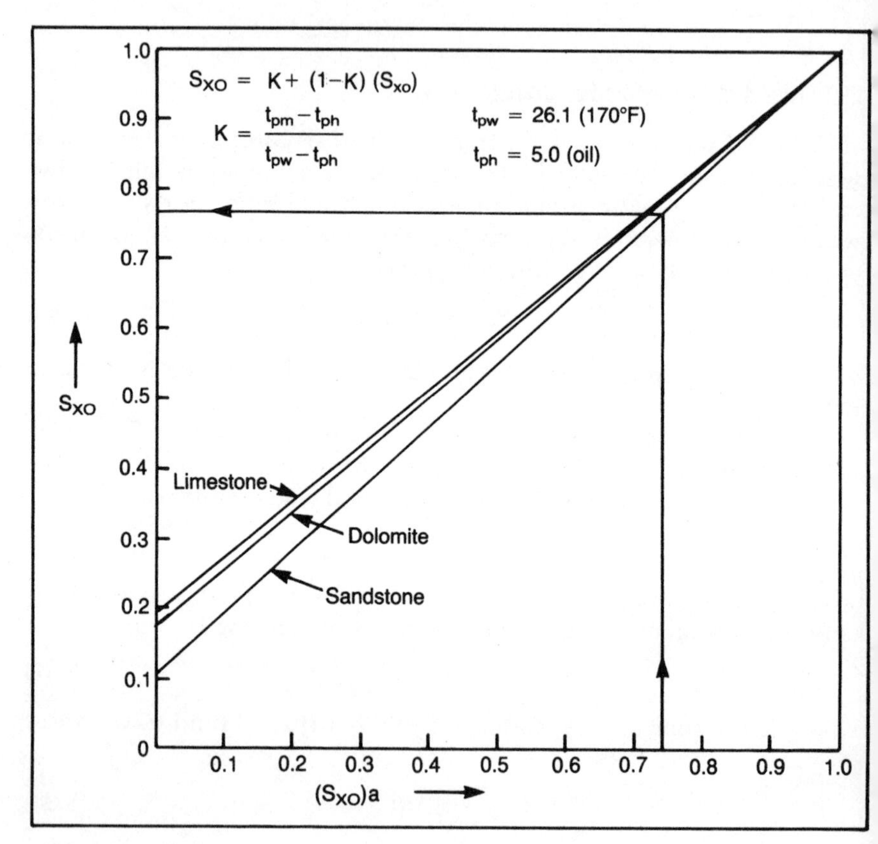

Fig. 5–53 Conversion of apparent flushed-zone water saturation $(S_{xo})_a$ to true saturation S_{xo}

averages 0.87, so a quick estimate of water saturation in sand A (see chapter 9) is

$$S_w = \sqrt{R_w/R_{wa}} = \sqrt{0.08/0.87} = 0.30$$

The Neutron-Density overlay shows the hydrocarbon in sand A is oil if the sand is clean. However, if it is slightly shaly, as is probably the case, correction for shale would result in Neutron-Density crossover in the upper part of the sand, indicating the presence of gas.

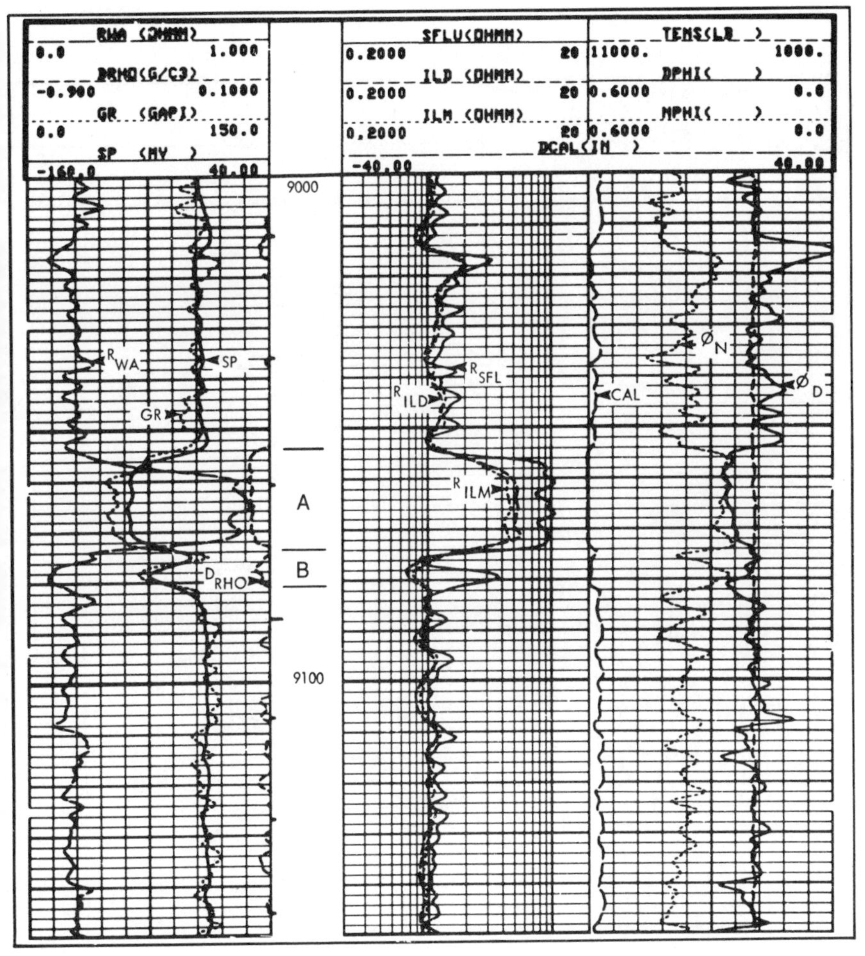

Fig. 5–54 Resistivity and Density-Neutron logs over the same interval as Fig. 5–51 (courtesy Schlumberger)

Fig. 5–55 shows the water-filled porosity, ϕ_{EP}, overlain with the total fluid-filled porosity ϕ obtained by averaging the Density and Neutron porosities. The former reads 0.20 in sand A, in agreement with the previous calculation, and the latter averages 0.27. The apparent flushed-zone water saturation is then

$$(S_{xo})_a = 0.20/0.27 = 0.74$$

The corrected S_{xo} value from Fig. 5–53 is 0.77. Moved hydrocarbon, as a fraction of the pore space, is

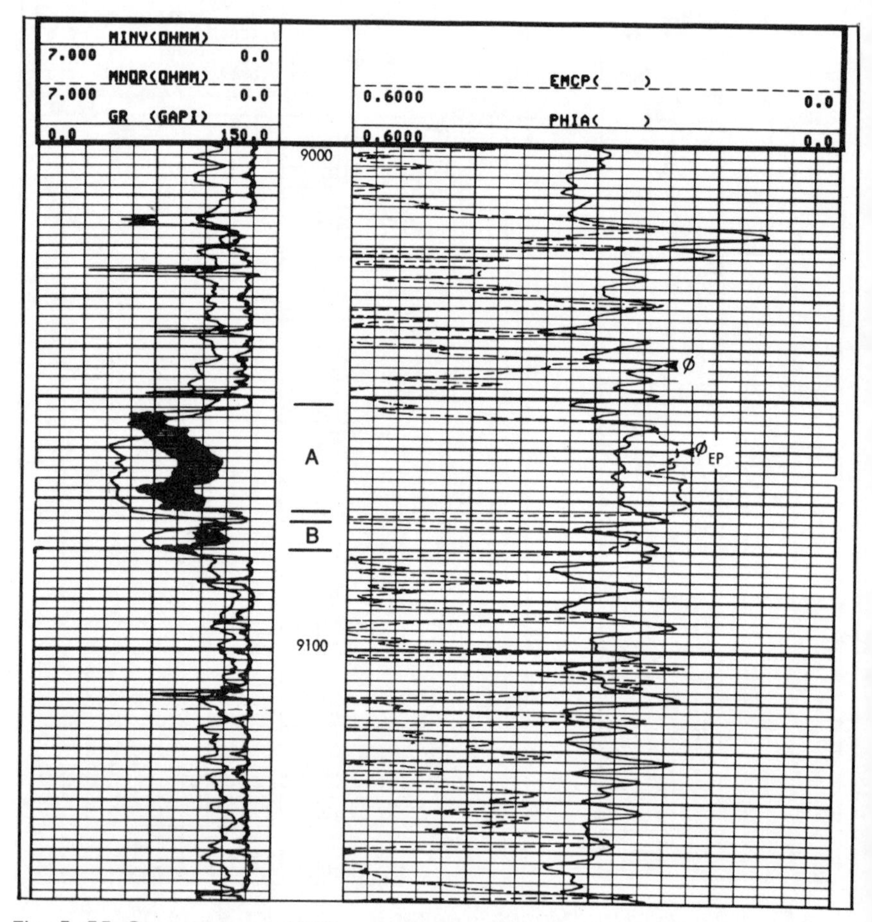

Fig. 5–55 Comparison of EPT porosity ϕ_{EP} with Density-Neutron porosity ϕ for same interval as Fig. 5–51 (courtesy Schlumberger)

$$S_{xo} - S_w = 0.77 - 0.30 = 0.47$$

Sand A should therefore be a good hydrocarbon producer. On the other hand for sand B, $\phi_{EP} \approx \phi$, which verifies it is water bearing.

Positive separation of the Microlog curves in sand A, as shown in Track 1, clearly indicates permeability in that zone, confirming its potential as a hydrocarbon producer. Sand B is also permeable. However, all other thin shaly sands over the recorded interval are indicated as impermeable.

The upper 4 ft of sand A was perforated and flowed 70 Mcfd and 30 bo/d, confirming the log analysis.

Example of Detection of Heavy Oil in Fresh Water Areas[53]

Fig. 5–56 is a composite log of a shallow well in Kern County, California, through a number of sands from which 12°–15° API gravity oil is produced. Formation water resistivities in the area vary widely, from 0.24 to 32 ohm-m at 75°F (25,000 to 150 ppm), so that determination of water saturation from the electrical logs is virtually impossible. The SP in Track 1 is of little help in reflecting R_w changes or even in delineating sand-shale boundaries.

Porosities of various sands, labeled A to E, are fairly uniform, averaging 38% as shown in Track 3. There is good agreement between Density-Neutron total fluid porosity ϕ and porosity ϕ_{core} from conventional cores. Track 3 also records ϕ_{EP}. It averages much less than ϕ, indicating considerable hydrocarbons in place.

S_{xo} values computed from the ϕ_{EP}/ϕ ratio and corrected as indicated by Fig. 5–53 are recorded in Track 2 along with residual oil saturation S_{oil}, measured in the cores. There is excellent agreement. For example, from 1,440–1,450 ft in zone E, both curves indicate 40% water saturation. This is also the estimated S_w value for the zone since little displacement of heavy oil by mud filtrate is expected.

With S_w equal to S_{xo}, water resistivity can be calculated from Archie's equation. Rearranging that equation gives

$$R_w = R_t \cdot (S_w \cdot \phi/c)^2 \tag{5.39}$$

In the interval 1,440–1,450 ft, $R_t = 250$ ohm-m, $\phi = 0.37$, $S_w = 0.4$, and $c = 0.9$ (for sands), which gives $R_w = 6.8$ ohm-m. A similar calculation for the lower part of zone D where R_t is 10 gives $R_w = 0.9$ ohm-m. This shows the rapid variability of R_w. Many sands in this area are steam flooded, which contributes to wide variation in R_w.

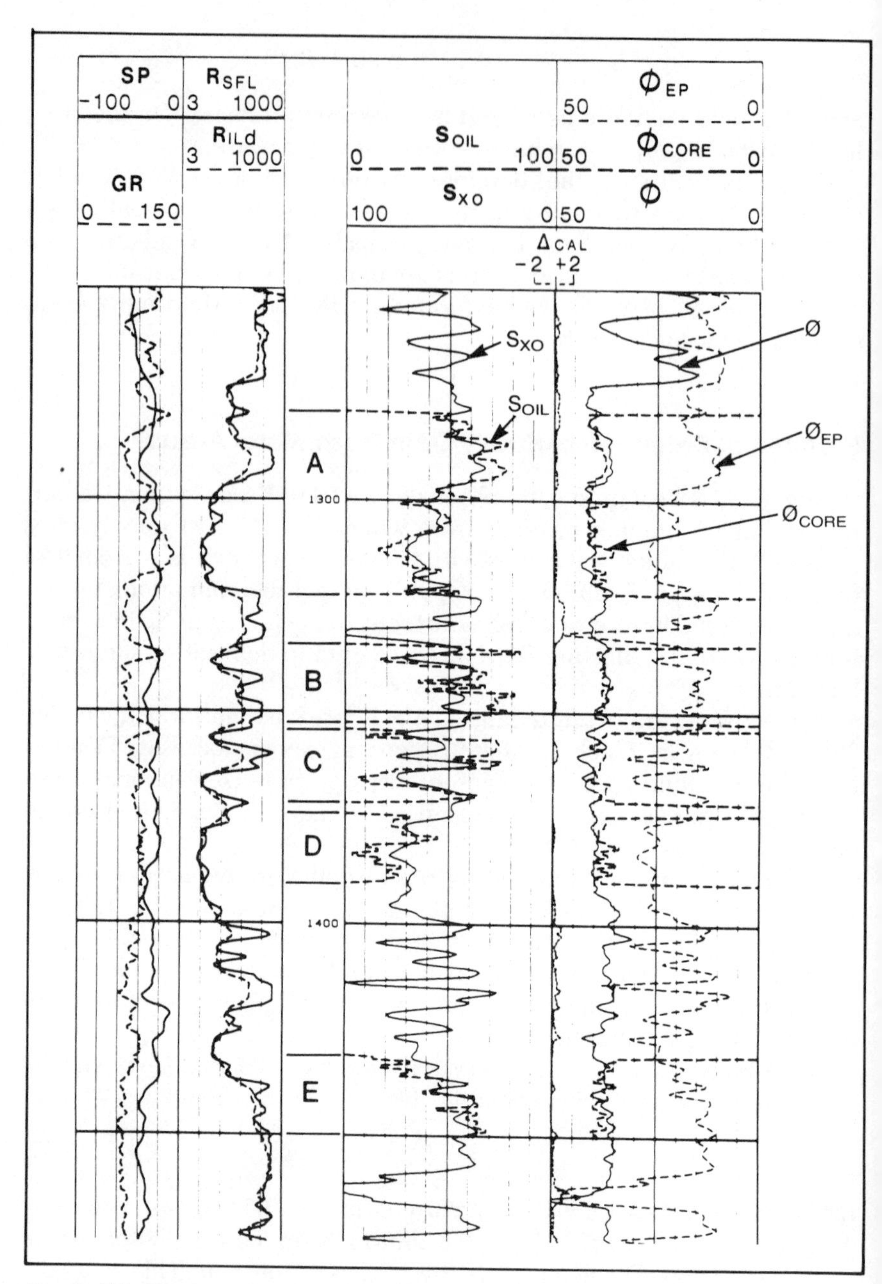

Fig. 5–56 EPT-derived porosities and water saturations in a shallow heavy-oil well (courtesy Schlumberger and SPWLA)

In cases where wells are logged through intervals in which steam has broken through, the steam appears to the EPT tool as hydrocarbon gas ($t_{pf} \approx 3$ ns/m) so that measured travel times are abnormally low. This results in calculated ϕ_{EP} and S_{xo} values likewise being unduly low. However, steam breakthrough also shows as a gas crossover on the Density-Neutron porosity overlay. When such occurs, the steam saturation can be estimated from the Neutron-Density response and the S_{xo} value from the EPT log can be corrected accordingly.[54]

Quick-Look Hydrocarbon Indication

Fig. 5–57 shows schematically how the combination of Induction, Density, Neutron, and EPT logs distinguish between fresh water, salt water, oil, and gas. Resistivity distinguishes fresh water from salt water, whereas the other curves do not. EPT distinguishes oil from fresh water while the other curves show only slight change. Finally, EPT and Neutron-Density together distinguish gas from oil, while the resistivity is not definitive.

Effect of Shale on the EPT Log

The EPT response in shales and shaly formations is not fully understood. The tool responds to water bound to clay surfaces (chapter 7) as well as to free water in clean pore space, but the response appears excessive. Figure 5–51, for example, shows t_{pl} in shale to be 22 ns/m. Correcting for attenuation (750 db/m) by Eq. 5.30 gives $t_{po} = 18.6$ ns/m. Assuming $t_{pm} = 7.6$ for shale (from Table 5–4, for a 50–50 matrix mixture of sand and dry colloids) and $t_{pw} = 26.1$, Eq. 5.32 gives for the calculated porosity of shale

$$(\phi_{EP})_{sh} = (18.6 - 7.6)/(26.1 - 7.6) = 0.59$$

However, Density-Neutron indicates a shale porosity of about 0.25 (Fig. 5–54), so the EPT reading is abnormally high.

A simplistic explanation is to assume the bound water in shale has effectively a higher dielectric constant than free water. Eq. 5–28 can be solved for t_{pf} if ϕ is known. That is

$$t_{pf} = [t_{po} - (1 - \phi)\, t_{pm}]/\phi \qquad (5.40)$$

Taking the value of 0.25 as the true shale porosity

$$(t_{pf})_{sh} = [18.6 - 0.75 \times 7.6]/0.25 = 52 \text{ ns/m}$$

The effective dielectric constant of the clay bound water in the zone considered is then given by Eq. 5.31 as

$$\epsilon_r = (t_{pf})^2_{sh}/11.1 = 52^2/11.1 = 244$$

This is roughly three times the dielectric constant of free water.

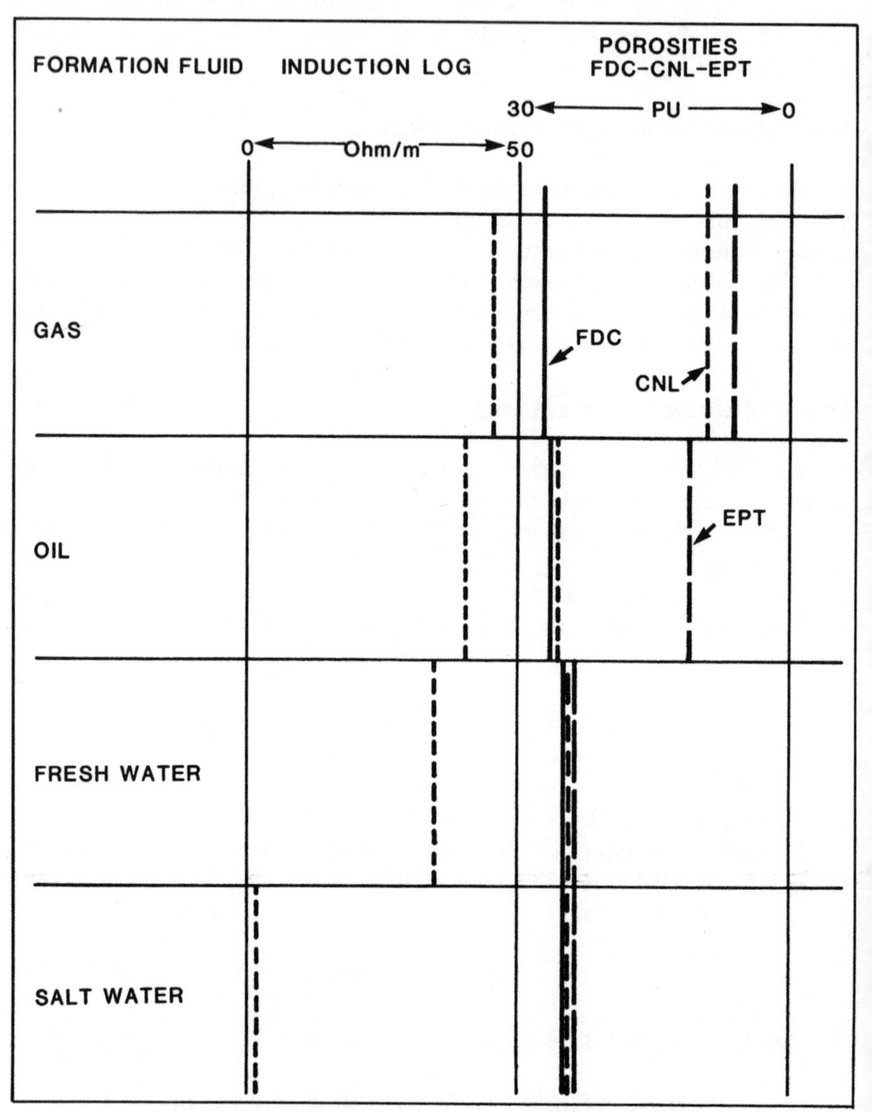

Fig. 5-57 Quick-look hydrocarbon indicator (after Schlumberger)

Laboratory measurements have shown that the dielectric constant of water-saturated porous rock indeed increases with clay content. The increase is astounding at frequencies less than 1,000 Hz where values up to 10^5 have been measured on shaly cores, but it is less pronounced with increasing frequency.[55] This behavior has been theoretically related to the concentration of thin, plate-like particles that signify clay content in the rock.[56,57] The platelet effect is predicted to be small at the 1.1 gHz operating frequency of the EPT, but logs indicate it is not negligible.

The anomalously high dielectric constant of shaly formations is clearly related to the substantial increase in signal attenuation that occurs simultaneously. If the two effects can be tied together in such a fashion that the specific surface area of entrained clay can be derived it would materially aid in the interpretation of shaly formations.

SUMMARY

COMPENSATED DENSITY LOG

- Measures formation density by sensing attenuation of gamma rays between source and pad-mounted detector applied against borehole wall. Auxiliary detector corrects for mud cake and rugosity. Density variation: 2.0–2.9 g/cc.
- Depth of penetration—4 in.; vertical bed resolution—3 ft; logging speed—1,800 ft/hr.
- Conversion of density to porosity is exact; requires knowledge of matrix and fluid densities.
- Accurate matrix densities must be input to obtain good porosity values at low porosity. When gas is present, correct fluid densities must be input for good porosities, particularly at high porosity.
- Advantages
  1. best porosity device in shallow, uncompacted formations
  2. effective porosity obtained, even with shale or clay present
  3. can be run in empty hole
- Disadvantages
  1. must be run slowly
  2. quality degraded in very rough hole; short, sharp washouts are particularly bad

3. heavy minerals (pyrite) cause densities to read too high

LITHO-DENSITY LOG

- Records additional photoelectric absorption curve, P_e. It distinguishes between sand, limestone, and dolomite, regardless of porosity if only one type of matrix present. If two types present, their relative amounts can be deduced from combination of ρ_b and P_e.
- Improves porosity determination in multimineral lithology.
- Aids in separating gas and matrix effect in tight formations.
- Disadvantage: P_e curve is useless in barite-loaded mud.

COMPENSATED NEUTRON LOG

- Measures formation porosity be sensing distance fast neutrons travel in a formation.
- Records ratio of pulse rates received at two detectors. Ratio converted to porosity with laboratory data.
- Depth of investigation—10 in.; vertical resolution—3 ft; logging speed—1,800 ft/hr.
- Better to interpret Neutron in conjunction with Density. For clean, liquid-bearing formation, correct porosity regardless of lithology is average value $\phi = (\phi_d + \phi_n)/2$.
- Gas-bearing formations normally show as a crossover of curves, with the Neutron reading lower porosity. Correct porosity for clean formations is $\phi = \sqrt{(\phi_d^2 + \phi_n^2)/2}$. Crossover is suppressed in dolomites and shaly formations.
- False indication of gas can be obtained when recording on a limestone matrix but lithology is sandstone.
- Overlay with Density shows matrix variations.
- Disadvantages: strongly influenced by lattice-held hydrogen and thermal neutron absorbers in clay. Not suited to empty holes (Sidewall Neutron better).

COMPENSATED SONIC LOG

- Measures rate of travel of compressional energy in formation; transit time variation: 40–150 μsec/ft.

- Borehole variation effects eliminated using a dual transmitter-receiver array.
- Vertical resolution—equals to span between receivers (2 ft typically); depth of penetration—1 in.; logging speed—5,000 ft/hr.
- Conversion of travel time to porosity simplest with empirical curves (Fig. 5–36) or $\phi = 0.63(1 - t_{ma}/t)$, where t_{ma} equals 44, 49, and 54 μsec/ft for dolomite, limestone, and sandstone, respectively.
- Best suited to well-compacted formations.
- Gas may or may not affect log; quantitative evaluation not possible.
- Secondary porosity in form of vugs or channels indicated by difference between Density-Neutron and Sonic porosities. Accurate values required.
- Advantages
  1. tolerates very rough holes
  2. combines well with Induction; can be run fast
  3. provides secondary porosity information
  4. ties in seismic time sections precisely with depth
- Disadvantages
  1. lithology must be known to compute porosity
  2. strongly affected by lack of compaction in shallow sands
  3. effect of gas unpredictable
  4. gas bubbles cause erratic operation

LONG-SPACING LOG

- Provides 8–10 ft and 10–12 ft spacing logs versus 3–5 ft for standard Sonic.
- Borehole compensation obtained by depth memorization.
- Provides best depth calibration of seismic sections.
- Use to avoid shale alteration and large hole effects that occur in shallow sand-shale sequences.
- Measures shear wave travel time. Chief application is determination of elastic moduli. Offsite processing required.

ELECTROMAGNETIC PROPAGATION LOG

- Measures travel times of microwaves in flushed zone.
- Gives water-filled porosity.
- Yields S_{xo} when compared with Density-Neutron.
- Determines hydrocarbon movability.
- Detects hydrocarbons in freshwater areas where electric logs are inadequate; works best if heavy oil.
- Delineates 100% water-bearing zones for determining R_w.
- Combination with Microlog defines producible intervals in great vertical detail.
- Vertical resolution—2 in.; depth of penetration—3 in.
- Mud cake or standoff greater than ⅜-in. thick deteriorates measurement.
- Shaliness causes anomalously high ϕ_{EP} readings.
- CNL-LDT logs mandatory for matrix identification in complex formation where lithology is variable. Combination length—80 ft; logging speed—1,800 ft/hr.

MICROLOG

- Indicates permeability by sensing presence of mud cake.
- Absolute values of permeability not obtainable.
- Vertical resolution—3 in. Depth of penetration—few inches.
- Mud-cake thicknesses estimated with little accuracy.

REFERENCES

[1]J. Tittman and J.S. Wahl, "The Physical Foundations of Formation Density Logging (Gamma-Gamma)," *Geophysics* (April 1965).

[2]R.P. Alger, L.L. Raymer, W.R. Hoyle, and M.P. Tixier, "Formation Density Log Applications in Liquid-Filled Holes," *Jour. Pet. Tech.* (March 1963).

[3]J.S. Wahl, J. Tittman, and C.W. Johnstone, "The Dual Spacing Formation Density Log," *Jour. Pet. Tech.* (December 1964).

[4]C. Clavier, A. Heim, and C. Scala, "Effect of Pyrite on Resistivity and Other Logging Measurements," *SPWLA Logging Symposium Transactions* (June 1976).

[5]B. Felder and C. Boyeldieu, "The Litho-Density Log," *Proc. 6th European Formation Evaluation Symposium* (London: March 1979).

[6]J.S. Gardner and J.L. Dumanoir, "Litho-Density Log Interpretation," *SPWLA Logging Symposium Transactions* (July 1980).

[7]D.C. McCall and J.S. Gardner, "Litho-Density Log Applications in the Michigan and Illinois Basins," *SPWLA Logging Symposium Transactions* (July 1982).

[8]J. Tittman, "Radiation Logging," U. of Kansas Conf. 1966, SPWLA Reprint Volume: *Gamma Ray, Neutron and Density Logging* (March 1978).

[9]J. Tittman, H. Sherman, W.A. Nagel, and R.P. Alger, "The Sidewall Epithermal Neutron Porosity Log," *Jour. Pet. Tech.* (October 1966).

[10]R.P. Alger, S. Locke, W.A. Nagel, and H. Sherman, "The Dual Spacing Neutron Log — CNL," *SPE 3565*, New Orleans (October 1971).

[11]L.S. Allen, C.W. Tittle, W.R. Mills, and R.L. Caldwell, "Dual-Spaced Neutron Logging for Porosity," *Geophysics*, Vol 32, No. 1 (1967).

[12]L.S. Allen, W.R. Mills, K.P. Desai, and R.L. Caldwell, "Some Features of Dual-Spaced Neutron Porosity Logging," *SPWLA Logging Symposium Transactions* (May 1972).

[13]H. Edmundson and L.L. Raymer, "Radioactive Logging Parameters for Common Minerals," *SPWLA Logging Symposium Transactions* (June 1979).

[14]H. Sherman and S. Locke, "Effect of Porosity on Depth of Investigation of Neutron and Density Sondes," *SPE 5510* (Dallas: September 1975).

[15]R. Gaymard and A. Poupon, "Response of Neutron and Formation Density Logs in Hydrocarbon-Bearing Formations," *The Log Analyst* (SPWLA), Vol 9, No. 5 (September-October 1968).

[16]F. Segesman and O. Liu, "The Excavation Effect," *SPWLA Logging Symposium Transactions* (May 1971).

[17]J. Suau, "An Improved Gas Correction Method for Density and Neutron Logs," *SPWLA Logging Symposium Transactions* (June 1981).

[18]R.R. Davis, J.E. Hall, Y.L. Boutemy, and C. Flaum, "A Dual Porosity CNL Logging System," *SPE 10296* (San Antonio: October 1981).

[19]H.D. Scott, C. Flaum, and H. Sherman, "Dual Porosity CNL Count Rate Processing," *SPE 11146* (New Orleans: September 1982).

[20]F.L. Paillet, "Predicting the Frequency Content of Acoustic Waveforms Obtained in Boreholes," *SPWLA Logging Symposium Transactions* (June 1981).

[21]F.P. Kokesh and R.B. Blizard, "Geometric Factors in Sonic Logging," *Geophysics*, Vol 24, No. 1 (February 1959).

[22]F.P. Kokesh, R.J. Schwartz, W.B. Wall, and R.L. Morris, "A New Approach to Sonic Logging and Other Acoustic Measurements," *Jour. Pet. Tech.* Vol 17, No. 3 (March 1965).

[23]K.B. Hartley, "Factors Affecting Sandstone Acoustic Compressional Velocities and an Examination of Empirical Correlations between Velocities and Porosities," *SPWLA Logging Symposium Transactions* (June 1981).

[24]W.G. Hicks and J.E. Berry, "Application of Continuous Velocity Logs to Determination of Fluid Saturation of Reservoir Rocks," *Geophysics*, Vol. 21, No. 3 (July 1956), pp. 739–754.

[25]M.R.J. Wyllie, A.R. Gregory, and G.H.F. Gardner, "Elastic Wave Velocities in Heterogeneous and Porous Media, "*Geophysics*, Vol 21, No. 1 (January 1956), pp. 41–70.

[26]M.R.J. Wyllie, A.R. Gregory, and G.H.F. Gardner, "An Experimental Investigation of Factors Affecting Elastic Wave Velocities in Porous Media," *Geophysics*, Vol. 23, No. 3 (July 1958), pp. 459–493.

[27]M.P. Tixier, R.P. Alger, and C.A. Doh, "Sonic Logging," *Jour. Pet. Tech.* Vol. 11, No. 5 (May 1959).

[28]L.L. Raymer, E.R. Hunt, and J.S. Gardner, "An Improved Sonic Transit Time-to-Porosity Transform," *SPWLA Logging Symposium Transactions* (July 1980).

[29]S.N. Domenico, "Effect of Water Saturation on Seismic Reflectivity of Sand Reservoirs Encased in Shale," *Geophysics* (December 1974), p. 759.

[30]S.N. Domenico, "Effect of Brine-Gas Mixture on Velocity in an Unconsolidated Sand Reservoir," *Geophysics* (October 1976), p. 882.

[31]D.H. Thomas, "Seismic Applications of Sonic Logs," *SPWLA Fifth European Logging Symposium* (Paris: October 1977).

[32]W.G. Hicks, "Lateral Velocity Variations Near Boreholes," *Geophysics*, Vol. 34, No. 3 (July 1959), pp. 451–461.

[33]E.R. Blakeman, "A Case Study of the Effect of Shale Alteration on Sonic Transit Time," *SPWLA Logging Symposium Transactions* (July 1982).

[34]J. Aron, J. Murray, and B. Seeman, "Formation Compressional and Shear Interval-Transit-Time Logging by Means of Long Spacing and Digital Techniques," *SPE 7446* (Houston: October 1978).

[35]C.C. Purdy, "Enhancement of Long Spaced Sonic Transit Time Data," *SPWLA Logging Symposium Transactions* (July 1982).

[36]R.W. Siegfried and J.P. Castagna, "Full Waveform Sonic Logging Techniques," *SPWLA Logging Symposium Transactions* (July 1982).

[37]J.D. Ingram, C.F. Morris, E.E. MacKnight, and T.W. Parks, "Shear Velocity Logs Using Direct Phase Determination," *SEG* (Los Angeles: 1982).

[38]H.D. Leslie and F. Mons, "Sonic Waveform Analysis: Applications," *SPWLA Logging Symposium Transactions* (July 1982).

[39]M.P. Tixier, C.W. Loveless, and R.A. Anderson, "Estimation of Formation Strength from the Mechanical Properties Log," *SPE 4532* (Dallas: September 1973).

[40]R.A. Anderson, D.S. Ingram, and A.M. Zanier, "Fracture-Pressure-Gradient Determination from Well Logs," *SPE 4135* (San Antonio: October 1972).

[41]G.R. Coates and S.A. Denoo, "Mechanical Properties Program Using Borehole Analysis and Mohr's Circle," *SPWLA Logging Symposium Transactions* (June 1981).

[42]G.R. Pickett, "Acoustic Character Logs and their Applications in Formation Evaluations," *Jour. Pet. Tech.* Vol. 15, No. 6 (June 1963).

[43]H.G. Doll, "The Microlog — A New Electrical Logging Method for Detailed Determination of Permeable Beds," *Pet. Trans. AIME*, Vol. 189 (1950), pp. 155–164.

[44]D.W. Hilchie, *Old Electrical Log Interpretation* (Golden, CO: Douglas W. Hilchie, Inc., 1979), pp. 103–117.

[45]W.C. Chew and S.C. Gianzero, "Theoretical Investigation of the Electromagnetic Wave Propagation Tool," *IEEE Transactions on Geoscience and Remote Sensing*, Vol. GE-19, No. 1 (January 1981), pp. 1–7.

[46]R. Freedman and J.P. Vogiatzis, "Theory of Microwave Dielectric Constant Logging Using the Electromagnetic Wave Propagation Method," *Geophysics*, Vol. 44, No. 5 (May 1979), pp. 969–986.

[47]R.N. Rau and R.P. Wharton, "Measurement of Core Electrical Parameters of VHF and Microwave Frequencies," *SPE 9380* (Dallas: September 21–24, 1980).

[48]J.P. Poley, J.J. Nooteboom, and P.J. de Waal, "Use of VHF Dielectric Measurements for Borehole Formation Analysis," *Log Analyst*, Vol. 19, No. 3 (May-June 1978), pp. 8–30.

[49]R.A. Meador and P.T. Cox, "Dielectric Constant Logging, A Salinity Independent Estimation of Formation Water Volume," *SPE 5504* (Dallas: September 1975).

[50]T.J. Calvert, R.N. Rau, and L.E. Wells, "Electromagnetic Propagation —A New Dimension in Logging," *SPE 6542* (Bakersfield, CA: April 1977).

[51]R.P. Wharton, G.A. Hazen, R.N. Rau, and D.L. Best, "Electromagnetic Propagation Logging: Advances in Technique and Interpretation," *SPE 9267* (Dallas: September 21–24, 1980).

[52]Wharton, Hazen, Rau, and Best, ibid.

[53]R. Freedman and D.R. Montague, "Electromagnetic Propagation Tool (EPT): Comparison of Log Derived and In-situ Oil Saturations in Shaly Fresh Water Sands," *SPE 9266* (Dallas: September 1980).

[54]R.P. Wharton and J.M. Delano Jr., "An EPT Interpretation Procedure and Application in Fresh Water, Shaly, Oil Sands," *SPWLA Logging Symposium Transactions* (June 1981).

[55]W.A. Hoyer and R.C. Rumble, "Dielectric Constant of Rocks as a Petrophysical Parameter," *SPWLA Logging Symposium Transactions* (June 1976).

[56]P.N. Sen, "The Dielectric and Conductivity Response of Sedimentary Rocks," *SPE 9379* (Dallas: September 1980).

[57]P.N. Sen, "Relation of Certain Geometrical Features to the Dielectric Anomaly of Rocks," *Geophysics*, Vol. 46, No. 12 (1981), p. 714.

Chapter 6

CLEAN FORMATION INTERPRETATION

Evaluation of clean formations is basically a question of inserting the right values of the parameters R_w; R_t and ϕ in the Archie water saturation equation

$$S_w = c\sqrt{R_w/R_t}/\phi \quad (6.1)$$

where
c = 0.9 for sands and 1.0 for carbonates

While apparently simple, difficulties can arise. One is determination of R_w when the water-bearing formations are not obvious, the SP is poor, and no sample or catalog values are available. Another is derivation of porosity when only a Density or Sonic log is run and appropriate values of matrix density or travel time are not known. These are common situations encountered by independent operators drilling spot wells in areas unfamiliar to them, particularly where formations are low in porosity. In cases of this nature porosity-resistivity crossplots can aid considerably in interpretation.

An even more fundamental question is whether the average values of cementation and saturation constants, built into Eq. 6.1, really apply to the specific formation being analyzed. These constants can vary considerably. One of the porosity-resistivity crossplots can help define cementation value.

Another aspect of importance to the geologist — and, to some extent, to the reservoir engineer — is rock typing and facies mapping in areas where lithology is quite variable and little core information is available. This occurs frequently in hard rock regions. Three methods of multimineral identification that can be used in these situations are presented later.

RESISTIVITY-POROSITY CROSSPLOTS

A porosity-resistivity crossplot is a convenient method of analyzing an interval where

- R_w is unknown but should be constant over the interval
- matrix density and/or velocity are not known
- at least a few water-bearing zones of different porosities in the interval are present
- formations of interest are clean

Two types of crossplots are the Hingle plot and the Pickett plot. With the former, values of R_w and matrix density or velocity applicable to the interval can be derived, and values of porosity and water saturation for individual levels can be read directly from the plot. With the latter, R_w can also be derived but not matrix density or velocity. Instead, the value of the cementation component, m, is obtained.

THE HINGLE PLOT[1,2]

Rearranging Eq. 6.1

$$\phi = c\sqrt{R_w/S_w^2} \cdot 1/\sqrt{R_t} \tag{6.2}$$

Assuming R_w and c are constant, a plot of ϕ vs $1/\sqrt{R_t}$ yields a straight line whose slope depends on S_w. Once the water-bearing line ($S_w = 1$) is established, other lines representing different values of S_w can readily be drawn on the chart and values of S_w for plotted points read by interpolation between the lines.

Fig. 6–1 is an example. Porosity increases linearly on the horizontal axis but resistivity decreases in very non-linear fashion on the vertical axis; it is scaled such that $1/\sqrt{R_t}$ is linear. Porosity for the plot is best obtained from Neutron-Density logs, and R_t is obtained from deep Induction or Laterolog readings corrected for invasion.

The procedure is to plot values of ϕ and R_t for all levels of interest in an interval, as shown on Fig. 6–1. The interval should be limited to a few hundred feet in depth over which R_w should be essentially constant. Those points falling at lowest resistivity for a given porosity represent the 100% water-bearing levels. An $S_w = 1$ line is established by drawing a straight line through the pivot point ($\phi = 0$, $R_t = \infty$) and the most northwesterly points on the plot. This line can also be labeled R_o.

The equation for the $S_w = 1$ line is, from Eq. 6.2

$$\phi = c\sqrt{R_w} \cdot 1/\sqrt{R_t} \tag{6.3}$$

R_w is determined by substituting in this equation values of ϕ and R_t for any point on the $S_w = 1$ line. Taking $\phi = 0.10$ and $R_t = 6.5$ as illustrated and placing $c = 1.0$, since the plot is for a carbonate section, we find $R_w = 0.065$ ohm-m.

This technique is one of the best for establishing R_w in an unknown area. It is equivalent to averaging R_w values calculated in all zones believed to be water bearing.

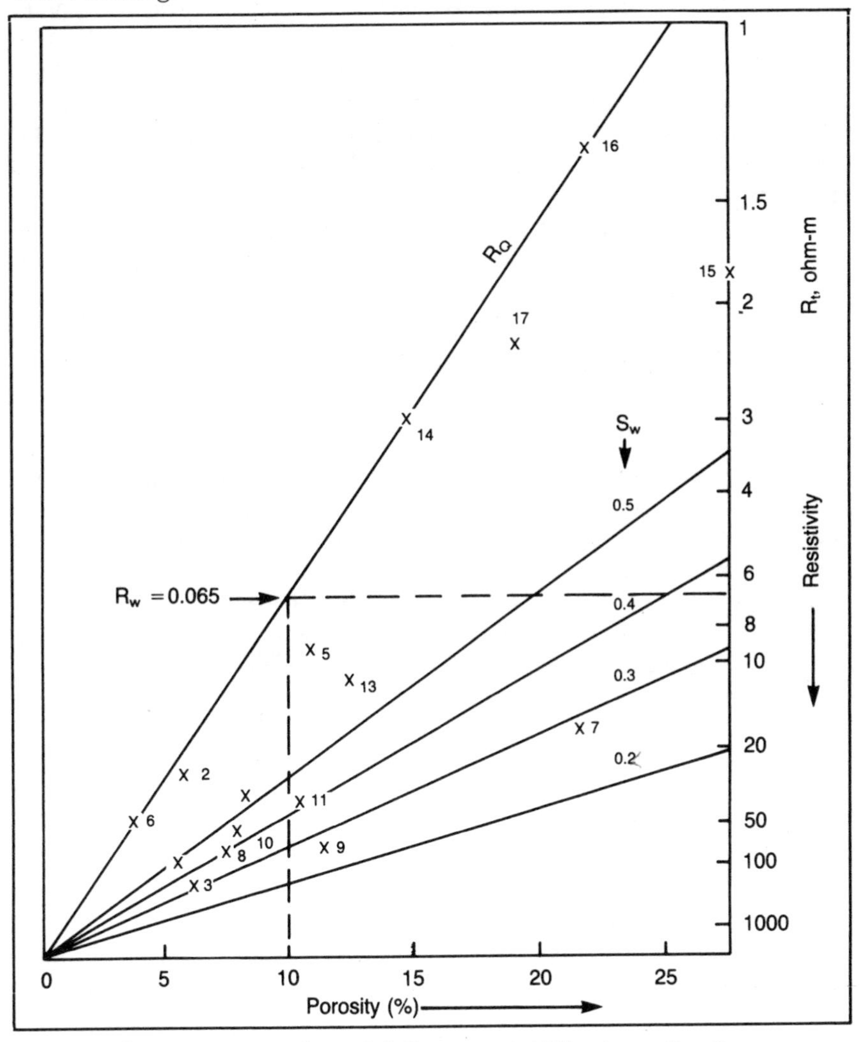

Fig. 6–1 Porosity-resistivity crossplot (Hingle method)

Having determined the R_o line, those lines representing values of S_w other than unity are established as follows. For a fixed porosity the Archie equation is

$$S_w = \sqrt{R_o/R_t} \tag{6.4}$$

The $S_w = 0.5$ line, for example, is therefore represented by points where $R_t = 4R_o$ for any given porosity. We pick a convenient porosity where it is easy to read R_o, in this case the point ($\phi = 17.5$, $R_o = 2.1$), and multiply R_o by 4 to obtain a point ($\phi = 17.5$, $R_t = 8.4$) for which $S_w = 0.5$. The $S_w = 0.5$ line is established by joining this point to the pivot point ($\phi = 0$, $R_t = \infty$) as illustrated. Other lines of constant S_w are drawn similarly. For example, $S_w = 0.3$ requires $R_t = 11.1\ R_o$ so that the point ($\phi = 17.5$, $R_t = 23$) fixes this line.

With a grid of constant S_w lines so established, the S_w value corresponding to any plotted point can be immediately estimated. For example, point 7 has an S_w of approximately 28%.

This technique allows quick evaluation of a large number of levels. Note that it is unnecessary to calculate R_w to obtain S_w values.

The Sonic-Resistivity Crossplot

If a Neutron-Density combination is run, porosities can be derived without knowledge of lithology and plotted directly on the crossplot. However, if only a Sonic log is available for porosity (the Induction-Sonic combination is still popular in some areas) and the lithology is unknown, porosities cannot be initially determined. A different procedure is then followed.

The horizontal axis of the plot is scaled in μsec/ft, increasing from left to right with the scale covering transit times, t, from those observed on the log down to possible matrix values. Fig. 6–2 is an example, with a scale from 50–110 μsec/ft.

Values of t and R_t for levels of interest are then plotted and an $S_w = 1$ line is drawn through the most northwesterly points. Extension of this line to $R = \infty$ determines the matrix travel time, in this case 55.5 μsec/ft, which is also the zero porosity point. The porosity corresponding to the travel time value at the right hand edge of the plot can then be determined by the Wyllie relation, Eq. 5.16. In this case for $t = 110$, $t_{ma} = 55.5$, and $t_f = 189$ (normal assumption for usual water), the corresponding porosity with no compaction correction is

$$\phi = (110 - 55.5)/(189 - 55.5) = 41\%$$

The horizontal axis is then scaled linearly in porosity between the $\phi = 0$ point and the $\phi = 41\%$ point. If the travel time in adjacent shales, t_{sh}, is greater than 100 μsec/ft, the porosity value so determined must be divided

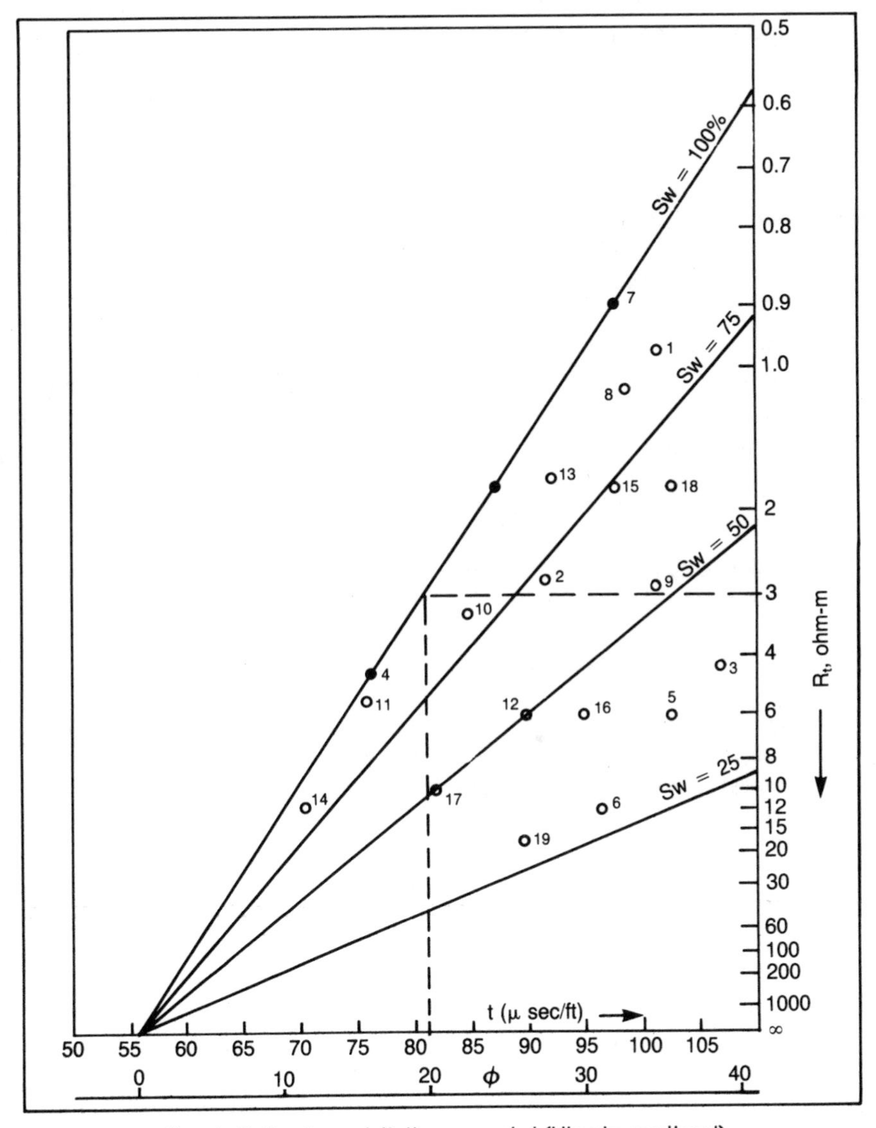

Fig. 6–2 Sonic-resistivity crossplot (Hingle method)

by the compaction correction factor, $B_{cp}(\approx t_{sh}/100)$, before establishing the porosity scale.

From there the procedure is the same as described for determining R_w, and estimating S_w values. Porosities as well as water saturations for the plotted levels can be read directly. In this case a matrix value of 55.5 μsec/ft indicates sandstone, so c = 0.9. Eq. 6.3 using the point ($\phi = 0.20$, $R_t = 3.0$) on the R_o line, gives $R_w = 0.15$ ohm-m.

The Density-Resistivity Crossplot

If only a Density log is available for porosity and matrix density is unknown, the procedure is analogous to that with Sonic logs. In this case the horizontal axis is scaled in grams per cubic centimeter, increasing from right to left and covering observed values up to possible matrix densities. A typical scale might be 2.3 g/cc at the right edge to 2.85 g/cc at the left. Plotting the points of interest and extending the $S_w = 1$ line to $R = \infty$ establishes the matrix density, at which point $\phi = 0$. Assuming ρ_{ma} is found to be 2.68 g/cc, for example, the right edge of the plot at $\rho = 2.3$ would correspond to a porosity given by Eq. 5.3 as

$$\phi = (2.68 - 2.30)/(2.68 - 1.0) = 22.6\%$$

The horizontal axis could then be scaled linearly between $\phi = 0$ and 22.6%. Procedure from there on follows that described for the Sonic case.

The Movable Oil Crossplot[3]

In cases where a Microlaterolog or MicroSFL log has been run, normally in conjunction with a deep Laterolog rather than a deep Induction R_t curve, the crossplot can be extended to determine movable oil. To do so the foregoing procedure is followed, plotting (ϕ, R_t) values, establishing the porosity scale (if necessary) and the S_w = constant lines, and determining R_w. It is particularly advisable to correct the LL_d resistivity readings for invasion before plotting them as R_t values since corrections can be as much as a factor of two. Fig. 6–3 is an example with plotted points shown as X.

Each level is then plotted again at its same porosity but at a resistivity value equal to R_{MLL} (or R_{MSFL}) for that level multiplied by the ratio R_w/R_{mf}, where R_{mf} is the value on the log heading corrected to the temperature at the zone of interest. This ratio normalizes the R_{xo} values from the microresistivity curve to what they would be if the water in the flushed zone had the same resistivity as that in the undisturbed region. On Fig. 6–3 the points so

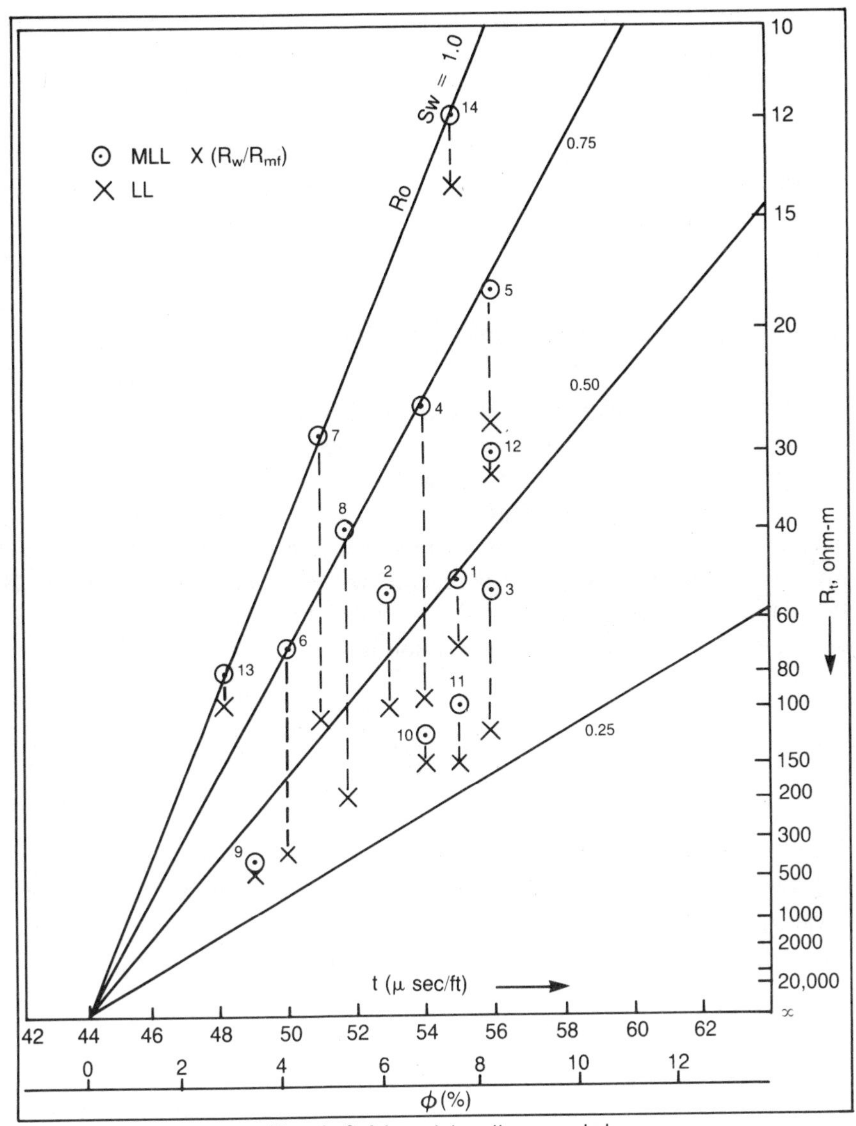

Fig. 6–3 Movable oil crossplot

obtained are shown as circles. The S_w values read for these points are actually S_{xo} values.

In water-bearing intervals the two points corresponding to a given level should both fall close to the $S_w = 1$ line. Levels 13 and 14 are examples. In hydrocarbon-bearing zones where the hydrocarbon is immobile, both points will again fall close together but at an S_w value less than unity. Points 9, 10, and 12 are cases in point. On the other hand where movable hydrocarbon is present, the points of a given level will separate significantly. The difference in indicated S_w values represents the movable oil (S_{xo}–S_w), as a fraction of the pore space. For example, point 8 indicates that movable oil constitutes approximately (75%–35%) or 40% of the pore space at that level.

Additional Remarks Concerning the Hingle Plot

First a word of caution. Two types of crossplot blanks with different divisions of the resistivity scale are found in service company chart books. One is labeled $F = 1/\phi^2$ or $m = 2$. This can be used for either carbonate or sandstone formations, providing Eq. 6.3 is utilized for calculating R_w with $c = 1.0$ for carbonates and 0.9 for sandstones. The resistivity scale can be divided or multiplied by 10 (or any factor) without changing the validity of the plot. The other plot is labeled $F = 0.62/\phi^{2.15}$, which is valid only for sandstones. In this case the basic relation is

$$S_w = \left[\frac{0.62}{\phi^{2.15}} \frac{R_w}{R_t}\right]^{1/2} \tag{6.5}$$

This equation, with $S_w = 1$, must be used in determining R_w rather than Eq. 6.3. Use of the $m = 2$ chart for all conditions is recommended; it is simpler.

The presence of gas can be detected (by crossover) and correct ϕ and S_w values obtained with the Density-Neutron log combination. With Density or Sonic logs alone, however, gas is not obvious, derived porosities will be too high, and S_w values will be too low.

Similarly, in the case of carbonates, lithology can vary between limestone and dolomite in the interval being analyzed if Density-Neutron logs are available but not if only Sonic or Density logs are run. In fact, even if the lithology varies between sand, limestone, and dolomite, the crossplot can be made with D-N porosities and a value of 0.95 utilized for c in calculating R_w. Derived water saturations will be reasonably correct. Crossplots made only with Density or Sonic logs will not be valid.

Where formations are shaly, the plot is unreliable except in limited areas such as the Gulf Coast where reasonable S_w values (but abnormally high effective porosity values) may be obtained by using Density-Neutron or Sonic porosity values. Use of Density porosities will give unduly high water saturations.

THE PICKETT PLOT[4,5]

The generalized form of the Archie equation, as indicated in Chapter 2, is

$$S_w^{\,n} = (a/\phi^m)(R_w/R_t) \tag{6.6}$$

which of course reduces to Eq. 6.1 when the saturation exponent, n, equals 2; the cementation exponent, m, equals 2; and the cementation constant, a, equals 1.0 for limestones and 0.81 for sandstones (equivalent to $c = 1$ or 0.9 since $c = \sqrt{a}$). Rearranging and taking logarithms

$$\log R_t = -m \log \phi + \log (aR_w) - n \log S_w \tag{6.7}$$

This equation shows that if a, R_w, n, and S_w are constant, a plot of log R_t vs log ϕ yields a straight line whose slope is $-m$. A plot of R_t vs ϕ on log-log paper is called a Pickett plot.

Fig. 6–4 is an example of a Pickett plot for a limestone reef, with points plotted for three wells penetrating the same interval. Guidelines for plotting are the same as for the Hingle plot except that porosity values must always be

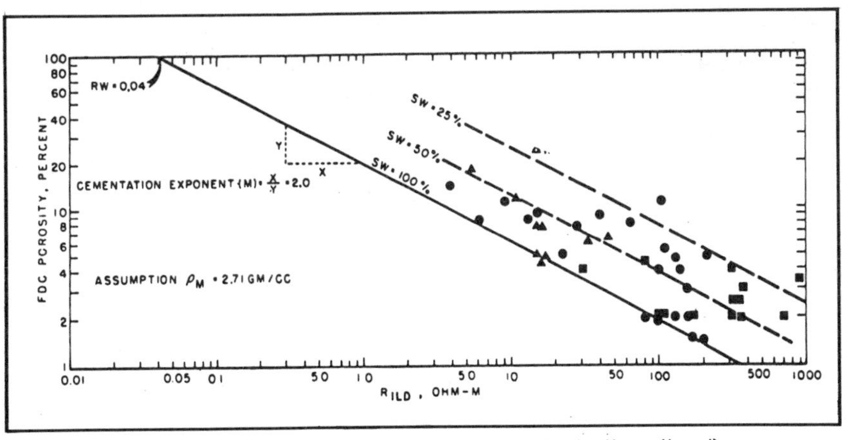

Fig. 6–4 Porosity-resistivity crossplot (Pickett method)

plotted on the porosity scale, not bulk densities or travel times. After the points are plotted, the R_o line ($S_w = 1$) is fixed by drawing it through the lowest resistivity points corresponding to different porosities — in this case the southwesterly points. The slope of the line, X/Y as indicated, gives the cementation exponent m. In this case it is 2.0.

Eq. 6.4 also shows that if the $S_w = 1$ line is extrapolated to $\phi = 1$ (at which point both log ϕ and log S_w are zero), the R_t value so intercepted is equal to aR_w. In this case, $aR_w = 0.040$. If R_w is unknown, a value for *a* must be assumed and R_w derived. In this case, since formations are limestone, *a* can be taken as 1.0 so that $R_w = 0.040$. However, if R_w is accurately known from other sources such as measurements on produced water, its value can be inserted and *a* derived. For example, if samples showed $R = 0.050$, *a* would be 0.80.

It is important that porosities cover as wide a range as possible if accurate values of R_w or *a* are to be determined. A range of at least a factor of two and preferably a factor of three is desirable.

As with the Hingle plot, it is not necessary to calculate R_w to derive S_w values for the various points. Lines representing water saturations less than 1.0 are drawn parallel to the $S_w = 1$ line but at higher resistivities. Eq. 6.4 applies, so that for any given porosity $R_t = 4R_o$ for $S_w = 0.5$, $R_t = 11R_o$ for $S_w = 0.3$, $R_t = 25R_o$ for $S_w = 0.2$, etc. With these lines water saturation can be quickly estimated for any plotted level.

Where only Density or Sonic logs are run, matrix density or travel times must be assumed in order to arrive at porosities for crossplotting. In this case $\rho_{ma} = 2.71$ g/cc is assumed. The matrix constants cannot readily be determined from the plot. If improper values are assumed, porosities will be in error but water saturation values will be correct (providing the lithology is constant). Remarks on the Hingle plot concerning gas and shaliness apply also to the Pickett plot.

The advantage of the Pickett plot over the Hingle plot is that it does not assume the standard values of *a* and *m* built into the Hingle plot. This is important in low-porosity formations where variations in *m* can alter S_w values considerably. Consequently, the Pickett plot is preferable in hard rock areas. However in such cases Density-Neutron porosities should be used to avoid the uncertainties in matrix density or travel time encountered when only Density or Sonic logs are utilized.

RANGE OF UNCERTAINTY IN CALCULATED WATER SATURATIONS

Two factors contribute to the uncertainty of calculated water saturations. One is deviation of the constants m, n, and c from the average values assumed in Eq. 6.1 ($m = n = 2$; $c = 0.9$ or 1.0). The other is possible errors in measurement of R_w, R_t, and ϕ.

Many case studies in the literature have reported values of m and n for specific formations. A considerable variation has been observed. Table 6–1 lists average values of m and n for various formations, as determined from core measurements in one laboratory.[6] For sandstone samples studied, m varies from 1.5 to 2.0 and n from 1.3 to 2.2. The spread is significant though average values are not far from 2.0.

There is good evidence, both experimental and theoretical, that the cementation exponent m is a function of grain shape.[7,8] The more angular and plate-like the shape, the higher the value of m. On these grounds m should increase with clay content since clay has flat platelet-type grain structure. Indeed this seems to be the case, with m ranging from approximately 1.6 for clean sand with fairly round grains to 2.2 for quite shaly sand.[9] The value of m has been correlated to both permeability[10] (in md) and porosity[11] with the relations $m = 1.28 + 2/(\log k + 2)$ and $m = 2.05 - \phi$. However, the determining factor really appears to be clay content, at least for sands. On general grounds the saturation exponent n should be essentially equal to m. The constant c would remain at 1.0, which it theoretically should be if the foregoing equations for m were adopted.

In the case of carbonates, the value of *m* is directly related to the fraction of pore space that is vugular — that consisting of separate vugs connected by intergranular porosity. Laboratory measurements have shown that *m* increases from 2.0 to 3.0 as the fraction of vugular pore space increases from 0 to 60%.[12] The effect is important at low porosities (5–10%) where water saturations calculated using the usual $m = 2$ value can be much too low.

Uncertainties in the measurements of R_w, R_t, and ϕ are in the range of ± 5–10%. When these are coupled with uncertainties of the same order of magnitude in m, n, and c, the net result is an average uncertainty in calculated water saturation in the range ±10–20%.[13] For example, $S_w = 0.40$ means $S_w = 0.40 \pm 0.04$ at best and perhaps ± 0.08 at worst.

Maximum precision in S_w is needed for reserve calculations or for decisions on whether to complete wells when S_w falls in the critical 45–65% range. Consequently, it is desirable to do everything possible to minimize uncertainty in S_w, such as measuring R_w on water samples, checking log porosities against core porosities, making borehole and bed-thickness corrections to the logs when necessary, cross-checking logs on adjacent wells, and deriving *m* and *a* values from Pickett plots. Even so, reducing the

TABLE 6-1 VALUES OF m AND n FOR VARIOUS FORMATIONS

	Lithology	Ave m	Ave n
Wilcox, Gulf Coast	SS	1.9	1.8
Sparta, So. La. (Opelousas)	SS	1.9	1.6
Cockfield, So. Louisiana	SS	1.8	2.1
Government Wells, So. Texas	SS	1.7	1.9
Frio, So. Texas	SS	1.8	1.8
Miocene, So. Texas	Cons. SS	1.95	2.1
	Uncons. SS	1.6	2.1
Travis Peak and Cotton Valley	HD. SS	1.8	1.7
Rodessa, East Texas	LS	2.0	1.6
Edwards, So. Texas	LS	2.0	2.8
Woodbine, East Texas	SS	2.0	2.5
Annona, No. Louisiana	Chalk	2.0	1.5
Nacatoch, Arkansas	SS	1.9	1.3
Ellenburger, W. Texas	LS and Dol.	2.0	3.8
Ordovician Simpson, W. Texas and New Mexico	SS	1.6	1.6
Pennsylvanian, W. Texas	LS	1.9	1.8
Permian, W. Texas	SS	1.8	1.9
Simpson, Kansas	SS	1.75	1.3
Pennsylvanian, Oklahoma	SS	1.8	1.8
Bartlesville, Kansas	SS	2.0	1.9
Mississippian, Illinois	LS	1.9	2.0
Mississippian, Illinois	SS	1.8	1.9
Pennsylvanian, Illinois	SS	1.8	2.0
Madison, No. Dakota	LS	1.9	1.7
Muddy, Nebraska	SS	1.7	2.0
Cretaceous, Saskatchewan, Canada	SS	1.6	1.6
Bradford, Pennsylvania	SS	2.0	1.6
Frio, Chocolate Bayou, Louisiana	SS	1.55–1.94	1.73–2.22
Frio, Agua Dulce, South Texas	SS	1.71	1.66
Frio, Edinburgh, South Texas	SS	1.82	1.47, 1.52
Frio, Hollow Tree, South Texas	SS	1.80, 1.87	1.64, 1.69
Jackson, Cole Sd., South Texas	SS	2.01	1.66
Navarro, Olmos, Delmonte, So. Texas	SS	1.89	1.49
Edwards Lime, Darst Creek Co.	LS	1.94, 2.02	2.04, 2.08
Viola, Bowie Field, No. Texas	LS	1.77	1.15
Lakota Sd. Crook Co., Wyoming	SS	1.52	1.28

Source: G.R. Coates and J.L. Dumanoir, "A New Approach to Log-Derived Permeability," *SPWLA Logging Symposium Transactions* (May 1973)

uncertainty in S_w below ±10% of the calculated value is difficult, especially if shaliness is a factor.

MULTIMINERAL IDENTIFICATION

In this section we consider how the relative amounts of different minerals making up clean complex formations may be determined by combining Density, Neutron, Sonic, and Spectral Gamma Ray data. The principal minerals of concern are sandstone, limestone, dolomite, and anhydrite. Perturbing components are shale and less-frequently encountered minerals such as gypsum, salt, polyhalite, and sulfur.

Identification of matrix makeup is particularly important in tight formations for several reasons. First, porosities may hover near cutoff values, about 5%, so the most accurate values obtainable from logs are desired. Dolomite and shale, for example, cause similar separations between limestone-based Neutron and Density porosity curves, but effective porosity is computed differently in the two cases. With dolomite the effective porosity is close to the average of ϕ_N and ϕ_D; with shale the effective porosity is closer to ϕ_D. The GR curve may not distinguish between the two situations because dolomites frequently are radioactive.

Second, tight formations often require acidizing or acid fracturing to stimulate production. Optimization of this operation requires knowledge of the rock matrix.

The third reason is geologic in nature. The trend of matrix development across a field may indicate preferential directions for offset wells. For example, dolomitization is often accompanied by increased porosity so that the direction of increasing dolomite content may be favorable for production.

There are three methods of combining information from the basic porosity tools to distinguish mineral composition. They are the M-N plot, the MID plot, and the Litho-Density-Neutron method. The first two of these utilize Density, Neutron, and Sonic data. The third uses Litho-Density and Neutron data; it can also make use of Spectral Gamma Ray information.

THE M-N PLOT

The M-N plot is a means of combining data from the three porosity devices in such a manner that effects due to porosity variation are almost eliminated and those due to matrix changes are maximized.[14] The quantities M and N are defined by

$$M = 0.01\ (t_f - t)/(\rho_b - \rho_f) \tag{6.8}$$

$$N = (\phi_{Nf} - \phi_N)/(\rho_b - \rho_f) \tag{6.9}$$

where t, ρ_b, and ϕ_N are log values of Sonic travel time (μsec/ft), bulk density (g/cc), and Neutron porosity (limestone units, fractional); and t_f, ρ_f, and ϕ_{Nf} are the corresponding values for pore fluid. The latter are taken as 189, 1.0, and 1.0, respectively, for fresh mud and 185, 1.1, and 1.0 for salt mud.

Porosity variations simultaneously increase or decrease the numerators and denominators of M and N so the parameters are almost independent of porosity. On the other hand matrix variations cause M and N to change. The values for single-mineral formations are obtained by inserting appropriate matrix parameters in Eqs. 6.8 and 6.9 in place of log values. The M-N plot of Fig. 6–5 is so generated. Open circles represent the fresh-mud case and black circles represent the salt-mud situation. In each case silica has two possible points, depending on the matrix velocity assumed, and dolomite has three slightly different points, depending on the porosity range. This particular plot applies only for the CNL Neutron. The matrix points will plot a little differently for other models of Neutron tools that have slightly different matrix dependency.

Two matrix triangles have been drawn for average fresh-mud conditions. They represent the common combinations of sandstone-calcite-dolomite and calcite-dolomite-anhydrite.

To utilize the chart, log values of t, ρ_b, and ϕ_N are read at a level of interest, M and N values are computed, and the corresponding point is plotted on the chart. If the formation consists of a binary mineral mixture, the point will fall on the line joining the two mineral points. If three minerals are present, it will fall within the triangle formed by the three mineral points. A point such as A could represent a mixture of either sandstone-calcite-dolomite, calcite-dolomite-anhydrite, or even a combination of all four minerals.

In principle, if the matrix is known to consist of only three components, the relative amounts of each can be determined by the position of the plotted point relative to the apex points of the appropriate triangle. However, the method is not accurate enough for this for several reasons:

- the basic silica-calcite-dolomite points are not widely separated
- their positions are somewhat dependent on assumed matrix velocity

- the location of any log-derived point is subject to the statistical fluctuations in Density and Neutron logs

In addition, the presence of shale, gypsum, secondary porosity, gas, salt, or sulfur shifts the points away from the basic triangles in the directions

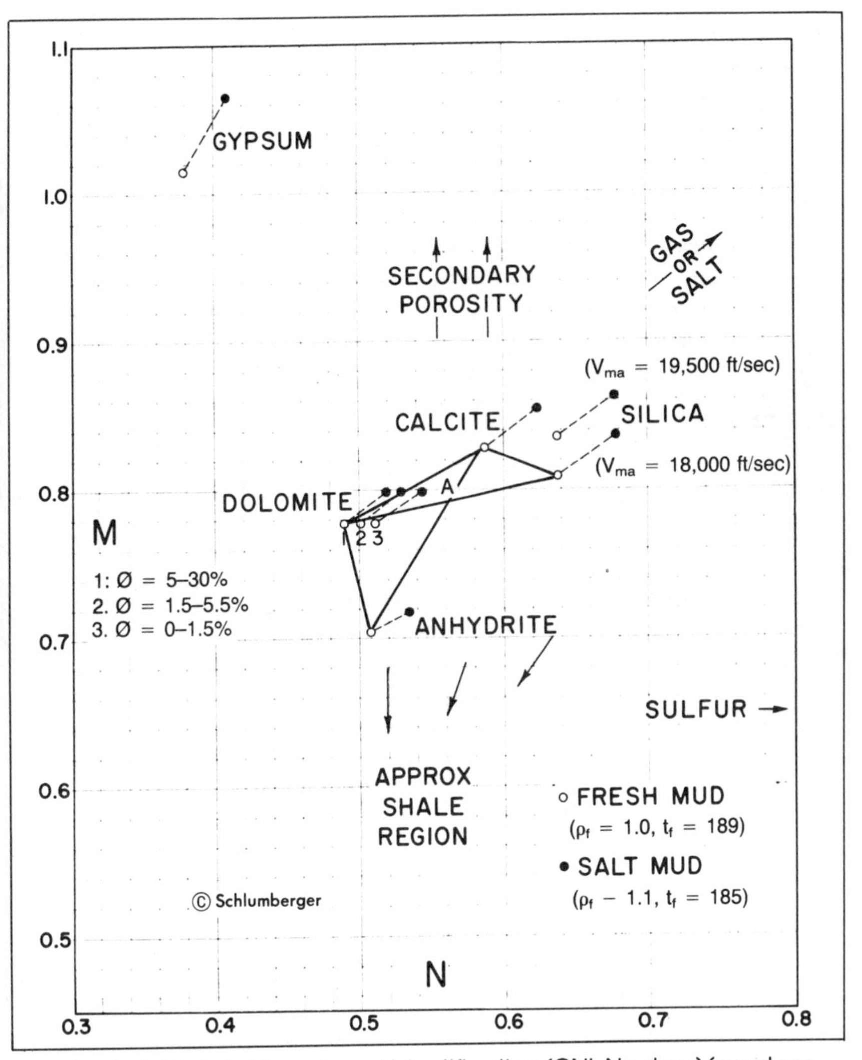

Fig. 6–5 M-N plot for mineral identification (CNL Neutron) (courtesy Schlumberger and SPWLA)

indicated. The plot is therefore best used on a statistical basis to indicate where a group of points from a given log interval tend to cluster.

Fig. 6–6 shows examples from the Permian basin area of West Texas where low-porosity, complex-carbonate sequences are common. (The matrix points are those applicable to the earlier SNP Sidewall Neutron tool.) Plot (*a*) indicates a section composed primarily of two binary mixtures, dolomite-calcite and dolomite-anhydrite, with minor secondary porosity. Plot (*b*) shows calcite-dolomite with no anhydrite along with considerable secondary porosity that increases with dolomite content. Plot (*c*), from a shallow zone, clearly indicates the presence of gypsum and some shaliness in a basically calcite-dolomite-anhydrite mixture. Plot (*d*) shows various amounts of shale in a sandstone-limestone mixture.

Creation of M-N plots is a laborious process by hand but is well adapted to log processing in a computer center. In that environment it is of considerable assistance in zoning logs for automatic interpretation. The same is true of MID and LDN plots.

THE MID PLOT

The MID (matrix identification) plot is another method of utilizing the same log data as for the M-N plot.[15] It is based on determining the apparent matrix density, $(\rho_{ma})_a$, and the apparent matrix travel time, $(t_{ma})_a$, for a level of interest; plotting the values on the MID plot of Fig. 6–7; and observing where the point falls relative to the positions of the single-mineral points. The latter are located at their known matrix densities and travel times.

To determine the apparent matrix density at a given depth level, values of ρ_b and ϕ_N are read from the appropriate Density-Neutron crossplot (Fig. 6–8). On this chart lines of constant $(\rho_{ma})_a$ have been created by interpolating between and extrapolating from the known matrix densities of sandstone, limestone, and dolomite. The $(\rho_{ma})_a$ value for any plotted point can be read from these lines. For example, $\rho_b = 2.45$ g/cc and $\phi_N = 0.20$ gives $(\rho_{ma})_a = 2.76$ g/cc.

Determination of apparent matrix velocity is similar. The travel time value, t, is read from the Sonic log and inserted along with ϕ_N in the appropriate Sonic-Neutron crossplot (Fig. 6–9). On this plot lines of constant $(t_{ma})_a$ have been created by interpolation and extrapolation from the known clean matrix values. For $t = 70$ μsec/ft and $\phi_N = 0.20$, $(t_{ma})_a$ is 46 μsec/ft.

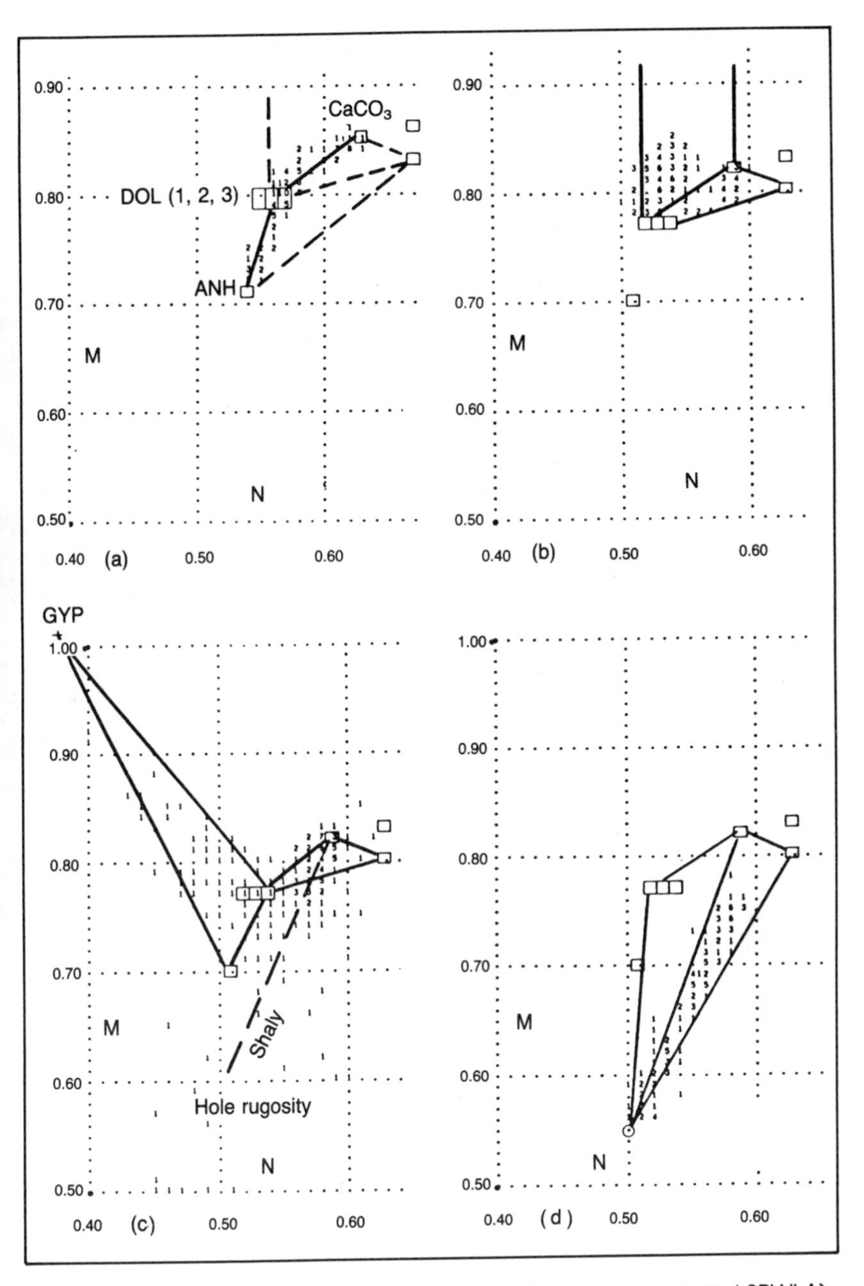

Fig. 6–6 Examples of M-N plots (courtesy Schlumberger and SPWLA)

The coordinate point $[(\rho_{ma})_a, (t_{ma})_a]$ so obtained is plotted on Fig. 6–7 and the matrix components are interpreted as for the M-N plot. Point B is the example case which indicates a calcite-dolomite mixture.

The same reservations on quantitative use of the chart apply as for the M-N plot. In particular, shale pushes points downward and salt and gas shift them upward and to the right. No single point is very significant by itself. Grouping of points indicates trends.

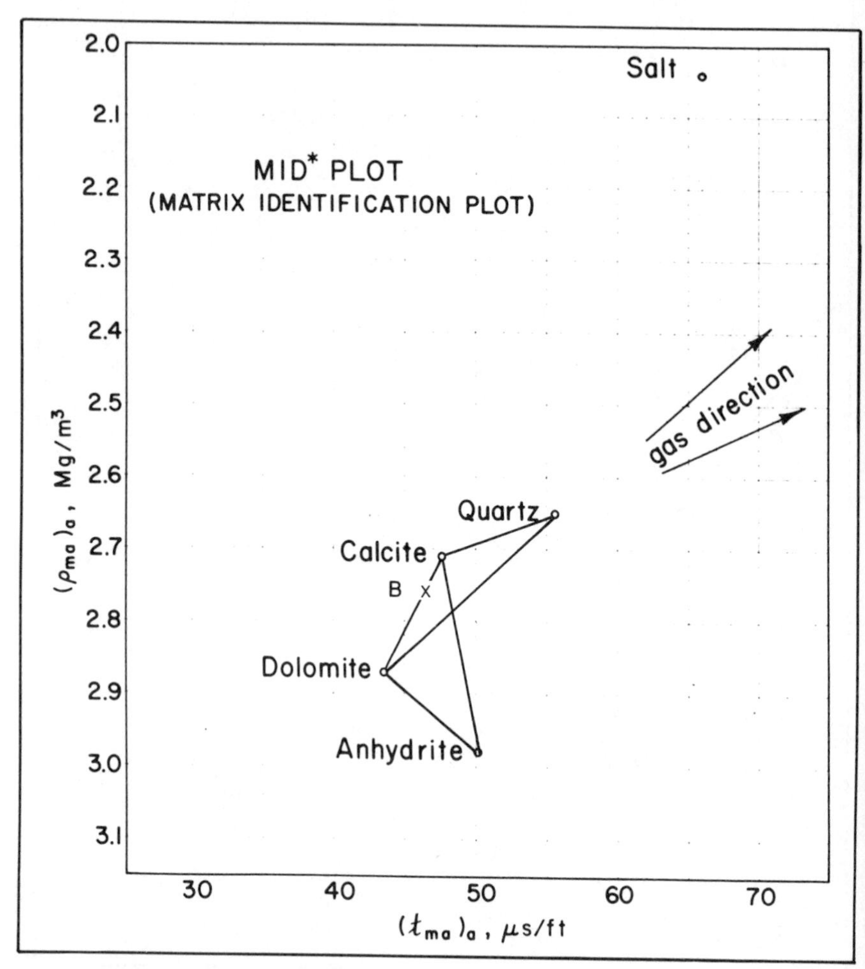

Fig. 6–7 MID plot for mineral identification (CNL Neutron) (courtesy Schlumberger and SPWLA)

The main advantages of the MID plot over the M-N plot are that no computations are required, the matrix points on the chart (Fig. 6–7) are not dependent on porosity or salinity, and the parameters plotted have a pseudophysical meaning. In contrast, M and N are abstract quantities. Use of both plots, however, can help resolve the effects of less common minerals—salt, gypsum, and sulfur. [16]

THE LITHO-DENSITY-NEUTRON METHOD

This is a relatively new method of mineral identification.[17] Its input is the bulk density, ρ_b, and the photoelectric absorption coefficient, P_e, from the Litho-Density log, and porosity, ϕ_N, from the Neutron log. The ρ_b and P_e values alone can be used to derive lithology (and porosity) when only two minerals are present, as described in chapter 5. With the addition of Neutron porosity, three minerals can be distinguished.

The method is based on a plot of the apparent matrix density, $(\rho_{ma})_a$, versus the apparent volumetric absorption index, $(U_{ma})_a$. The crossplot is shown in Fig. 6–10.

The volumetric photoelectric absorption index, U, of a given formation is derived from its ρ_b and P_e values as

$$U = P_e(\rho_b + 0.1883)/1.0704 \quad (6.10)$$

This coefficient is additive for different components in a mixture so that for a formation of porosity ϕ

$$U = \phi \cdot U_f + (1 - \phi)U_{ma} \quad (6.11)$$

where U_f is the absorption index for pore fluid and U_{ma} is the absorption index for the matrix. Values of these parameters are listed in Table 5–2.

For a formation of unknown matrix, rearranging Eq. 6.11 gives the apparent matrix absorption index as

$$(U_{ma})_a = (U - U_f \cdot \phi)/(1 - \phi) \quad (6.12)$$

The procedure for a level of interest is to read the log values of ϕ_N, ρ_b, and P_e and insert the first two in Fig. 6–8 (or equivalent) to find apparent matrix density, $(\rho_{ma})_a$, and apparent total porosity, ϕ_{ta}. For example, $\phi_N = 0.20$ and $\rho_b = 2.52$ give $(\rho_{ma})_a = 2.8$ and $\phi_{ta} = 0.16$.

Next, the values of P_e, ρ_b, and ϕ_{ta} are inserted in the nomogram of Fig. 6–11 to find $(U_{ma})_a$. This nomogram solves Eqs. 6.10 and 6.12. The example indicated with $P_e = 3.65$ gives $(U_{ma})_a = 10.9$.

The values of $(\rho_{ma})_a$ and $(U_{ma})_a$ so obtained are inserted in the plot of Fig. 6–10. Approximate percentages of limestone, sandstone, and dolomite can be read from the grid lines. The example, point C indicates almost equal amounts of dolomite and calcite with possibly a small amount of quartz.

Fig. 6–12 shows example plots from various zones in wells of mixed lithology and low porosity. The groupings of points clearly define different lithologies. Some intervals contain gas, as indicated by points shifted upward from their true locations. Gas causes $(\rho_{ma})_a$ to decrease without significantly altering $(U_{ma})_a$.

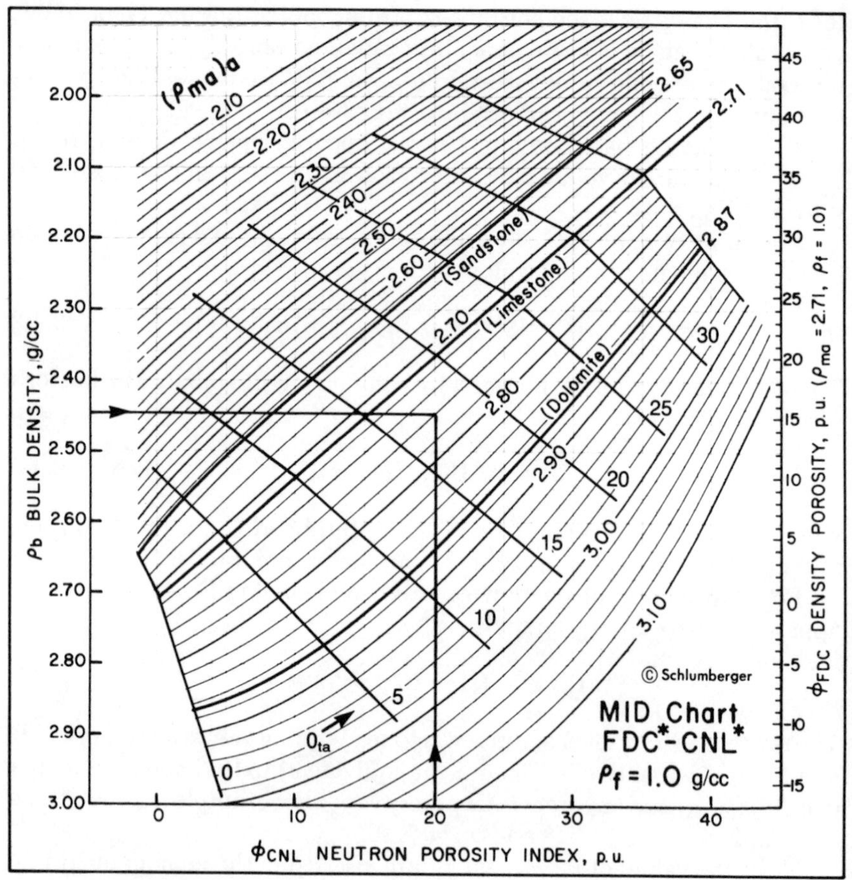

Fig. 6–8 Chart for determining apparent matrix density (courtesy Schlumberger)

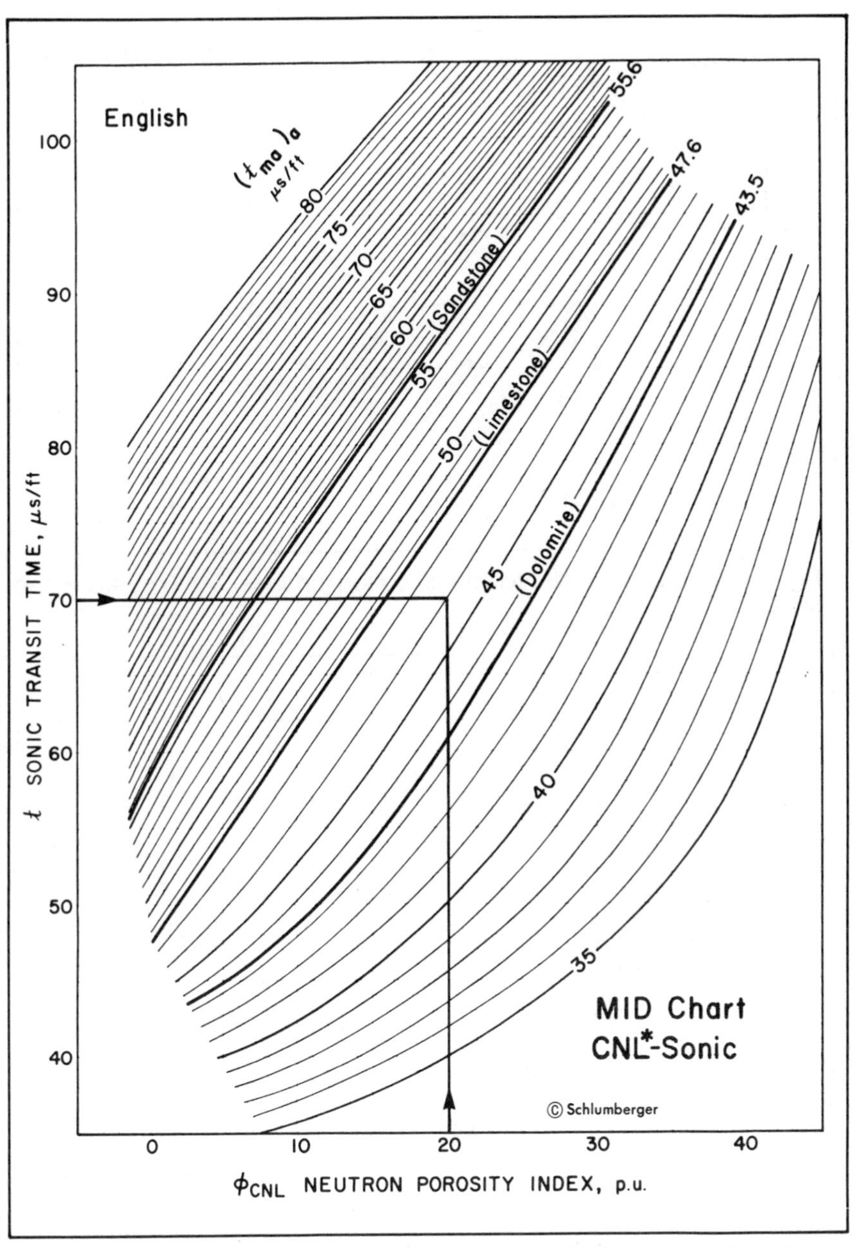

Fig. 6–9 Chart for determining apparent matrix travel time (courtesy Schlumberger)

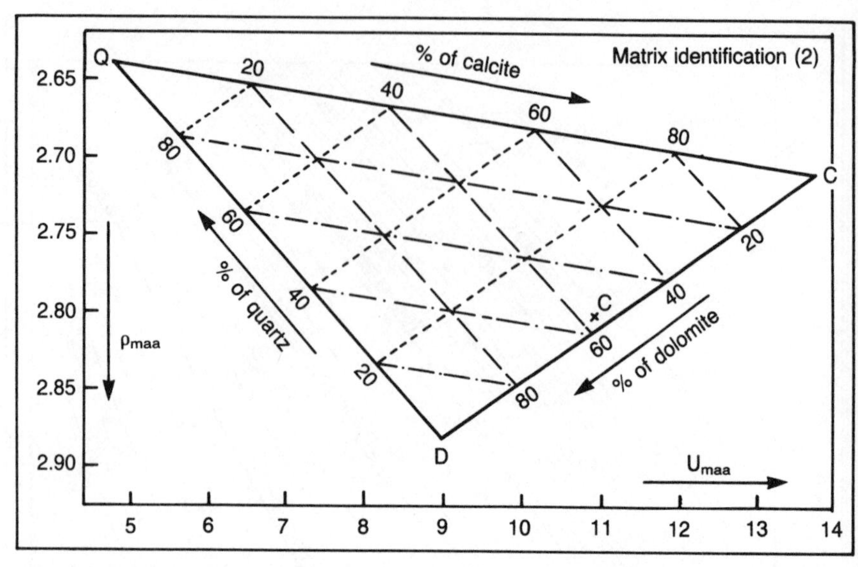

Fig. 6–10 ρ-U plot for mineral identification (courtesy Schlumberger and SPWLA)

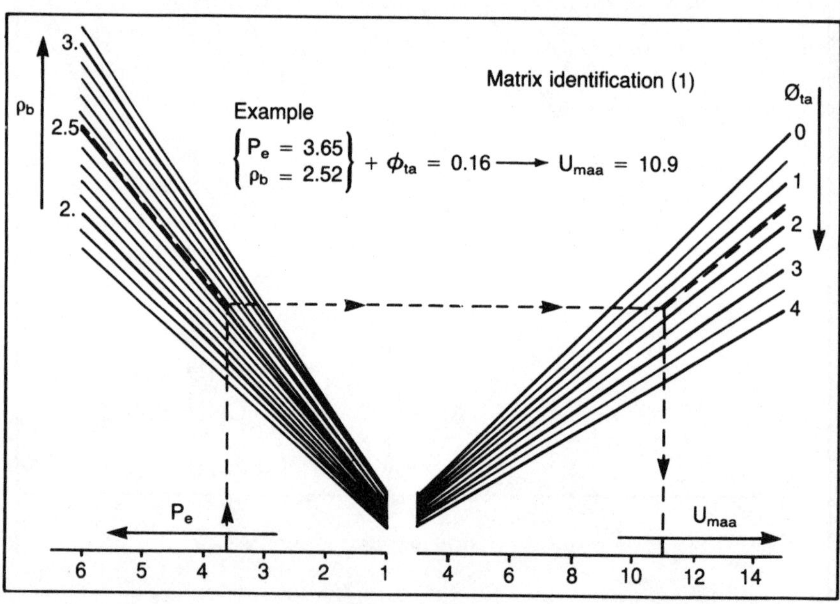

Fig. 6–11 Nomogram for determining apparent matrix photoelectric absorption coefficient (courtesy Schlumberger and SPWLA)

As with the M-N and MID plots, the presence of clay or shale is a very perturbing factor. Fig. 6–13 shows approximate locations of kaolinite, illite, and chlorite points as well as that of feldspar, which frequently

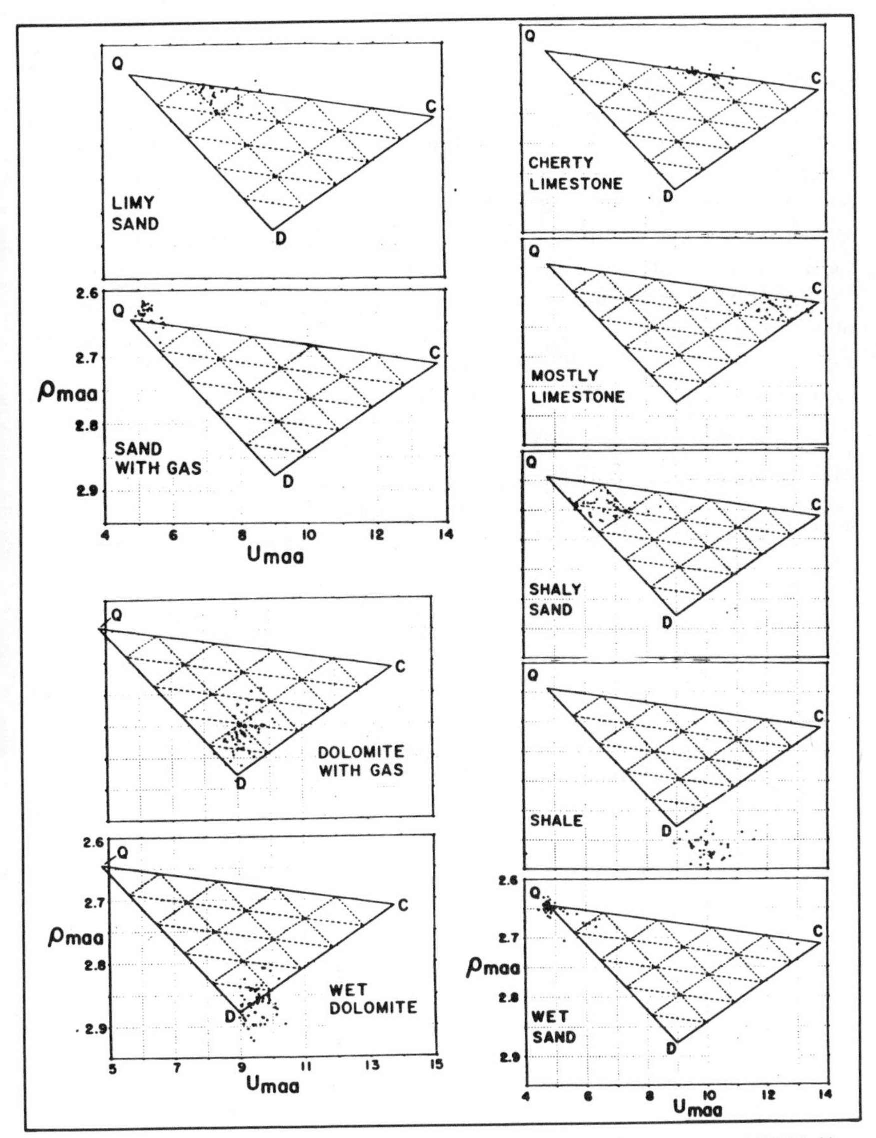

Fig. 6–12 Examples of ρ-U plots (courtesy Schlumberger and SPWLA)

accompanies clays. The locations of anhydrite and salt points are also shown.

Clay or shale inclusion shifts points toward the lower right, causing quartz formations to appear excessively dolomitic and calcite-dolomite intervals to appear anhydritic. The ambiguity can be resolved to some extent with Spectral Gamma Ray data using Fig. 6–14. This is an empirical plot of percent potassium versus ppm thorium on which the approximate locations of (Th, K) points for formations containing 100% low-potassium clay such as kaolinite (Cl_1), 100% high-potassium clay such as illite (Cl_2), and 100% feldspar (Fel) are indicated. Log-derived points are plotted on this chart, and the percentages of the two different types of clay and feldspar are estimated. Corrections can then be made to lithologies indicated on the $(\rho_{ma})_a - (U_{ma})_a$ plot.

Fig. 6–15 is an example of a shaly carbonate zone. At first glance the ρ-U plot indicates primarily a calcite-dolomite-anhydrite mixture. However, the thorium-potassium plot shows the majority of levels are shaly and contain a mixture of high-potassium clay and feldspar. When this is taken into

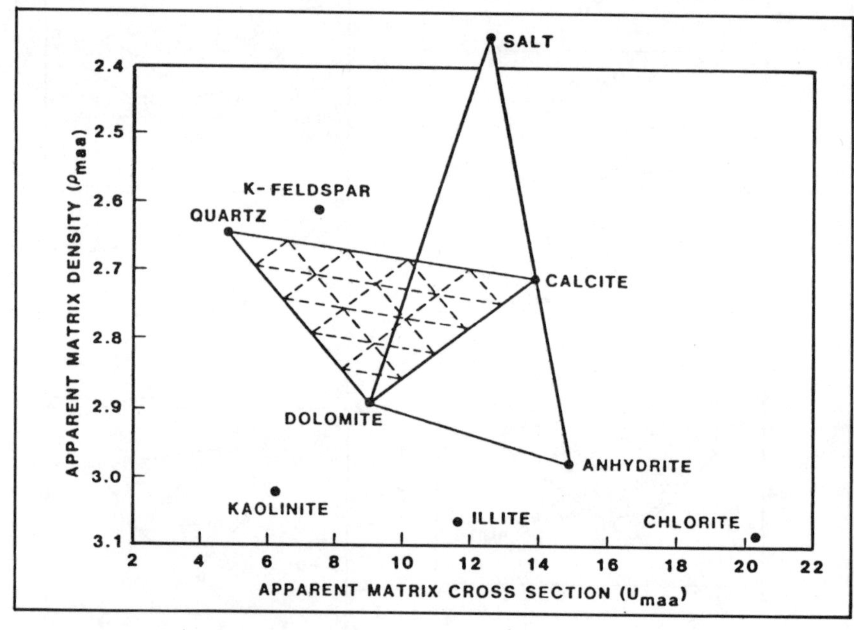

Fig. 6–13 ρ-U plot showing clay and evaporite locations (courtesy Schlumberger and SPWLA)

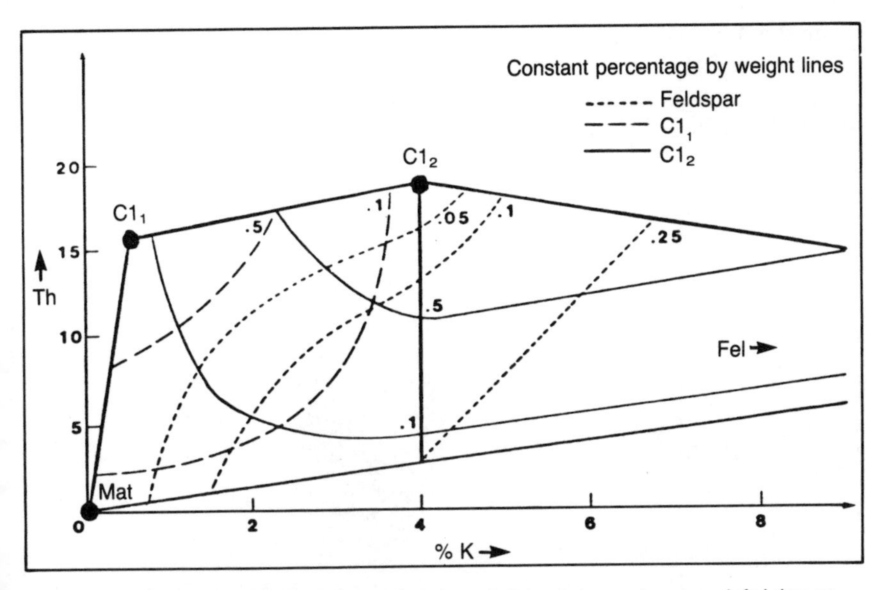

Fig. 6–14 Thorium-potassium plot for determining clay and feldspar content (courtesy Schlumberger and SPWLA)

account, it indicates the matrix is composed of variable amounts of calcite, clay, and feldspar with little dolomite and no anhydrite.

TRENDS IN MULTIMINERAL IDENTIFICATION

We can expect to see further refinement of the Litho-Density-Neutron method. The ρ-U plot is better than the M-N or MID plots because the basic quartz-calcite-dolomite-anhydrite points are more widely separated and there is no uncertainty in matrix travel time to contend with.

Even with the addition of Spectral GR information, however, there is not enough independent data to distinguish the various matrix components unambiguously. The induced Gamma Spectroscopy tool,* a new logging tool currently being tested, will do much to close the gap. This tool generates bursts of neutrons that interact with the formation nuclei and cause them to emit gamma rays of characteristic energies. By detecting these gamma rays and cataloging their energies, the relative amounts of hydrogen, calcium, silicon, chlorine, sulfur, iron, carbon, and oxygen can be determined and,

* Termed GST (Gamma Spectroscopy Tool) by Schlumberger and C/O by Dresser-Atlas.

from them, percentages of sand, limestone, clay, and other components can be estimated.[18,19]

At present, existing tools in this nature are being directed toward locating hydrocarbons behind casing, where the greatest need exists; but there is

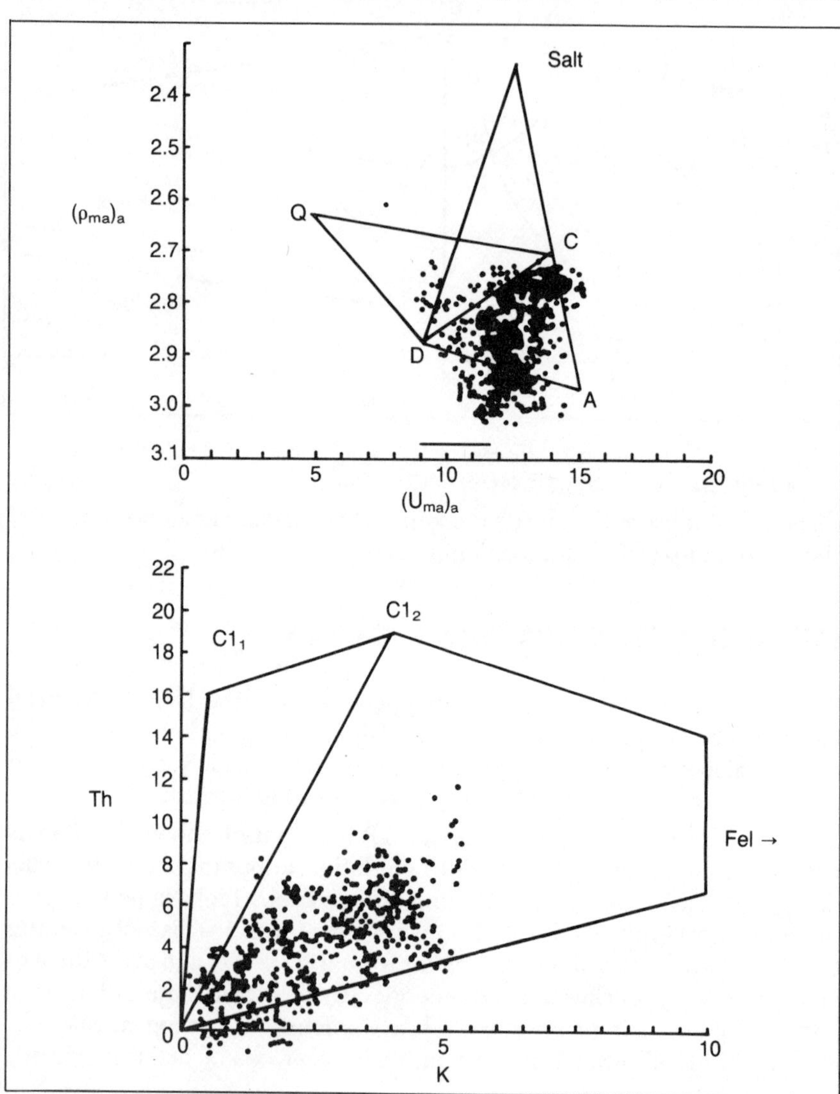

Fig. 6–15 Mineral identification using both ρ-U and Th-K plots (courtesy Schlumberger, © SPE-AIME)

no reason they cannot be used for mineral identification in open hole. Combination of the Litho-Density, Spectral Gamma Ray, and induced Gamma Spectroscopy logs will resolve most mineral components in the near future.

REFERENCES

[1]A.T. Hingle, "The Use of Logs in Exploration Problems," *SEG paper*, (Los Angeles: 1959).

[2]W.H. Fertl, "Hingle Crossplot Speeds Long-Interval Evaluation," *OGJ* (January 1979), pp. 113–118.

[3]R.H. Lindley, "Use of Differential Sonic-Resistivity Plots to Find Movable Oil in Permian Formations," *J. Pet. Tech.* Vol. 13, No. 8 (August 1961).

[4]G.R. Pickett, "A Comparison of Current Techniques for Determination of Water Saturation from Logs," *SPE paper* (Denver: 1966).

[5]W.H. Lang Jr., "Porosity-Resistivity Cross-Plotting," *SPWLA Logging Symposium Transactions* (May 1972).

[6]G.R. Coates and J.L. Dumanoir, "A New Approach to Log-Derived Permeability," *SPWLA Logging Symposium Transactions* (May 1973).

[7]E.R. Atkins and G.H. Smith, "The Influence of Particle Shape on the Formation Resistivity Factor of Sandstones and Shales," *SPE 1560-G* (Denver: October 1960).

[8]P.N. Sen, "The Dielectric and Conductivity Response of Sedimentary Rocks," *SPE 9379* (Dallas: September 1980).

[9]A.E. Bussian, "A Generalized Archie Equation," *SPWLA Logging Symposium Transactions* (July 1982).

[10]J. Clemenceau Raiga, "The Cementation Exponent in the Formation Factor Porosity Relation: The Effect of Permeability," *SPWLA Logging Symposium Transactions* (June 1977).

[11]D.K. Sethi, "Some Considerations About the Formation Resistivity Factor — Porosity Relations," *SPWLA Logging Symposium Transactions* (June 1979).

[12]F.J. Lucia, "Petrophysical Parameters Estimated from Visual Descriptions of Carbonate Rocks: A Field Classification of Carbonate Pore Space," *Jour. Pet. Tech.* (March 1983), pp. 628–637.

[13]C. Khelil, "Analysis of Errors in Logging Parameters and Their Effects on Calculating Water Saturation," *SPWLA Logging Symposium Transactions* (May 1971).

[14]J.A. Burke, R.L. Campbell Jr., and A.W. Schmidt, "The Litho-Porosity Crossplot," *The Log Analyst* (November–December 1969).

[15]C. Clavier and D.H. Rust, "The MID Plot: A New Lithology Technique," *The Log Analyst* (November–December 1976).

[16]J.V. Crues Jr., "Lithology Crossplots: Applications in an Evaporite Basin—the Maverick Basin of Southwest Texas," *SPWLA Logging Symposium Transactions* (June 1977).

[17]J.S. Gardner and J.L. Dumanoir, "Litho-Density Log Interpretation," *SPWLA Logging Symposium Transactions* (July 1980).

[18]P. Westaway, R. Herzog, and R.E. Plasek, "The Gamma Spectrometer Tool–Inelastic and Capture Gamma Ray Spectroscopy for Reservoir Analysis," *SPE 9461* (Dallas: September 1980).

[19]W.A. Gilchrist Jr., J.A. Quirein, Y.L. Boutemy, and J.R. Tabanou, "Application of Gamma Ray Spectroscopy to Formation Evaluation," *SPWLA Logging Symposium Transactions* (July 1982).

Chapter 7

SHALY FORMATION INTERPRETATION

The presence of shale in reservoir rock is an extremely perturbing factor in formation evaluation. On the one hand it complicates the determination of hydrocarbons in place; on the other hand it affects the ability of the reservoir to produce those hydrocarbons. Most sands contain some shale or clay. The effect of this is to

- reduce the effective porosity, often signficantly
- lower the permeability, sometimes drastically
- alter the resistivity from that predicted by Archie's equation

Clay, which is a major component of shale, consists of extremely fine particles that have very high surface area and are therefore capable of binding a substantial fraction of pore water to their surfaces. This water contributes to the electrical conductivity of the sand but not its hydraulic conductivity. It cannot be displaced by hydrocarbons and will not flow. For this reason we define *effective porosity* as the pore space occupied by only nonclay-bound fluid and *total porosity* as that occupied by both clay-bound and nonclay-bound fluid.

A shaly hydrocarbon-bearing formation can exhibit a resistivity little different from that of a nearby clean water sand or that of adjacent shales. This means shaly pay sands can be difficult to find on resistivity logs and, even if they can, application of the standard Archie equation can give water saturations that are too pessimistic.

A case in point is shown in Fig. 7–1.[1] At first glance none of the sands would be picked as productive on the Induction log. A close examination and shaly sand analysis, however, shows that the interval marked PERF contains an estimated 14% hydrocarbons by volume. After perforation, the zone produced approximately 161,000 bbl oil, 61,000 Mcf gas, and 114,000 bbl water in the first two years. This production could easily have been overlooked.

Too much shale in a reservoir rock will kill its production through excessive reduction in permeability. However, a modest amount of shale, if disseminated in the pores, can be beneficial in trapping interstitial water and permitting commercial hydrocarbon production from zones of abnormally high water saturation. The foregoing example is a good illustration of this situation. If the perforated zone had been a clean, coarse sand of the calculated water saturation (55%), it would have produced at perhaps a 4:1

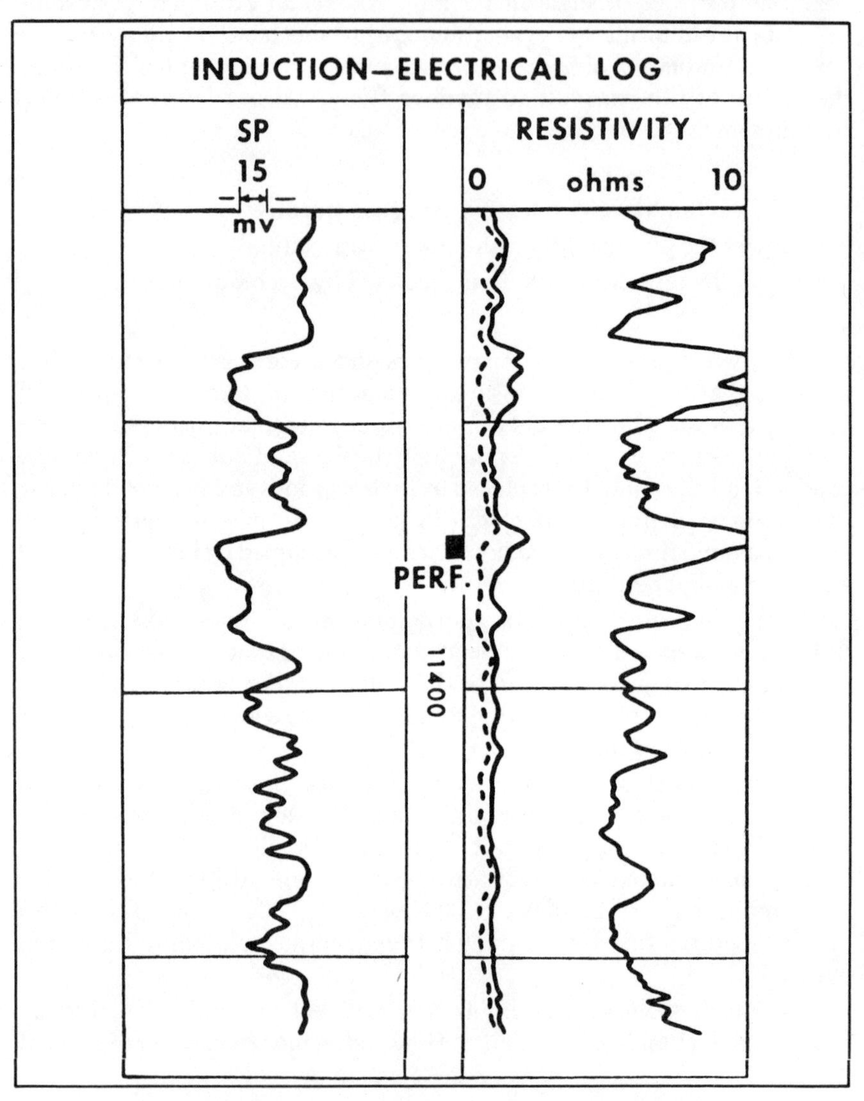

Fig. 7–1 Low-resistivity pay sand (courtesy Schlumberger)

water-oil ratio, though at greater rates. However, the fine clay particles in the pore space increased the irreducible water saturation and allowed production at the actual 0.7:1 water-oil ratio.

The effect of shaliness on electrical conductivity is illustrated by Fig. 7–2, a plot of the conductivity of a water-saturated sandstone, C_o, as a function of the conductivity of the saturating water, C_w.

If the sand is clean, the plot will be a straight line passing through the origin with slope 1/F as predicted by the formation factor relation, Eq. 2.1.* Writing it in terms of conductivity (the inverse of resistivity)

$$C_o = C_w/F \approx \phi^2 \cdot C_w \quad (7.1)$$

However, if some of the rock matrix is replaced by shale, maintaining the same effective porosity, the line will be displaced upward and the straight portion will interrupt the C_o axis at some value, C_{excess}. This is the *excess conductivity* contributed by the shaliness. It follows that use of the Archie saturation equation

$$S_w = c\sqrt{C_t/C_w}/\phi \quad (7.2)$$

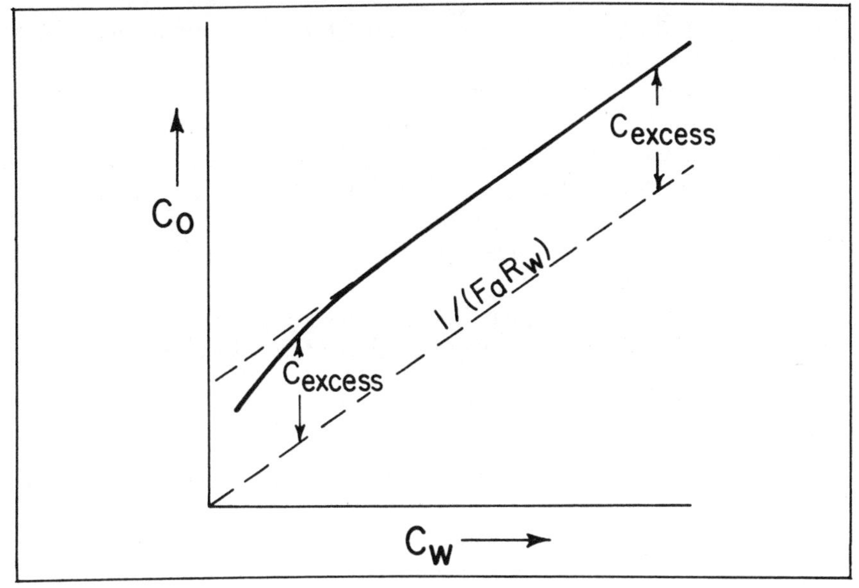

Fig. 7–2 Excess conductivity contributed by clay (courtesy Schlumberger)

*This statement is not entirely true. Even clean sand grains have a small surface conductance that gives C_0 a finite value when $C_w = 0$. This effect is important when logging freshwater wells but is insignificant in normal hydrocarbon logging situations.

will give saturations that are too large because C_t will be abnormally high for a given effective porosity, ϕ. A modified water saturation equation that includes a shaliness term must be used.

THE NATURE OF SHALE

Shale is a mixture of clay minerals and silt laid down in a very low-energy environment, principally by settlement from still water. Silt consists of fine particles, mostly silica, with small amounts of carbonates and other nonclay minerals. The solids of a typical shale may consist of about 50% clay, 25% silica, 10% feldspar, 10% carbonates, 3% iron oxide, 1% organic material, and 1% other material. The shale may also contain 2–40% water by volume. It is the clay component of the shale that affects logs in abnormal ways.

Clay is comprised of crystalline clay minerals.[2] These are hydrous aluminum silicates of the general formula $X(Al_2O_3) \cdot Y(SiO_2) \cdot Z(OH)$, which contain small amounts of other elements such as magnesium, potassium, iron, and titanium. Clay of detrital nature is a weathering product of preexisting rock, so its composition is quite variable, depending on the environment and conditions of temperature, humidity, and acidity under which it was formed.

Clay particles have a layered platelet structure. The crystalline platelets are very thin, 5–10 Å, but may extend to about 10,000 Å in length or width. They are stacked one above the other with spacings between them of 20–100 Å. The clay particles are therefore extremely small — about 2μ in maximum dimension.* This is 10 to 100 times smaller than average sand grains. Therefore, there is ample space in sandstone pores for clay to exist, and indeed it is found there.

Clay minerals are classified into specific groups according to their crystal-structure. Those of concern in sedimentary rocks are montmorillonite (a form of smectite), illite, kaolinite, chlorite, and mixed-layer minerals. Table 7–1 lists properties of these clay groups that are important in formation evaluation.

The first data column gives an important parameter, the cation exchange capacity (CEC). Note that montmorillonite and illite have much larger values than chlorite and kaolinite.

The second column lists the porosity that the CNL Neutron log would read theoretically in a 100%-dry clay formation because of hydrogen bound in the crystal lattice.[3] This hydrogen does not contribute to conductivity.

---

*One angstrom unit (Å) equals 10^{-8}cm; one micron (μ) equals 10^{-4}cm.

TABLE 7-1 CLAY PROPERTIES OF CONCERN IN LOGGING

Clay Type	CEC meq/g	ϕ_{CNL}	ρ(av), g/cc	Minor Constituents	Spectral GR Components (av) K, %	U, ppm	Th, ppm
Montmorillonite	0.8–1.5	0.24	2.45	Ca, Mg, Fe	0.16	2–5	14–24
Illite	0.1–0.4	0.24	2.65	K, Mg, Fe, Ti	4.5	1.5	<2
Chlorite	0–0.1	0.51	2.8	Mg, Fe	–	–	–
Kaolinite	0.03–0.06	0.36	2.65	–	0.42	1.5–3	6–19

Montmorillonite and illite have smaller values than chlorite and kaolinite—the opposite of CEC.

The next column lists the average dry clay density. It varies both with hydrogen concentration and with content of minor-constituent heavy minerals such as iron (succeeding column). There is a range on the order of $\pm 10\%$ in both ϕ_{CNL} and ρ values since clays vary widely in detailed composition.

The final three columns list average concentrations of naturally radioactive components in the clay.[4] Of interest is the high potassium concentration of illite and the high thorium content of montmorillonite.

Montmorillonite is somewhat unique in that it swells in contact with water. Water intrudes between the platelets and forces them apart. The fresher the water, the greater the swelling. In fact, lattice spacing increases as $(21 + 11/\sqrt{C})$ Å, where C is the water salinity in moles/liter (1 mole/liter $\approx$ 60,000 ppm).[5]

Another feature of montmorillonite is that it undergoes diagenesis to illite at the higher subsurface temperatures. This frees water and contributes to overpressuring of adjacent sands.

SHALE OR CLAY DISTRIBUTION IN SHALY SANDS

Most logging tools average formation response over 2- to 4-ft vertical intervals. In these "unresolvable" intervals shale or clay may be disposed in the sand in three ways or in combinations thereof: laminated, dispersed, and structural (Fig. 7–3).[6,7]

Laminated

In this form, thin shale laminations — fractions of an inch to many inches in thickness — are interspersed with clean sand. The effective porosity and the permeability of the shale are essentially zero so that the overall porosity and permeability (horizontal) of the averaged interval is reduced in proportion to the fractional volume of shale. For example, 40% shale will theoretically reduce effective porosity and permeability to 60% of the clean sand values.

It would appear that laminated shale over 50% by volume could be tolerated in an otherwise porous sand. However, it is unlikely that thin shale laminations extend uniformly very far from the wellbore. The interlayered sand laminations may pinch out nearby. Consequently, 30–40% laminated shale is the maximum amount normally tolerable for production.

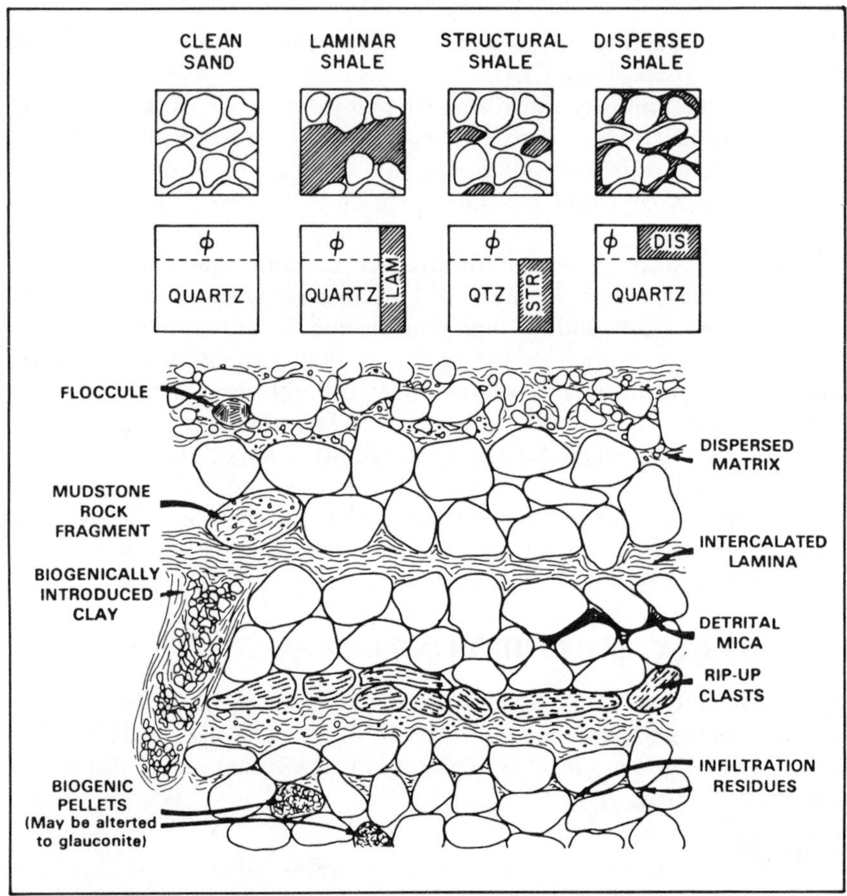

Fig. 7–3 Forms of shale distribution in sediments (after Wilson, Schlumberger)

The clay in laminated shale is of detrital origin. Since such clay is derived from diverse rock types and soils, it is generally mixtures of two or more clay minerals. Following deposition, the shale and sand laminations may tend to be homogenized by the reworking of organisms and infiltration of clay particles into the sand by water movement. Nevertheless, one would expect shale laminations to have somewhat the same composition as nearby thick shale beds.

Dispersed

In this form clay, not shale, is disseminated in the pore space of the sand. It replaces pore fluid. This type of distribution is very damaging because a relatively small amount of clay can choke pores and reduce effective porosity and particularly permeability to nonproducible values. Maximum tolerable clay content is approximately 40% of the sand pore space or about 15% by volume.

Much of the dispersed clay is of authigenic origin. That is, it literally grows in place after deposition of the sand as a result of chemical interaction between the pore fluid and minor constituents of the sand such as feldspar. Almost all sandstones contain some of this clay.

Authigenic clay has been characterized into discrete-particle, pore-lining, and pore-bridging types through observations of scanning electron microscope (SEM) pictures (Fig. 7–4).[8] The discrete-particle type consists mainly of kaolinite, which builds up as isolated booklets that lower porosity or permeability only a little. The pore-lining type coats the grains with whiskers, forming micropores that trap a good deal of pore water and significantly lower permeability. The pore-bridging type chokes the pore space with a mass of tendrils that significantly lowers effective porosity and drastically reduces permeability. Fig. 7–5 shows the effect on permeability. Pore-lining clay lowers permeability one order of magnitude; pore-bridging clay lowers it yet another order.

Because of their in situ origin, authigenic clays tend to be purer, more crystalline, and more likely to consist of a single mineral than detrital clays. Their composition may differ radically from that of detrital clay in nearby shale beds. In addition these clays are not subjected to overburden pressure, so a given quantity of clay (particularly montmorillonite) may trap more water than it would in a compacted shale. Thus, it is risky in log analysis to assume that the clay in a shaly sand has the same characteristics as clay in a nearby shale, although we are forced to do so at present.

Structural

In this form clay grains, which may be aggregates of clay particles or mudstone clasts, take the place of sand grains. Porosity and permeability of the sand is affected very little. Consequently, this type of clay is least objectionable, but it does not occur frequently.

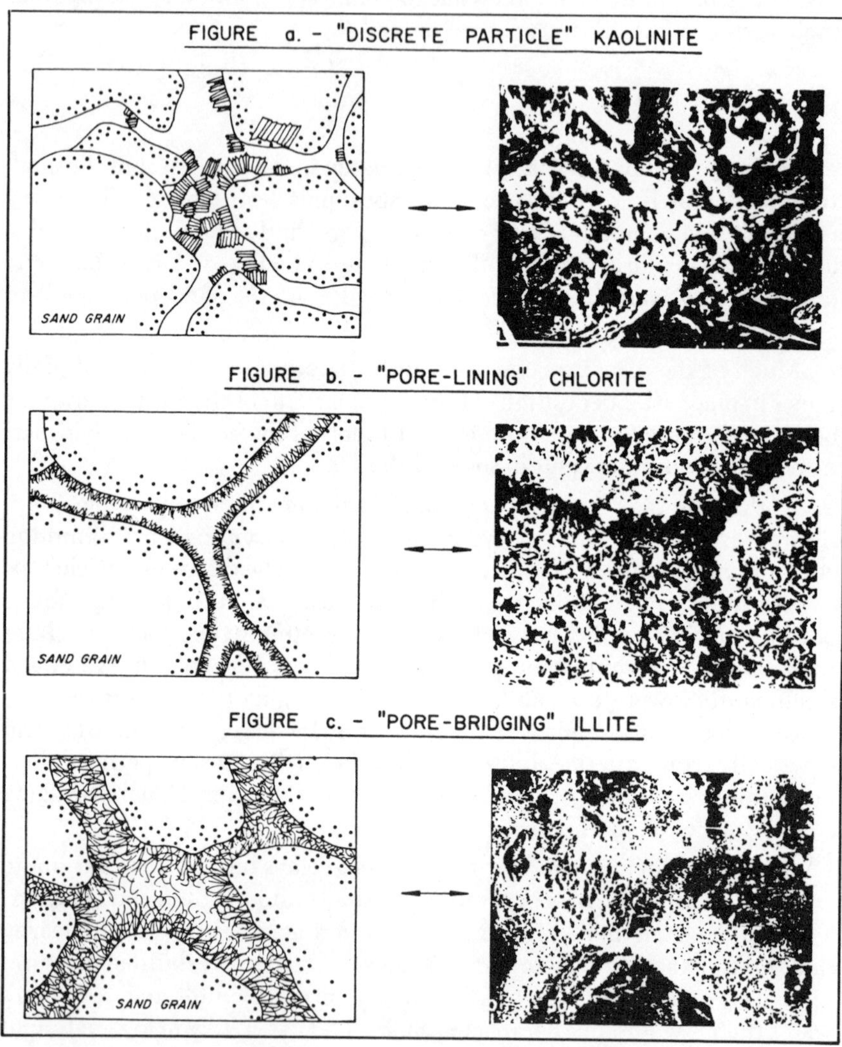

Fig. 7–4 Forms of authigenic clay in sandstone pore space (after Neashan, © SPE-AIME)

Given a certain overall fraction of clay or shale by volume, its particular distribution in the averaging intervals of Induction, Density, and Neutron tools does not affect their responses too much, although the Sonic log responds differently to laminated and dispersed clay. Water saturation and effective porosity can be determined without being too concerned about the distribution.

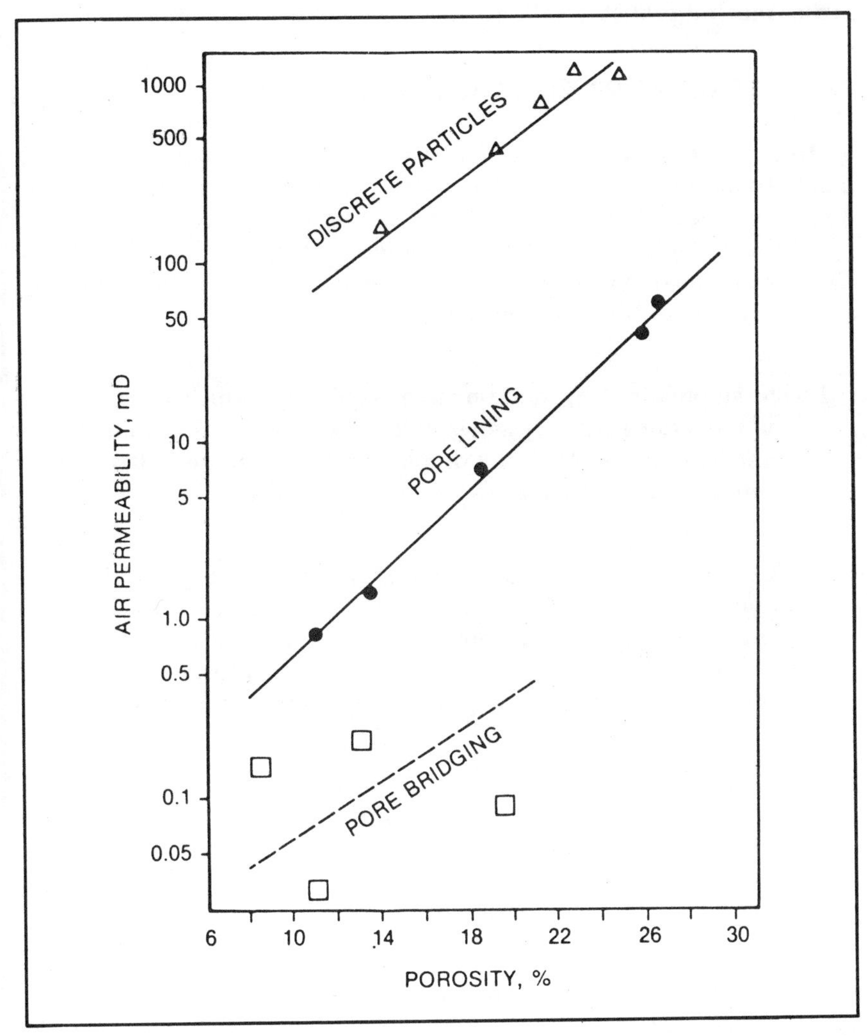

Fig. 7–5 Effect of authigenic clay type on permeability (after Neashan, © SPE-AIME)

However, the particular disposition strongly affects the producibility of the formation. Dispersed clay is much worse than laminated shale. For example, a zone may calculate 15% effective porosity with 50% water saturation. Such a zone could be a 30% porosity sand interlayered 50-50 with shale laminations. Or it could be a 30% porosity sand with pore space half-filled with authigenic clay. The former would produce at a much higher rate but probably with greater water cut than the latter.

SHALY SAND INTERPRETATION MODELS

Interpretation of shaly sands is still evolving. Over the years a large number of different models for calculating water saturation have been used. A comparative evaluation can be found elsewhere.[9]

Three methods once employed extensively are still used to some extent. They are oriented to logging tools available at the time and therefore may still be needed on older logs (see page 261):

1. The automatic compensation method (1950s) in which Sonic porosity and Induction resistivity are used directly in the Archie equation with compensating effects. It is a simple approach that works best in medium- to high-porosity sands with dispersed clay.

2. The dispersed model (1960s) using Sonic and Density porosities. The former reads essentially total porosity and the latter reads effective porosity in dispersed-clay sands so that the difference is indicative of the degree of shaliness. The method is directed toward sands with authigenic clay but has also given good results with laminated shale.

3. The Simandoux model (1970s) which uses porosity from Density-Neutron and shale fraction determined from GR, SP, or other shale indicators. This method has been the backbone of the service companies' shaly sand interpretation programs for the last ten years. It is applicable to dispersed or laminated shale.

At the present time a transformation to shaly sand models based on cation exchange capacity, CEC, rather than on shale fraction, V_{sh}, is underway. Two versions are in existence: the Waxman-Smits and the Dual-Water.

CATION EXCHANGE CAPACITY

The most important property of clay in log evaluation is its *cation exchange capacity*. This is the source of the excess conductivity depicted in Fig. 7–2.

Crystalline clay platelets are negatively charged as a result of ion substitutions in the lattice and broken bonds at the edges. Charge-balancing cations, typically Na^+, reside on the surface of dry clay. When the clay is in contact with a saline solution, these Na^+ cations are held in suspension close to the clay surface and, as a result, repel the Cl^- anions in the solution from the clay surface.

The current picture of the Na^+ ion and H_2O molecule concentration near the clay surface is shown in Fig. 7–6.[10] Directly on the surface of the clay is a monolayer of adsorbed water. Next to it is a layer containing hydrated Na^+ ions sufficient to balance the negative platelet charge.

The concentration of sodium cations can be measured by chemical means and is termed the cation exchange capacity. It is called such because the Na^+ cations are exchanged for Ba^{++} cations—for example, when a barium chloride $BaCl_2$ solution is flushed through the clay. (This is one way CEC is measured.) CEC is expressed in milliequivalents per gram of dry clay. (One meq = 6×10^{20} atoms.) Table 7–1 shows that CEC is high for

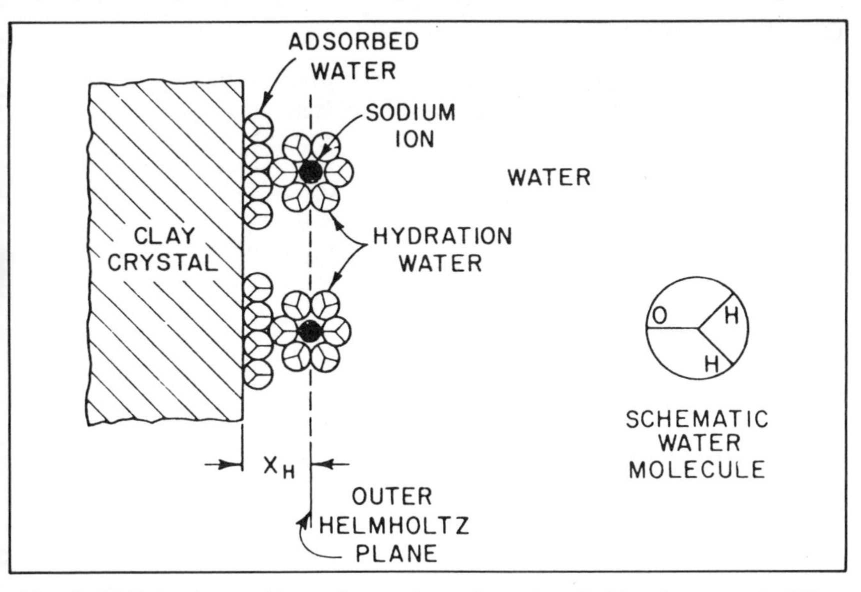

Fig. 7–6 Water bound to a clay surface (courtesy Schlumberger, © SPE-AIME)

montmorillonite, intermediate for illite, and low for chlorite and kaolinite. The high end of the range for each type probably applies to laminated clay and the low end applies to dispersed clay because the former undergoes severe mechanical stressing during compaction that creates broken bonds.

Cation exchange capacity may also be expressed as milliequivalents per unit volume of pore fluid, Q, given by

$$Q = CEC \cdot \rho(1 - \phi)/\phi \text{ meq/cc} \quad (7.3)$$

where ρ is the density in g/cc of the dry clay particles and ϕ is the porosity of the clay.

Relation of CEC to Surface Area

Measurements have shown that cation exchange capacity is essentially a reflection of the specific surface area of a clay regardless of type.[11] This is depicted in Fig. 7–7, which shows that a single value of approximately 450 m^2 of surface area per meq of cations applies to the clays of interest. (This implies that the negative charge density at the clay surface is the same for all clays.) Surface areas of clays, varying from 800 m^2/g for montmorillonite to 20 m^2/g for kaolinite, are hundreds of times greater than those of sands, which vary from 0.01 to 5 m^2/g.

Clay-Bound Water

The model of Fig. 7–6 predicts a very important clay property, i.e., there is a layer of water next to the clay surface that is essentially immovable.

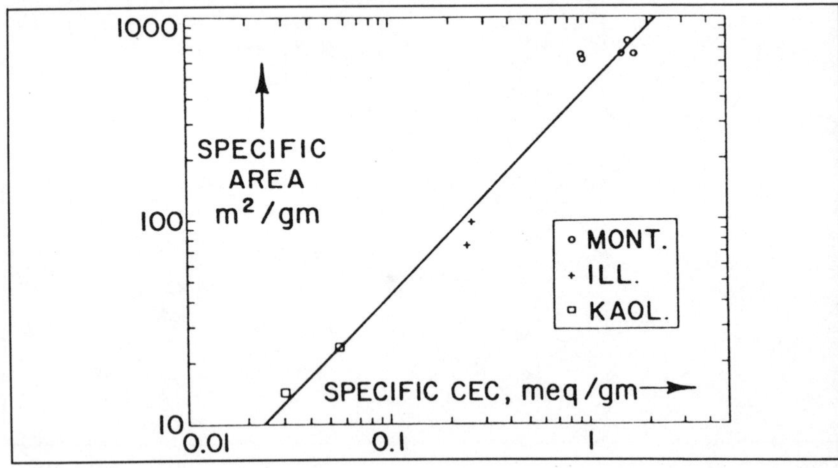

Fig. 7–7 Relation of area to CEL for API standard clays (after Patchett and SPWLA)

The electrostatic binding force is so strong that the water cannot be squeezed out even by tremendous overburden pressures. By the same token, oil migrating into a shaly sand will not replace this water.

The amount of clay-bound or anion-free water has been measured on shaly sand cores and the results are given, in grams of water per meq of exchangeable cations, as[12]

$$W = 0.22 + 0.084/\sqrt{C} \tag{7.4}$$

where C is the concentration of sodium chloride (moles/liter) in the water in equilibrium. Assuming the density of the bound water is close to 1.0, Eq. 7.4 also represents the volume of bound water in cc/meq. Dividing W by the specific surface area of clay, 450×10^4 cm^2/meq, gives the approximate thickness of the bound layer as $4.9 + 1.9\sqrt{C}$ Å.

For a formation that has a cation exchange capacity of Q meq/cc of pore fluid, this leads to the fundamental relation that the fraction of pore water, S_b, that is bound to the clay is

$$S_b = W \cdot Q \tag{7.5}$$

Q values of producible shaly sands range up to about 1.0 meq/cc. W is approximately 0.3 cc/meq, which means that up to about 30% of the pore water can be bound in such sands. Shaly sands with Q values higher than 1.0 are generally too tight to produce.

A shaly water-bearing sand therefore contains two types of water: bound water tied to the clay and free water in the remaining pore space. The latter is really not all free, since it includes irreducible water associated with the sand grains, but is equivalent to the water in a clean sand.

Shaly Sand Partitioning

Fig. 7–8 shows the partitioning of a hydrocarbon-bearing shaly sand as envisioned with this model. The rock matrix is comprised of normal sand particles, silt particles, and dry clay particles. The fluid is composed of bound water, free water, and hydrocarbons.

The total porosity (bound water + free water + hydrocarbons) is designated as ϕ_t, and the free or effective porosity is ϕ_e. The latter is given by

$$\phi_e = \phi_t(1 - S_b) \tag{7.6}$$

The volumetric fraction of hydrocarbons is

$$\phi_h = \phi_t(1 - S_{wt}) \tag{7.7}$$

where S_{wt} is the fraction of total pore space containing water. This is a difficult quantity to determine in shaly sand interpretation. The charge-balancing Na^+ cations (also called counterions) that are associated with the clay give rise to electrical conductivity. This takes the form of cations migrating from one exchange site to another when an electric field is imposed. Externally it manifests itself as the excess conductivity illustrated in Fig. 7–2.

The manner in which the counterion conductivity is considered to act in the pore space affects the calculation of the total sand conductivity and the evaluation of water saturation. Two models are in current use: the Waxman-Smits and the Dual Water.

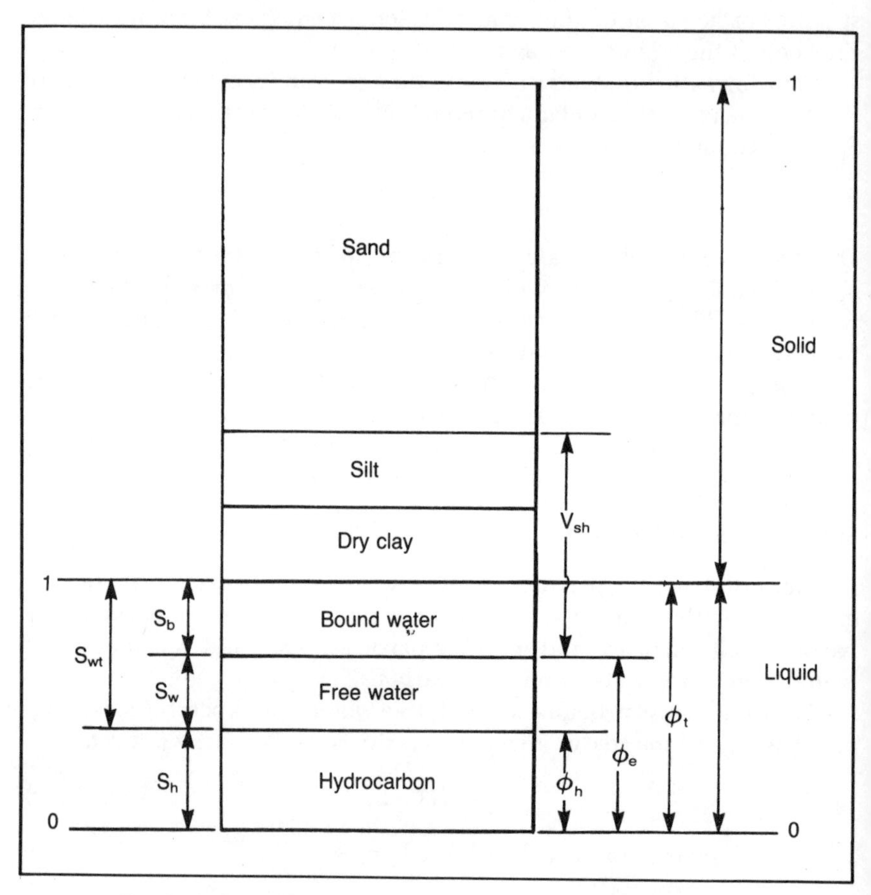

Fig. 7–8 Partitioning of a shaly sand in Dual-Water model

Waxman-Smits Model (W-S)

In this conception the cation conduction and the conduction of the normal sodium chloride electrolyte are assumed to act independently in the pore space, providing parallel conduction paths.[13] That is, the effective water conductivity at 100% water saturation is taken as

$$C_{we} = C_w + BQ \tag{7.8}$$

where C_w is the conductivity of the free pore water in mho/m and B is the specific counterion conductivity in mho/m per meq/cc.

When hydrocarbons enter the pore space and displace free water, the counterions are more concentrated in the remaining water; the effective water conductivity becomes

$$C_{we} = C_w + BQ/S_{wt} \tag{7.9}$$

Archie's principle leads directly to the following expression for the conductivity of the hydrocarbon-bearing shaly sand

$$C_t = (S_{wt} \cdot \phi_t)^2(C_w + BQ/S_{wt})^* \tag{7.10}$$

This equation can be solved for S_{wt} if the other quantities are known. C_t, ϕ_t, and C_w are obtainable from the resistivity and porosity logs, leaving B and Q to be determined.

The value of B has been measured by applying Eq. 7.10 to shaly sand cores with known Q values saturated with water of different salinities. B is 3.83 mho/m per meq/cc at 77°F and at water salinities greater than 30,000 ppm. At lower salinities it increases somewhat. It is also extremely temperature dependent, rising to 25 at 340°F.[14]

Application of Eq. 7.10 then hinges on the determination of Q. This matter will be deferred until later.

One major objection has been raised to the W-S model. It predicts, through Eq. 7.8, that water sands of constant C_w but increasing shaliness will have increasing effective water conductivities to the point that shales should appear to contain quite saline water. There is a good deal of evidence to the contrary. The Dual-Water model was devised to circumvent this constraint.

---

*Strictly speaking, this equation should be written

$$C_t = S_{wt}^{\ n} \cdot \phi_t^{\ m}(C_w + BQ/S_{wt})/a$$

For simplicity the saturation and cementation exponents *n* and *m* have been taken as 2 and the cementation constant *a* as 1.

Dual-Water Model (D-W)

The D-W model represents that the counterion conduction will be restricted to the bound water where the counterions reside and that the normal electrolyte conduction will be confined to the free water, inasmuch as the Cl^- anions are excluded from the bound portion.[15] There is therefore a mixture of two waters: *bound water* of conductivity C_b occupying a fraction of pore space equal to S_b and *free water* of conductivity C_w occupying the remaining pore space. The effective water conductivity for 100% saturation is then

$$C_{we} = C_w(1 - S_b) + C_b \cdot S_b \tag{7.11}$$

Hydrocarbons displace free water. When this happens, relative amounts of bound and free water are changed, giving an effective water conductivity

$$C_{we} = C_w(1 - S_b/S_{wt}) + C_b \cdot S_b/S_{wt} \tag{7.12}$$

The conductivity of the hydrocarbon-bearing sand is then

$$C_t = (S_{wt} \cdot \phi_t)^2[C_w(1 - S_b/S_{wt}) + C_b \cdot S_b/S_{wt}] \tag{7.13}$$

Since the counterion conductivity BQ is restricted to the fractional pore space WQ, the D-W model further predicts that the bound water conductivity C_b will be their quotient

$$C_b = B/W \tag{7.14}$$

Using this relation and Eq. 7.5 for S_b, the conductivity equation becomes

$$C_t = (S_{wt} \times \phi_t)^2[C_w(1 - WQ/S_{wt}) + BQ/S_{wt}] \tag{7.15}$$

This is the D-W relation for water saturation. It has one more term than the equivalent W-S relation, Eq. 7.10. Both equations, of course, reduce to the clean-formation expression, Eq. 7.2 with $c = 1$, when $Q = 0$.

The quantity B has also been determined by applying Eq. 7.15 to the same core measurements as in the W-S case. The value derived is 2.05 mho/m per meq/cc at 77°F instead of 3.83. With this value and $W = 0.3$ (Eq. 7.4), we arrive at the important result, through Eq. 7.14, that the bound water conductivity is

$$C_b = 6.8 \text{ mho/m at } 77°\text{F} \tag{7.16}$$

Consequently, the D-W model predicts that shales, which contain only bound water, should have water conductivities essentially independent of salinities in adjacent water sands and dependent only on temperature (through B).

W-S vs D-W Relations

On the basis of available shaly sand information, it is impossible to make an impartial choice between the W-S and D-W saturation relations because of the lack of data wherein the counterion conductivity B and the cementation exponent m have been determined independently of the saturation model. If the same values of B, m, and n are used in both equations, the D-W relation will typically give water saturation values 10% higher than the W-S relation. If the different model-dependent B values indicated above are used, answers will be even farther apart. However, use of a higher cementation exponent in the W-S relation, such as 2.2 instead of 2.0 (which is theoretically justified), will bring the answers back together.

The present situation is that both shaly sand models are being used successfully with judicious choice of parameters. Correct determination of Q is the overriding factor in either case.

SHALE POROSITY AND CONDUCTIVITY

The Dual-Water concept leads to an insight into the porosities and conductivities of such formations.

When initially laid down, shales — muds at that stage — may contain as much as 80% water by volume. As the mud is overlain, two effects occur simultaneously. The clay, being plastic when mixed with water and having very small particle size, is forced into pore space (between silt particles, for example) and displaces the water there. At the same time, loosely bound water held between the clay platelets (particularly with montmorillonite clay) is squeezed out. The expelled water migrates vertically to adjacent sands and thence laterally to a relief aquifer or fault. This compaction process generally takes place in the first 3,000 ft of depth, with most of it in the first 1,000 ft.

A compacted shale is therefore left primarily with only bound water that cannot be expelled with further overburden increase because the electrostatic binding forces are too strong. For such a shale

$$S_b = 1 \tag{7.17}$$

which means, from Eq. 7.5

$$Q_{sh} = 1/W \tag{7.18}$$

Since shales are laid down in waters with moderate salinities, the value of W is approximately 0.30, giving $Q_{sh} = 3.3$. Consequently, all well-compacted shales should have Q values of approximately 3.3, regardless of the particular type and content of constituent clay. The latter two factors will, however, determine the porosity of the shale.

Shale Porosity

Consider a shale in which the fraction of solids that is dry clay is Y and the cation exchange capacity of the clay is CEC_{cl} meq/g. Assuming that grain densities of dry clay and silt are the same, the CEC of the dry shale is

$$CEC_{sh} = CEC_{cl} \cdot Y \tag{7.19}$$

Applying Eq. 7.3 with $Q = 3.3$ and $\rho = 2.65$, leads to a prediction of shale porosity as

$$\phi_{sh} = 0.80\ (CEC_{cl} \cdot Y)/(1 + 0.80\ CEC_{cl} \cdot Y) \tag{7.20}$$

Thus, porosities of compacted shales are primarily determined by the content and CEC value of constituent clay and not by overburden pressure.

Porosities calculated from Eq. 7.20 are listed in Table 7–2 for the range of Y values normally found in shales.

TABLE 7-2 CALCULATED SHALE POROSITIES

Clay Type	Average CEC_{cl} (meq/gm)	ϕ_{sh}		
		Y = 0.35	Y = 0.55	Y = 0.75
Montmorillonite	1.00	0.22	0.30	0.37
Illite	0.25	0.07	0.10	0.13
Kaolinite/chlorite	0.04	0.01	0.015	0.02

Table 7–2 shows that shale porosities can vary widely. The greater the fraction of clay and the greater its CEC, the higher the shale porosity. Allowing for the range of CEC values given in Table 7–1, montmorillonite shales will have porosities in the range of 20–50%, illite shales will have porosities in the range of 5–20%, and kaolinite/chlorite shales will have porosities in the range of 1–5%. This explains why shale porosities do not necessarily decrease uniformly with depth and why high shale porosities can be found at great depths.

The Density log will read shale porosities correctly, provided the grain and bound-water densities of the shale are the same as those for sands, normally 2.65 and 1.0 g/cc. There is some evidence that shale values are typically higher, perhaps 2.85 and 1.2 g/cc, in which case the Density porosity will be a little too low. In any case it should be a clue to the type of clay in a shale.

The Neutron log will show substantially higher porosities. It will add to the actual porosity the contribution from the hydrogen bound in the crystal lattice. Using the ϕ_{CNL} values shown in Table 7–2, CNL shale porosities will be as listed in Table 7–3.

TABLE 7-3 TYPICAL NEUTRON POROSITIES OF SHALES

		$(\phi_{sh})_{CNL}$		
Clay Type	CEC_{cl} (meq/gm)	Y=0.35	Y=0.55	Y=0.75
Montmorillonite	1.00	0.28	0.39	0.48
Illite	0.25	0.15	0.22	0.29
Chlorite	0.04	0.19	0.30	0.40
Kaolinite	0.04	0.14	0.22	0.29

It is apparent that Neutron porosities in shales are not clay distinctive. They are all in the range of 14–50%. In fact, Neutron tools using thermal detectors (e.g., CNL and equivalent) may read even higher porosities if there are significant quantities of thermal neutron absorbers such as boron, lithium, cadmium, gadolinium, and samarium in the shales.

Although compacted shales can have high water contents, their effective porosities are zero because all of the water is tightly bound to the clay surface. By the same token permeabilities are virtually nil, in the range of 10^{-3}–10^{-6}md.

Shale Conductivity

Since compacted shale contains only bound water, its conductivity is, by Archie's principle

$$C_{sh} = \phi_{sh}^2 \cdot C_b \tag{7.21}$$

where ϕ_{sh} is given by Eq. 7.20 and C_b is given by Eq. 7.14.

The Dual-Water model therefore predicts that shale conductivities should vary in accordance with the CEC capacity and content of clay, which determines ϕ_{sh}, and with temperature, which largely determines C_b. Confirmation of this prediction is illustrated in Fig. 7–9, which shows calculated versus measured shale conductivities for a wide range of wells. Published data listed log-derived conductivities along with CEC_{cl} and Y values from core analysis.[16] Porosities were calculated from the CEC_{cl} and Y values using Eq. 7.20. Published temperatures permitted calculation of C_b using the temperature dependency[17]

$$C_b = 6.8(1 + 0.0545t - 1.127\ 10^{-4}\ t^2) \qquad (7.22)$$

where t = (°C − 25). The cementation exponent was allowed to vary from

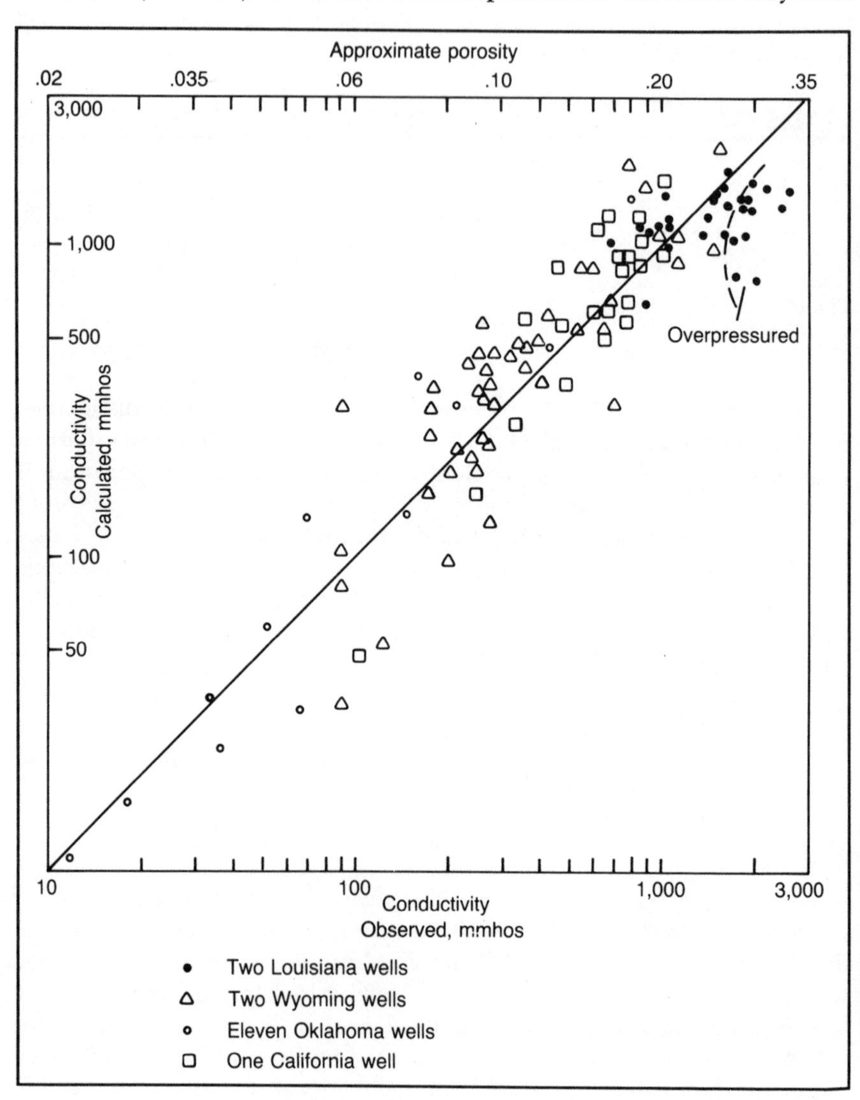

Fig. 7–9 Calculated vs observed shale conductivities (D-W model)

1.8–2.2 in accordance with $m = 1.8 + 0.6\ CEC_{sh}$, since increased clay surface area implies increased tortuosity. Conductivities so calculated are compared to measured shale conductivities in Fig. 7–9.

The agreement between calculated and measured values is remarkable, considering that the comparison covers 16 wells from Louisiana to California with depths ranging from 2,000–15,000 ft, CEC_{sh} values ranging from 0.02–0.7, and temperatures ranging from 100–275°F. Some of the spread between measured and calculated values can certainly be attributed to logging tool averaging. CEC_{sh} values for cores taken within 1–2 ft of each other varied by a factor of 1.4, which corresponds to factor-of-two variations in calculated conductivities.

The D-W model therefore explains why shale porosities can vary all the way from 2% to 35% and resistivities can vary from 0.3 to 100 ohm-m. This being the case, it should work well for shaly sands.

APPLICATION OF THE DUAL-WATER METHOD TO SHALY SANDS

For practical application we shall use the following form of the D-W saturation relation, derived from Eq. 7.13 by replacing conductivities by resistivities ($C = 1/R$) and rearranging terms

$$S_{wt}^{2} - S_{wt} \cdot S_b(1 - R_w/R_b) = R_w/(R_t \cdot \phi_t^{2}) \qquad (7.23)$$

The second term of this relation applies the shale correction. If it is omitted, the expression reverts to the familiar Archie relation. To apply the equation, the parameters S_b, ϕ_t, R_w, and R_b must be determined. First consider S_b, the bound water fraction in the shaly sand.

From Eqs. 7.5 and 7.18, S_b may be written as the following ratio (also called normalized Q[18])

$$S_b = Q/Q_{sh} \approx 0.3\,Q \qquad (7.24)$$

To determine S_b accurately requires a direct measurement of Q; a Q-log is sorely needed. Unfortunately, no such log is currently available, although measurements on cores can be readily made in the laboratory or even at the wellsite.[19,20] Consequently, we are forced into indirect methods using shale indicators.

In terms of V_{sh}, the volumetric fraction of shale (including its bound water), the effectively porosity, ϕ_e, can be written

$$\phi_e = \phi_t - V_{sh} \cdot \phi_{tsh} \qquad (7.25)$$

where ϕ_t is the total porosity of the shaly sand and ϕ_{tsh} is the total porosity of the shale fraction in the sand. Equating this expression to that of Eq. 9.6 gives

$$S_b = V_{sh} \cdot \phi_{tsh}/\phi_t \tag{7.26}$$

Determination of S_b therefore reduces to obtaining V_{sh} from available shale indicators. This is a key result.

Evaluation of V_{sh}

No single logging measurement accurately measures V_{sh}. Consequently, V_{sh} is usually estimated from several shale indicators and the lowest value is used.[21,22] The two best indicators are the Density-Neutron difference and the GR log. A fallback indicator that is less reliable is the SP log. All techniques assume that the shale in a shaly sand is the same as that in adjacent shales. This is a reasonable premise for sands with shale laminations, but it is very questionable for sand with dispersed clay. Nevertheless, there is no alternative.

1. V_{sh} from the Density-Neutron Difference

Because of the lattice-bound hydrogen in clay, a gas-free shaly sand will always read a higher Neutron porosity than Density porosity, as illustrated by the differences in Tables 7–2 and 7–3. The larger the fraction of shale, the greater the difference. The effect is linear, so the shale fraction is given by

$$(V_{sh})_{ND} = (\phi_n - \phi_d)/(\phi_{nsh} - \phi_{dsh}) \tag{7.27}$$

where the numerator represents the difference in Neutron and Density porosities in the shaly sand and the denominator represents the difference in nearby shale. The latter will typically be 0.15 to 0.30, depending on the amount and type of clay in the shale.

This method cannot be used when gas is present or suspected since gas distorts the ϕ_n and ϕ_d values.

2. V_{sh} from The Gamma Ray Log

Gamma Ray deflection increases with shale content of a formation. Consequently, an index of the degree of shaliness of a sand is obtained by linearly interpolating between the clean sand level and the shale level

$$I_{sh} = (GR - GR_{cl})/(GR_{sh} - GR_{cl}) \tag{7.28}$$

where

GR = reading in the sand of interest, APIU
GR_{cl} = average reading in nearby clean sands, APIU
GR_{sh} = average reading in nearby 100% shales, APIU

I_{sh} will vary from zero in a clean sand to 1.0 in shale.

Estimation of the clean sand and 100% shale levels, as illustrated in Fig. 7–10 is not always easy. There may be few clean sands and the shales may vary considerably in activity, so a good deal of judgment is required. An occasional abnormally high shale reading should be ignored.

The fractional volume of shale, V_{sh}, will be equal to the shale index, I_{sh}, if the density of the formation does not vary with shale content. This is the situation when thin shale laminations are intermixed with clean sand layers of the same bulk density. In this case the straight-line relationship of Fig. 7–11, converting I_{sh} to V_{sh}, applies.

On the other hand when increasing clay content is accompanied by a substantial increase in bulk density, as it is when authigenic clay grows in the pores of originally clean high-porosity sands, then the curved line of Fig. 7–11 transforming I_{sh} to V_{sh} applies.[23] Many cases will fall between the two extremes. Consequently, a more generally applicable relation that might be applied when a Density log accompanies the Gamma Ray is

$$(V_{sh}) = I_{sh} \cdot (\rho/\rho_{sh})^3 \qquad (7.29)$$

where ρ is the density of the formation of interest and ρ_{sh} is the density of nearby shale. The exponent 3 is an educated guess; it has never been determined precisely.

Where spectral Gamma Ray logs are run, improvement may be effected in certain areas by eliminating the U component and determining $(V_{sh})_{GR}$ using only the Th + K components. If feldspars or micaceous formations are prominent, the potassium component should be eliminated or subdued.[24] One field study reported excellent correlation between $(V_{sh})_{GR}$ and CEC values measured on cores and poor correlation between $(V_{sh})_{ND}$ and the same CEC values.[25] This was attributed to the GR responding primarily to montmorillonate and illite, with high uranium and potassium contents respectively, along with these clays having high CEC values. On the other hand the Neutron-Density separation gives greatest weight to kaolinite and chlorite, which have low CEC values. This is an argument in favor of the Gamma Ray as the better CEC indicator.

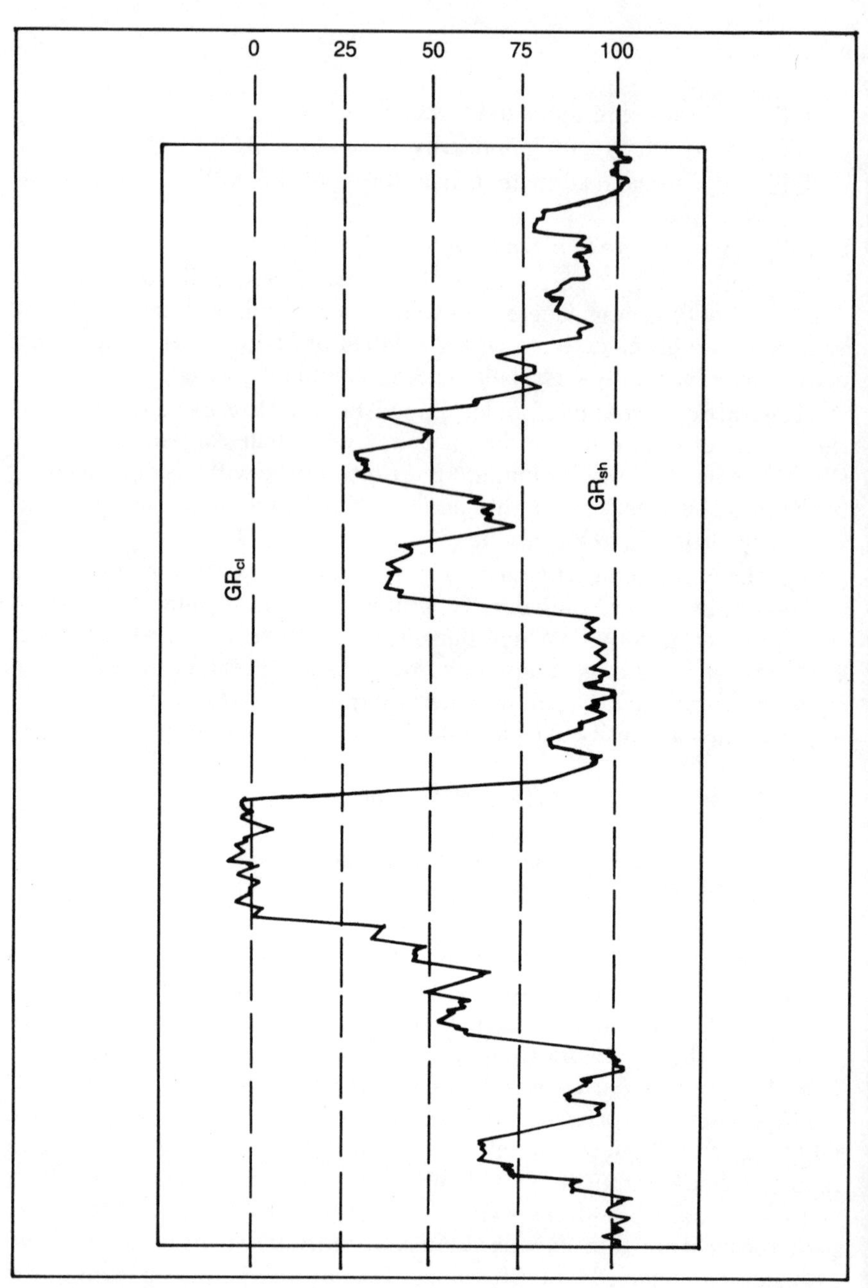

Fig. 7–10 Determination of shale indication, I_{sh}, from the GR curve

3. V_{sh} from the SP Log
In similar fashion, V_{sh} can be calculated from the SP log as

$$(V_{sh})_{SP} = (SP - SP_{cl})/(SP_{sh} - SP_{cl}) \tag{7.30}$$

where the numerator is the difference in millivolts between the SP level in the zone of interest and the clean formation level and the denominator is the difference between the shale and clean levels (the SSP). This relation is valid only under certain conditions, as pointed out in Chapter 3.

With several V_{sh} values so determined, standard procedure is to pick the lowest value as the correct one, excluding the crossplot value when gas is indicated. The reason is that most side effects cause calculated V_{sh} values to

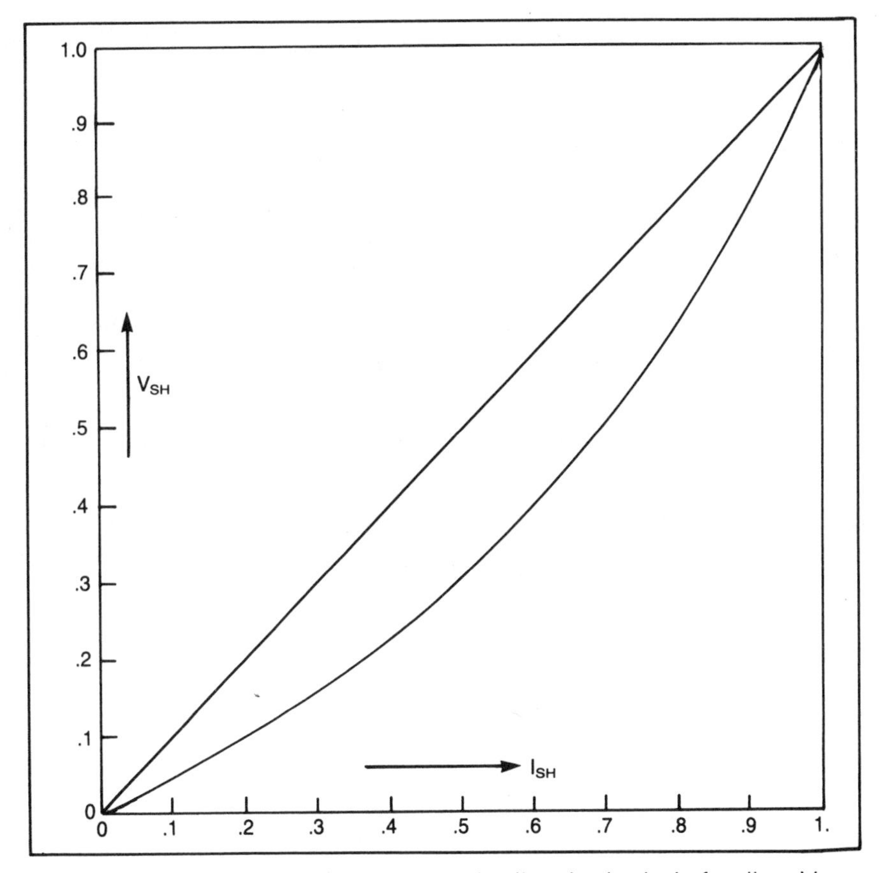

Fig. 7–11 Conversion of GR shale indication, I_{sh}, to shale fraction, V_{sh} (courtesy Schlumberger, © SPE-AIME)

be too high. Heavy minerals or neutron absorbers in the shaly sand will cause $(V_{sh})_{ND}$ to be too large. Hole enlargements in the shales will cause $(V_{sh})_{GR}$ to be too great; it is particularly important to correct the GR readings when caving is severe and the mud weight is high before computing V_{sh}. Hydrocarbons in the shaly sand will often cause $(V_{sh})_{SP}$ to be too high. Consequently, the lowest value is picked, but even it is not likely to be very accurate.

Determination of Effective Porosity

The next step is to determine the effective porosity, ϕ_e, of the shaly sand. The Density and Neutron porosities are first corrected for shale as follows

$$\phi_{dc} = \phi_d - V_{sh} \cdot \phi_{dsh} \tag{7.31}$$

$$\phi_{nc} = \phi_n - V_{sh} \cdot \phi_{nsh} \tag{7.32}$$

If no gas is present, the corrected porosities should be close together. The effective porosity can be taken as the average

$$\phi_e = (\phi_{dc} + \phi_{nc})/2 \tag{7.33}$$

If gas is present, it will show up as a crossover or enhanced crossover of the corrected porosities, ϕ_{nc} being significantly less than ϕ_{dc}. This is the main reason for proceeding in this fashion. With gas the effective porosity may be taken as

$$\phi_e = \sqrt{(\phi_{dc}^2 + \phi_{nc}^2)/2} \tag{7.34}$$

The effect of these calculations is illustrated in the crossplot of Fig. 7–12, which applies to sand or limestone provided the porosity values input correspond to the matrix chosen. On such a plot, clean formation points fall along the 45° line and shaly formation points fall to the right of the line. Gas-bearing formations will plot to the left if not too shaly. The shale point, S, may fall anywhere in the indicated shale zone, depending on the type and content of clay in the shale. Point P represents a gas-free shaly formation point. Correcting for shale translates this point to P_1 (parallel to the line OS), and averaging porosities at P_1 gives the effective porosity, P_2.

If the same sand contained gas, it would show up as some point such as P_3 on the plot. Correcting for shale moves that point to P_4, accentuating the gas effect. Correcting for gas via Eq. 7.34 is equivalent to translating point P_4 to the 45° line in a direction parallel to the gas correction line. This brings P_4 back essentially to P_2.

Determination of ϕ_{tsh}, the total porosity of the shale, is required next. Unfortunately, there is no accurate method of measuring this quantity. Dry clay densities may vary all the way from 2.4–3.0 g/cc, so the Density porosity (based on 2.65 g/cc) may be too high or too low; (most often it will be too low.) The Neutron porosity will always be too high. Therefore, a common equation is

$$\phi_{tsh} = \delta\, \phi_{dsh} + (1 - \delta)\, \phi_{nsh} \tag{7.35}$$

where δ is a constant between 0.5 and 1.0, depending on local experience.

The total porosity, ϕ_t, and the bound-water fraction, S_b, for the shaly sand are then

$$\phi_t = \phi_e + V_{sh} \cdot \phi_{tsh} \tag{7.36}$$

$$S_b = V_{sh} \cdot \phi_{tsh}/\phi_t \tag{7.37}$$

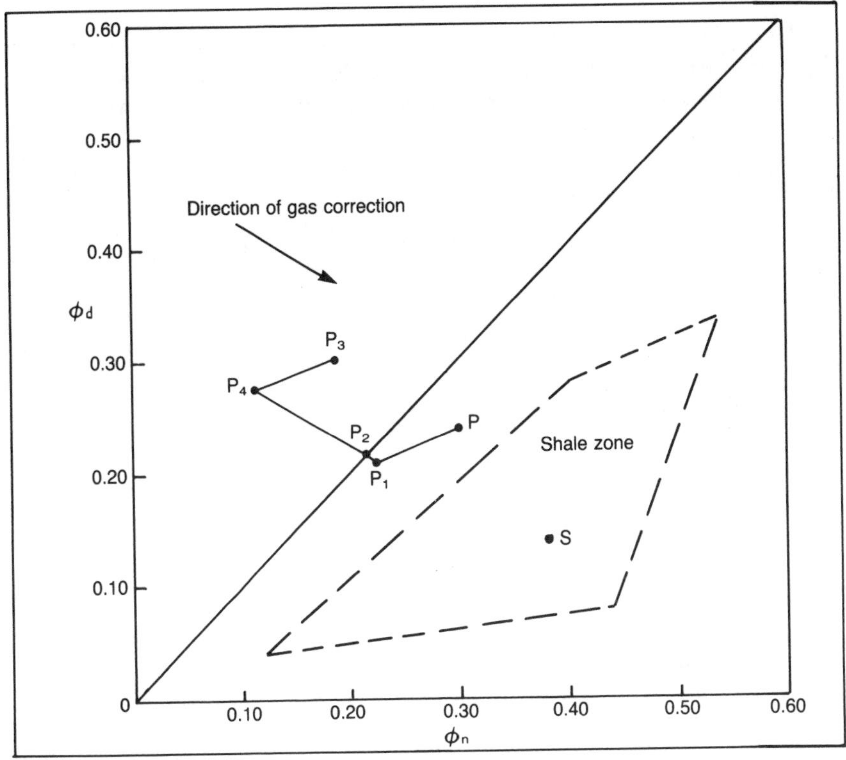

Fig. 7–12 Correction for shale and gas effects

Evaluation of Water Resistivities

It remains to determine the free and bound-water resistivities in order to calculate water saturation. The free-water resistivity, R_w, is best obtained from a nearby clean-water sand. For such a sand, $S_b = 0$ and $S_{wt} = 1$ so that Eq. 7.23 reduces to

$$R_w = R_{cl} \cdot \phi_{cl}^2 \tag{7.38}$$

where R_{cl} and ϕ_{cl} are the observed resistivity and porosity of the clean sand. An alternate method is to obtain R_w from the SP, as described in Chapter 3.

Likewise, the bound-water resistivity, R_b, is best determined from a nearby shale. For shale, $S_b = 1$ and $S_{wt} = 1$ so that Eq. 7.23 gives

$$R_b = R_{sh} \cdot \phi_{tsh}^2 \tag{7.39}$$

where R_{sh} and ϕ_{tsh} are the resistivity and total porosity of the shale.

Determination of Water Saturation

All of the factors required for water saturation determination through Eq. 7.23 are thereby determined. To solve that equation, it is useful first to calculate the apparent water resistivity of the shaly sand, which is

$$R_{wa} = R_t \cdot \phi_t^2 \tag{7.40}$$

Eq. 7.23 then can be written

$$S_{wt}^2 - S_{wt} \cdot S_b (1 - R_w/R_b) = R_w/R_{wa} \tag{7.41}$$

from which the total water saturation is

$$S_{wt} = b + \sqrt{b^2 + (R_w/R_{wa})} \tag{7.42}$$

where

$$b = S_b (1 - R_w/R_b)/2 \tag{7.43}$$

It is a relatively simple matter to obtain S_{wt} values, provided an R_{wa} log is recorded during logging with a Density-Neutron porosity input to the R_{wa} computer equivalent to Eq. 7.35. In this case R_w, R_b, and R_{wa} for the zone of interest can be read directly from that log. Once S_b is determined, its value and those of R_w/R_b and R_w/R_{wa} can be entered into the charts of Fig. 7–13 and S_{wt} can be read as indicated by the dashed lines.

Total water saturations so calculated will be higher than effective water saturations that have been historically used. If desired, the latter can also be computed as

$$S_{we} = (S_{wt} - S_b)/(1 - S_b) \tag{7.44}$$

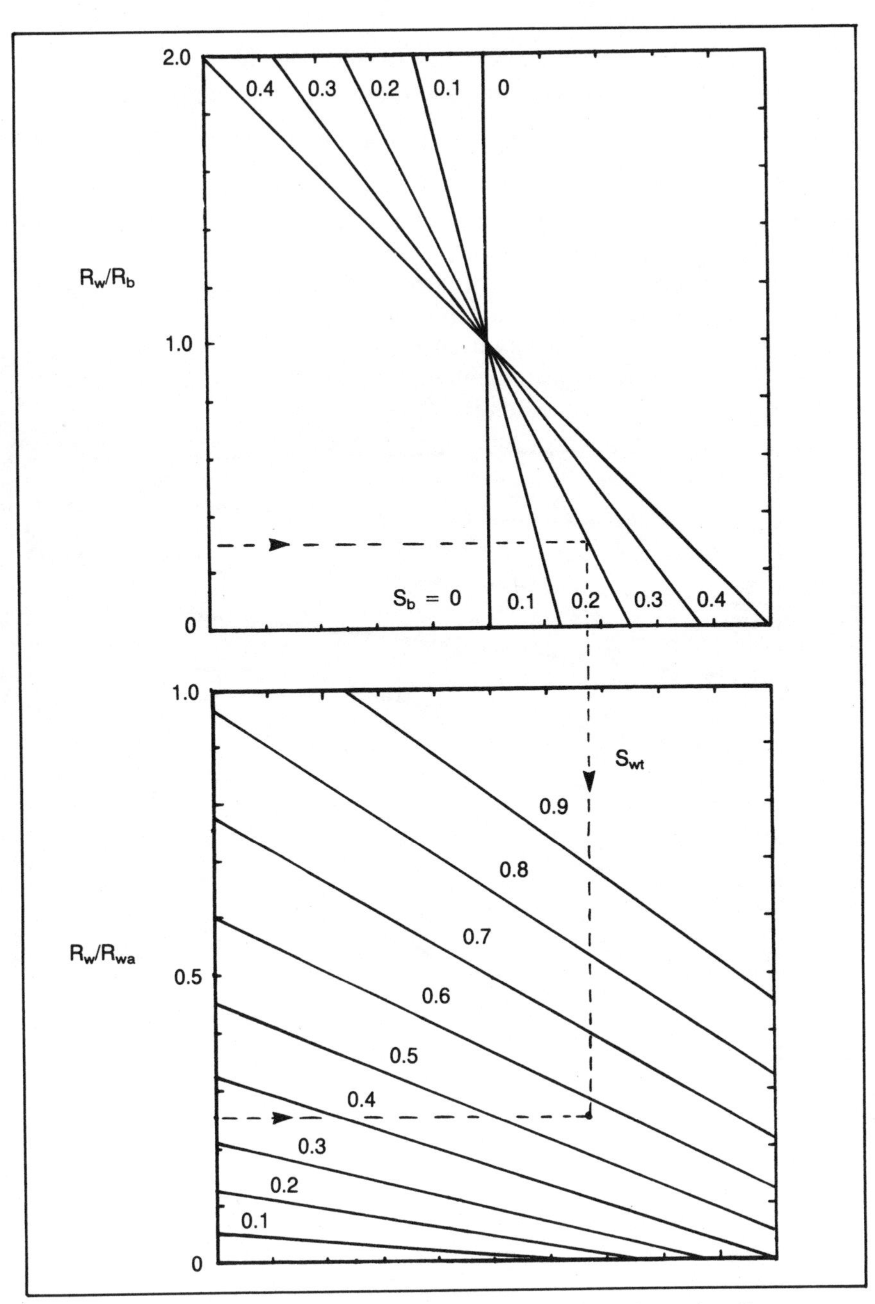

Fig. 7–13 Graphical determination of total water saturation

Fig. 7–14 provides a graphical solution to this equation.

Finally, hydrocarbon content as a fraction of total volume is

$$\phi_h = \phi_t(1 - S_{wt}) \tag{7.45}$$

In spite of all of the equations, shaly sand interpretation is as much of an art as a science. The most critical parameter, V_{sh}, is difficult to pick in many cases. The relation for total shale porosity, ϕ_{tsh}, is by no means firmly established. There may be a lack of clean sands to establish R_w. The applicable cementation and saturation exponents m and n may differ from 2.0, although the value of the cementation constant a is immaterial when R_w, R_b, and R_{wa} are determined as indicated. In short, there is no substitute for experience in the region of interest.

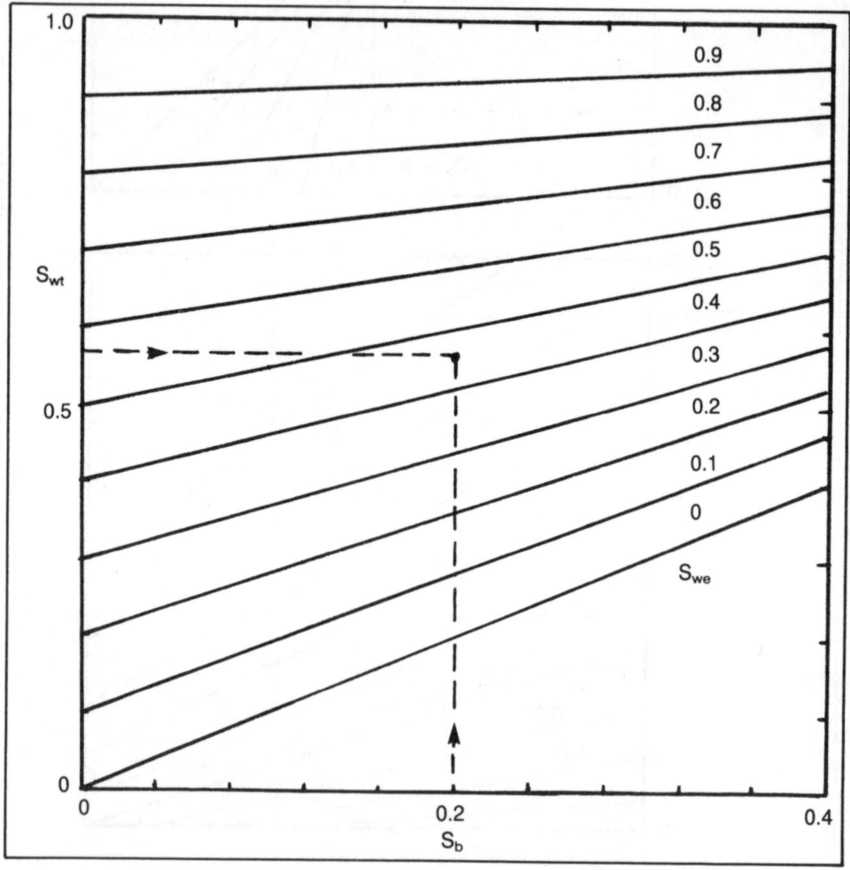

Fig. 7–14 Graphical determination of effective water saturation

One final point: Values of ϕ_d and ϕ_n indicated on the log must be corrected to the appropriate matrix before calculations are performed. That is, if the logs have been recorded on limestone matrix but the shaly formations of interest are believed to be sandstone, the log values should be converted to sand matrix using Fig. 5–15 or 5–21. Likewise, significant environmental corrections should be made before computation. Those that might be required are invasion corrections to R_t values, borehole size corrections to GR readings, and temperature/pressure adjustments to CNL values.

SUMMARY OF DUAL WATER INTERPRETATION

1. Read the resistivities, porosities, GR, and SP values in the sand of interest, in a nearby shale, and in a nearby clean sand. Correct porosity values to the appropriate matrix if necessary.

2. Calculate V_{sh}

$$(V_{sh})_{ND} = (\phi_n - \phi_d)/(\phi_{nsh} - \phi_{dsh})$$

$$(I_{sh})_{GR} = (GR - GR_{cl})/(GR_{sh} - GR_{cl})$$

Convert I_{sh} to V_{sh} using Fig. 7–11.

$$(V_{sh})_{SP} = (SP - SP_{cl})/(SP_{sh} - SP_{cl})$$

Choose the minimum value. Omit $(V_{sh})_{ND}$ if gas is indicated.

3. Correct the porosities for shaliness

$$\phi_{dc} = \phi_d - V_{sh} \cdot \phi_{dsh}$$

$$\phi_{nc} = \phi_n - V_{sh} \cdot \phi_{nsh}$$

Look for gas indication ($\phi_{nc} < \phi_{dc}$).

4. Calculate the effective porosity of the shaly sand

$$\text{No gas: } \phi_e = (\phi_{dc} + \phi_{nc})/2$$

$$\text{With gas: } \phi_e = \sqrt{(\phi_{dc}^2 + \phi_{nc}^2)/2}$$

5. Determine the total porosity of the nearby shale

$$\phi_{tsh} = \delta\,\phi_{dsh} + (1 - \delta)\phi_{nsh}$$

where

$$\delta = 0.5 \text{ to } 1.0$$

6. Determine the total porosity and the bound-water fraction of the sand

$$\phi_t = \phi_e + V_{sh} \cdot \phi_{tsh}$$

$$S_b = V_{sh} \cdot \phi_{tsh}/\phi_t$$

7. Determine the free-water resistivity from a nearby clean sand

$$R_w = R_{cl} \cdot \phi_{cl}^2$$

8. Determine the bound-water resistivity from a nearby shale

$$R_b = R_{sh} \cdot \phi_{tsh}^2$$

9. Determine the apparent water resistivity in the shaly sand

$$R_{wa} = R_t \cdot \phi_t^2$$

10. Determine the total water saturation corrected for shale

$$S_{wt} = b + \sqrt{b^2 + (R_w/R_{wa})}$$

where

$$b = S_b\,(1 - R_w/R_b)/2$$

11. Determine the effective water saturation of the shaly sand

$$S_{we} = (S_{wt} - S_b)/(1 - S_b)$$

12. Determine the volumetric fraction of the hydrocarbon

$$\phi_h = \phi_t\,(1 - S_{wt})$$

Example

Fig. 7–15 is a composite log from an offshore Louisiana well through a massive sand-shale series. The sand of interest is from 8,505–8,545 ft. Above and below this interval (not shown) are a number of clean gas sands with pronounced Neutron-Density crossovers (as much as 30 porosity units) and Induction resistivities as high as 30 ohm-m. What does the shaly sand of interest contain?

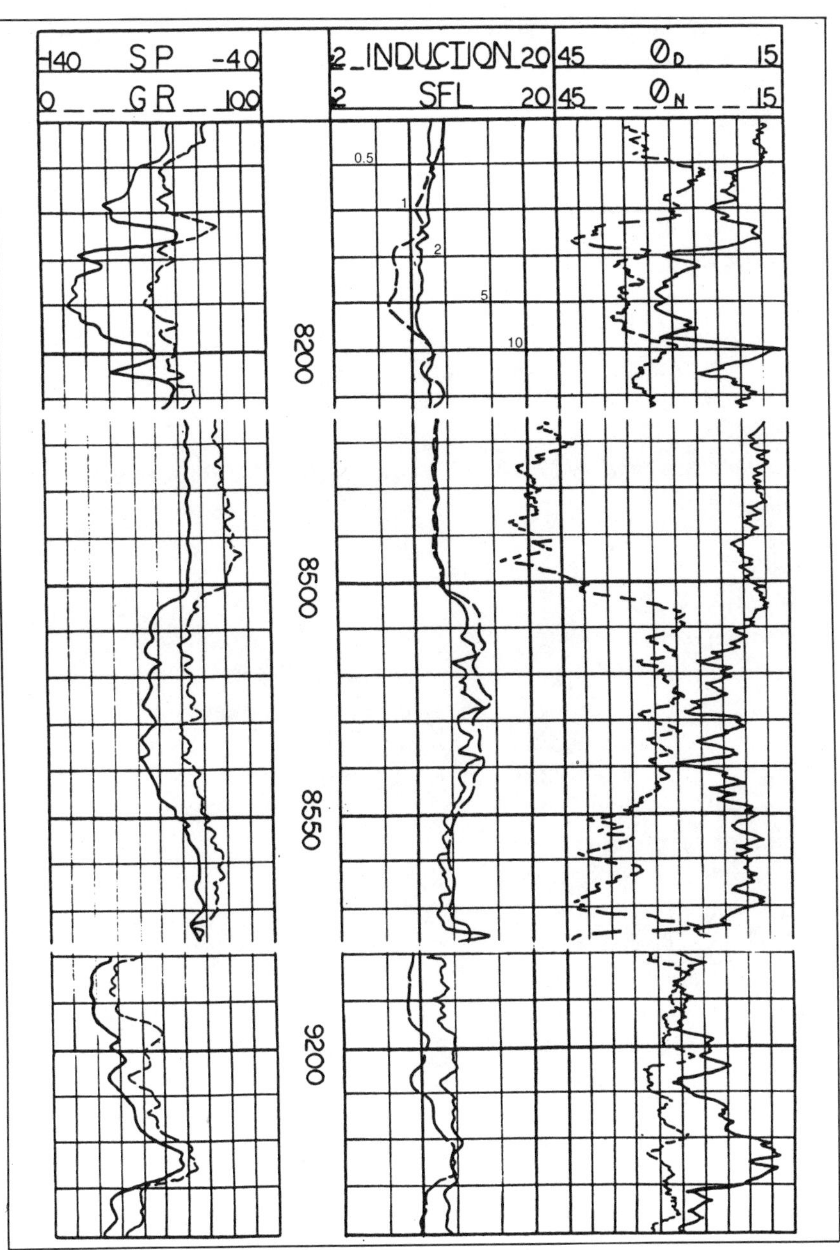

Fig. 7–15 Shaly sand example–offshore Louisiana

Applying the above step-by-step procedure, gives the following results.

In the sand from 8,510–8,540 ft

$R_t = 3$
$\phi_d = 0.26$
$\phi_n = 0.33$
$GR = 63$
$SP = -95$

In nearby shale 8,470–8,500 ft

$R_{Sh} = 1.2$
$\phi_{dSh} = 0.20$
$\phi_{nSh} = 0.50$
$GR_{Sh} = 87$
$SP_{Sh} = -75$

In nearby clean sands 8,178–8,193 ft and 9,180–9,193 ft (averages)

$R_{Cl} = 0.65$
$\phi_{Cl} = 0.34$
$GR_{Cl} = 36$
$SP_{Cl} = -122$

Calculation of V_{Sh}

$(V_{Sh})_{ND}$ is not applicable because gas is suspected.

$(I_{Sh})_{GR} = (63 - 36)/(87 - 36) = 0.53$
$(V_{Sh})_{GR} = 0.34$ from Fig. 7–11 (curved line)
$(V_{Sh})_{SP} = (-95 + 122)/(-75 + 122) = 0.57$
Choose $V_{Sh} = 0.34$

Calculation of effective porosity

$\phi_{dc} = 0.26 - 0.34 \times 0.20 = 0.19$
$\phi_{nc} = 0.33 - 0.34 \times 0.50 = 0.16$
Gas is indicated since $\phi_{nc} < \phi_{dc}$
$\phi_e = \sqrt{(0.19^2 + 0.16^2)/2} = 0.17$

Calculation of water saturation

$\phi_{tSh} = 0.7 \times 0.20 + 0.3 \times 0.50 = 0.29$ $(\delta = 0.7)$
$\phi_t = 0.17 + 0.34 \times 0.29 = 0.27$
$S_b = 0.34 \times 0.29/0.27 = 0.36$
$R_w = 0.65 \times 0.34^2 = 0.075$
$R_b = 1.2 \times 0.29^2 = 0.10$
$R_{wa} = 3 \times 0.27^2 = 0.22$
$b = 0.36(1 - 0.075/0.10)/2 = 0.045$
$S_{wt} = 0.045 + \sqrt{0.045^2 + 0.075/0.22} = 0.63$

$S_{we} = (0.63 - 0.36)/(1 - 0.36) = 0.42$
$\phi_h = 0.27(1 - 0.63) = 0.10$

The conclusion is that the zone of interest does contain gas and has an effective porosity of 17% and a water saturation in the effective pore space of 42%. Since that water is not tied to entrained clay but is associated only with the clean sand fraction, it is probable that the zone would produce considerable water with any gas. Consequently, this interval was not perforated. The clean gas zones below the interval (not shown) calculated approximated 10% water saturation and produced dry gas on production.

SUMMARY OF EARLIER SHALY SAND INTERPRETATION METHODS

Automatic Compensation method, with only Resistivity and Sonic logs (1950s)

Shale causes R_t to read too low and ϕ_s to read too high, compensating each other in the water saturation equation. However, observed porosity must be corrected for shale to obtain effective porosity. Relations are

$$S_w = 0.9\sqrt{R_w/R_t}/\phi_s \qquad \text{(a)}$$

$$\phi_e = \phi_s - V_{sh} \cdot \phi_{ssh} \qquad \text{(b)}$$

where

ϕ_s = porosity from Sonic without shale correction
R_t = resistivity from deep Induction
V_{sh} = lowest of GR and SP indicators
ϕ_{ssh} = porosity from Sonic in adjacent shale

This method is still used in the Gulf Coast to obtain quick answers.

Where Density-Neutron is run in place of Sonic, ϕ_s in Eq. (a) is replaced by uncorrected D-N porosity given by

$$\phi_{dn} = \sqrt{(\phi_d^2 + \phi_n^2)/2} \qquad \text{(c)}$$

where

ϕ_d = porosity from Density without shale correction
ϕ_n = porosity from Neutron without shale correction

Effective porosity, ϕ_e, is determined as in the Simandoux method.

Dispersed clay method, with Resistivity, Sonic, and Density logs (1960s)[26]

The Sonic log sees dispersed clay in pore water as a slurry and gives a porosity equal to the sum of their volumetric fractions. The Density senses only the water-filled porosity, which is less. The fraction of the clean-sand-intergranular space occupied by clay, termed q, is

$$q = (\phi_s - \phi_d)/\phi_s \tag{d}$$

Water saturation is given by

$$S_w = \left[\sqrt{\frac{0.8}{\phi_s^2} \cdot \frac{R_w}{R_t} + \left(\frac{q}{2}\right)^2} - \frac{q}{2} \right] /(1 - q) \tag{e}$$

Effective porosity is

$$\phi_e = \phi_d - V_{sh} \cdot \phi_{dsh} \tag{f}$$

The method is not reliable in gas sands; ϕ_d can be greater than ϕ_s, giving negative q values that are meaningless. It is also inappropriate for carbonates that have little dispersed clay.

Permeability will in general be too low for commercial production if $q > 2\,\phi_e$ or $q > 0.4$.

Simandoux method with Resistivity, Density and Neutron logs (1970s)[27]
Water saturation is

$$S_w = \frac{c \cdot R_w}{\phi_e^2} \left[\sqrt{\frac{5\,\phi_e^2}{R_w \cdot R_t} + \left(\frac{V_{sh}}{R_{sh}}\right)^2} - \frac{V_{sh}}{R_{sh}} \right] \tag{g}$$

where

c = 0.40 for sands; 0.45 for carbonates
V_{sh} = lowest of the various shale indicators
R_w = formation-water resistivity
R_t = deep resistivity (corrected for invasion)
R_{sh} = deep resistivity reading in adjacent shale
ϕ_e = effective porosity

Effective porosity is

$$\phi_e = \sqrt{(\phi_{dc}^2 + \phi_{nc}^2)/2} \tag{h}$$

where ϕ_{dc} and ϕ_{nc} represent Density and Neutron porosities corrected for shale given by

$$\phi_{dc} = \phi_d - V_{sh} \cdot \phi_{dsh} \quad \text{(i)}$$

$$\phi_{nc} = \phi_n - V_{sh} \cdot \phi_{nsh} \quad \text{(j)}$$

where ϕ_d and ϕ_n are Density and Neutron porosities read in the zone of interest and ϕ_{dsh} and ϕ_{nsh} corresponding values in adjacent shales.

Eq. (g) may be solved with Figs. 7–16 and 7–17. Values of A and B are found with the nomogram of Fig. 7–16. A and B are then inserted in Fig. 7–17 to obtain S_w.

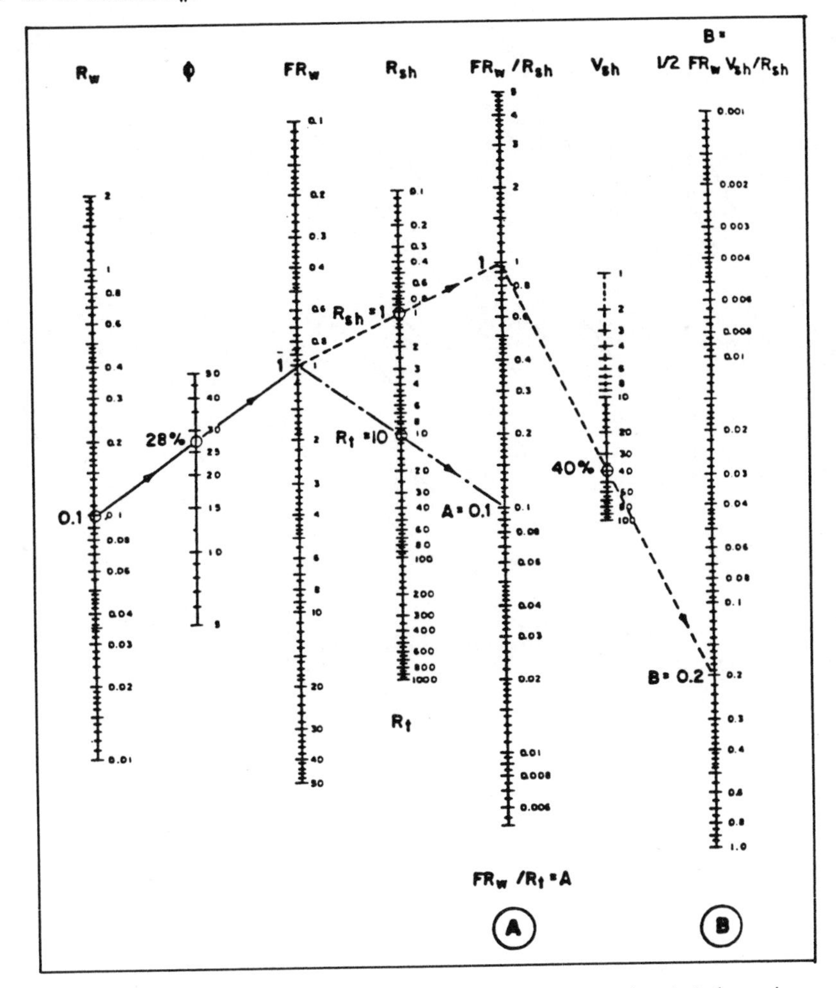

Fig. 7–16 Water saturation. Simandoux method, chart 1 (courtesy Dresser)

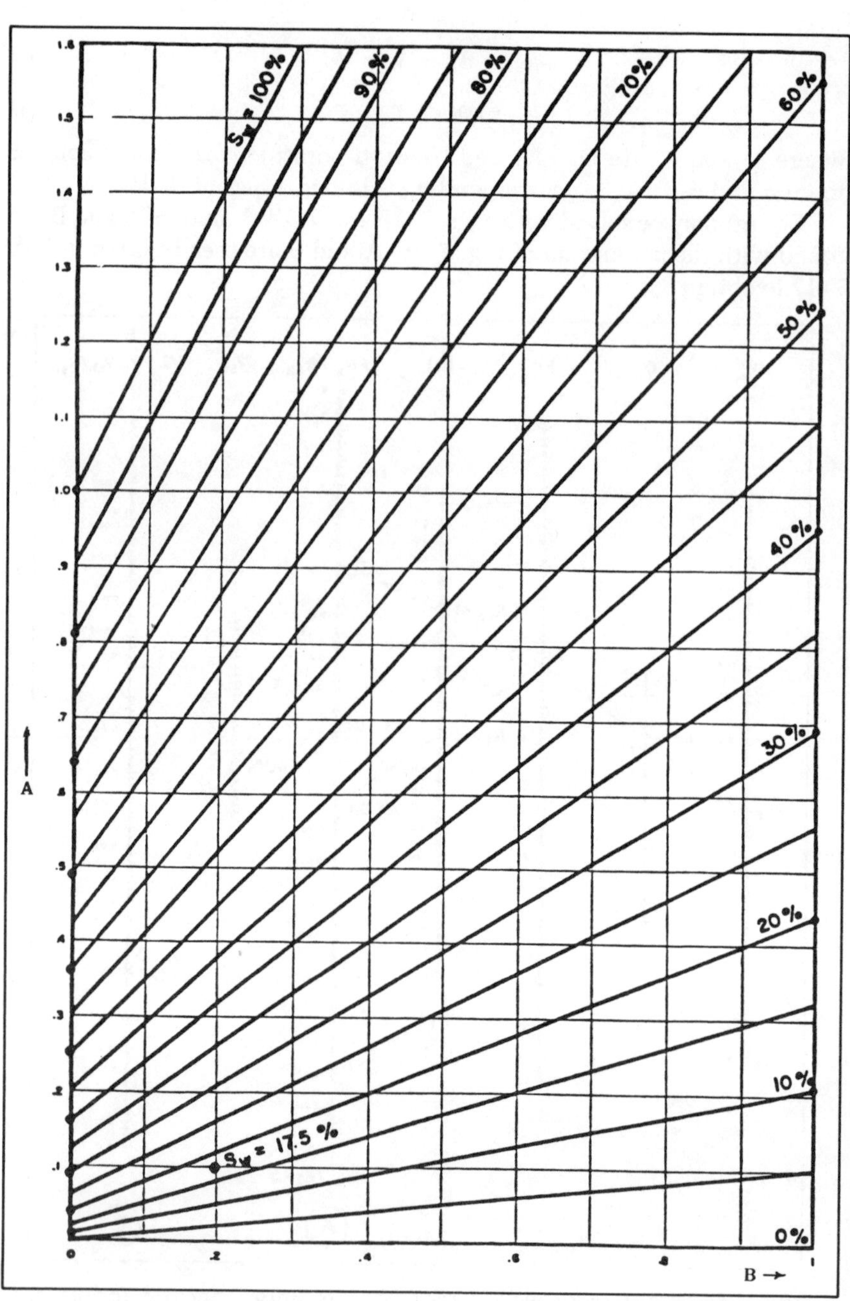

Fig. 7–17 Water saturation, Simandoux method, chart 2 (courtesy Dresser)

Dresser uses the S_w relation as indicated. Schlumberger uses the same expression with R_w replaced by $R_w(1 - V_{sh})$.

REFERENCES

[1]M.P. Tixier, R.L. Morris, and J.G. Connell, "Log Evaluation of Low Resistivity Pay Sands in the Gulf Coast," *SPWLA Logging Symposium Transactions* (1968).

[2]R.E. Grim, *Applied Clay Mineralogy* (New York: McGraw-Hill, 1962).

[3]H. Edmundson and L.L. Raymer, "Radioactive Logging Parameters for Common Minerals," *SPWLA Logging Symposium Transactions* (June 1979).

[4]N. Ruhovets and W.H. Fertl, "Digital Shaly Sand Analysis Based on Waxman-Smits Model and Log-Derived Clay Typing," *SPWLA/SAID Logging Symposium Transactions* (France: 1981).

[5]H.J. Hill, O.J. Shirley, and G.E. Klein, "Boundwater in Shaly Sands — Its Relation to Q_v and Other Formation Properties," edited by M.H. Waxman and E.C. Thomas, *The Log Analyst*, Vol. 20, No. 3 (May–June 1979).

[6]Schlumberger, "Log Interpretation—Principles," (1972).

[7]M.D. Wilson and E.D. Pittman, "Authigenic Clays in Sandstones: Recognition and Influence on Reservoir Properties and Paleonvironmental Analysis," *Jour. Sed. Petrology*, Vol. 47, No. 1 (March 1977), pp. 3–31.

[8]J.W. Neasham, "The Morphology of Dispersed Clay in Sandstone Reservoirs and Its Effects on Sandstone Shaliness, Pore Space and Fluid Flow Properties," *SPE 6858* (Denver: October 1977).

[9]W.H. Fertl and G.W. Hammack, "A Comparative Look at Water Saturation Computations in Shaly Pay Sands," *SPWLA Logging Symposium Transactions* (May 1971).

[10]C. Clavier, G. Coates, and J. Dumanoir, "The Theoretical and Experimental Bases for the 'Dual Water' Model for the Interpretation of Shaly Sands" (Denver: October 9–12, 1977).

[11]J.G. Patchett, "An Investigation of Shale Conductivity," *SPWLA Logging Symposium Transactions* (June 1975).

[12]Hill, Shirley, and Klein, Ibid.

[13]M.H. Waxman and L.J.M. Smits, "Electrical Conductivities in Oil-Bearing Shaly Sands," *Soc. Pet. Eng. J.* (June 1968), pp. 107–122.

[14]M.H. Waxman and E.C. Thomas, "Electrical Conductivities in Oil-Bearing Shaly Sands–I. The Relation Between Hydrocarbon Saturation and Resistivity Index; II. The Temperature Coefficient of Electrical Conductivity," *SPE Journal*, No. 14 (February 1974), pp. 213–225.

[15]Clavier, Coates, and Dumanoir, Ibid.

[16]Patchett, Ibid.

[17]Waxman and Thomas, Ibid.

[18]I. Juhasz, "Normalized Q_v—The Key to Shaly Sand Evaluation Using The Waxman-Smits Equation in the Absence of Core Data," *SPWLA Logging Symposium Transactions* (June 1981).

[19]A.E. Worthington, "An Automated Method for the Measurement of Cation Exchange Capacity of Rocks," *Geophysics*, No. 38 (February 1973), pp. 140–153.

[20]E.C. Thomas, "The Determination of Q_v from Membrane Potential Measurements on Shaly Sands," *J. Pet. Tech.* (September 1976), pp. 1087–1096.

[21]A. Poupon and R. Gaymard, "The Evaluation of Clay Content from Logs," *SPWLA Logging Symposium Transactions* (May 1970).

[22]R.C. Ransom, "Methods Based on Density and Neutron Well Logging Responses to Distinguish Characteristics of Shaly Sandstone Reservoir Rocks," *The Log Analyst*, Vol 18, No. 3 (1977), p. 47.

[23]C. Clavier et al., *J. Pet. Tech.* (June 1971).

[24]G. Marett, P. Chevalier, P. Souhaite, and J. Suau, "Shaly Sand Evaluation Using Gamma Ray Spectrometry Applied to the North Sea Jurassic," *SPWLA Logging Symposium Transactions* (June 1976).

[25]W.L. Johnson and W.A. Linke, "Some Practical Applications to Improve Formation Evaluation of Sandstones in the Mackenzie Delta," *CWLS Logging Symposium Transactions* (June 1976).

[26]R.P. Alger, L.L. Raymer, W.R. Hoyle, and M.P. Tixier, "Formation Density Log Applications in Liquid-Filled Holes," *J. Pet. Tech.* (March 1963).

[27]P. Simandoux, "Dielectric Measurements in Porous Media and Application to Shaly Formations," Revue de l'Institut Francais du Petrole, Supplementary Issue, 1963, pp. 193–215; English translation in *SPWLA* Reprint Volume *Shaly Sand* (July 1982).

Chapter 8

PREDICTION OF PRODUCIBILITY

Standard resistivity and porosity logs provide good answers for the quantity of oil or gas in situ but not on the producibility of those hydrocarbons. Prediction of the latter requires knowledge of reservoir pressure, formation permeability, and irreducible water saturation. None of these parameters is precisely determined by standard logs, although under certain circumstances approximate values can be deduced.

The wire-line tool that can be of considerable assistance in predicting productivity is the Multiple Formation Tester. It measures formation pressures, allows calculation of permeability, and retrieves a sample of reservoir fluid for analysis. With these parameters a reasonable, though not infallible, estimate of production rate and type can be made.

Before discussing the Formation Tester, we need to set forth basic flow relations, review some pertinent aspects of irreducible water saturation and permeability, and consider the limitations in estimation of permeability from logs.

FLOW RELATIONS

The rate of production of oil, q_o, in stock tank barrels per day (stb/d) from a homogenous formation under radial flow is given to a close approximation by[1]

$$q_o = 1.0 \cdot 10^{-3} k_o \cdot h (P_r - P_f)/(\mu_o \cdot B) \tag{8.1}$$

where

k_o = effective permeability of the formation to oil, md
h = height of the producing interval, ft
P_r = shut in reservoir pressure, psia
P_f = pressure in the wellbore at the producing level, psia
μ_o = oil viscosity at reservoir temperature, cp
B = formation volume factor, res bbl/stb

For average oil (30°API, GOR = 500) the quantity μB is approximately unity so that

$$q_o \approx 1.0 \cdot 10^{-3} k_o \cdot h (P_r - P_f) \tag{8.2}$$

Correspondingly, the rate of production of gas, $q_{g(scfd)}$, under similar conditions is closely

$$q_g = 0.10\, k_g \cdot h\, (P_r^2 - P_f^2)/(\mu_g \cdot Z \cdot T) \tag{8.3}$$

where

k_g = effective permeability of the formation to gas, md
Z = gas deviation factor at reservoir temperature and pressure
T = formation temperature, °R

For average gas conditions the product $\mu_g ZT$ is approximately 5 so that

$$q_g \approx 0.020\, k_g \cdot h\, (P_r^2 - P_f^2) \tag{8.4}$$

Of the four parameters required to estimate production rates from Eqs. 8.2 or 8.4, the most important is effective permeability, k_o or k_g. It is discussed in detail below.

Producing thickness, h, is readily determined from logs, principally the deep Induction curve. Where thin producing beds are interlayered with impermeable shale streaks, a Microlog in conjunction with the Induction will provide the most accurate producing thickness, sometimes called *sand count*.

Reservoir pressure, P_r, is often assumed to be the hydrostatic pressure of salt water at the depth of interest, that is, 0.46 psi/ft × vertical depth in ft. However, this can be a substantial overestimate in old fields where zones may be depleted or a considerable underestimate where producing intervals are overpressured. Consequently, a direct measurement is desirable.

The bottom-hole flowing pressure, P_f, cannot be determined by measurement at the time of logging. It is a function of both reservoir parameters (pressure, water-oil ratio, gas-oil ratio, bubble point pressure) and well parameters (depth, perforation efficiency, tubing size, choke size and placement, flow-line size and length, separator pressure). In effect, when the well is opened the reservoir will increase its flow rate until the reservoir pressure, which is the driving force, is balanced by the back pressures consisting of the hydrostatic head of the fluid in the tubing, separator pressure, and friction pressure drops in perforations, tubing, chokes, and flow line. A systems analysis encompassing all of these parameters is necessary to optimize flow rate and to determine corresponding bottom-hole pressure, P_f. Wellsite programs to perform such an analysis are becoming available from logging and testing service companies.

Barring a systems analysis, the reservoir potential may be characterized simply by its specific productivity index, SPI, defined as its production rate per foot of producing interval (h = 1) and per psi of pressure drawdown ($P_r - P_f = 1$). For oil, Eq. 8.2 gives

$$SPI \approx 1 \times 10^{-3} k_o \quad \text{b/d per psi/ft} \tag{8.5}$$

That is, the SPI of an average oil reservoir is its permeability in darcies, which is an easy relation to remember.

For gas it is more common to characterize the reservoir by its absolute open-flow potential (AOFP), obtained by placing $P_f = 0$, in which case the specific AOFP per foot of producing interval is

$$SAOFP \approx 0.020\, k_g \cdot P_r^2 \quad \text{scfd/ft} \tag{8.6}$$

It is clear from these relations that effective permeability is the key parameter in production prediction. We now consider exactly what this is.

ABSOLUTE, RELATIVE, AND EFFECTIVE PERMEABILITIES

Absolute permeability, k, is a property of the rock reflecting its flowability.[2] It is the value that applies with a single fluid phase (water, oil, or gas) in the pore space. Where two or more fluid phases coexist in the pores, which is always the case in hydrocarbon-bearing rock, the presence of one phase hinders the flow of the other phases. The reduction factor is termed the relative permeability, k_r, which has values between one and zero. In other words, the effective permeability to a given phase is the product of the absolute permeability of the rock times the relative permeability of the phase.

Fig. 8–1 shows the relative permeability of water and oil in a typical rock as a function of water saturation of the rock.[3] The left side of this plot represents the situation existing in the undisturbed zone of an oil-bearing reservoir well above the water table. Water saturation is at its irreducible value, S_{wi}. No water will flow, so the relative permeability to water, k_{rw}, is zero. Oil will flow virtually unhindered because the water exists only on the grain surfaces, at grain contacts, and in very fine pores, leaving all major passageways open for oil flow. Thus, the relative permeability of oil, k_{ro}, is close to unity.

At the other extreme, the right-hand side of the plot applies to the invaded portion of an oil-bearing zone where residual oil occupies 10–40% of the pore space and water occupies the remainder. The residual oil is immobile so that k_{ro} is zero. However, water will not flow unhindered because the residual oil is left as isolated globules occupying a number of the medium-to-large pore spaces. These substantially reduce the number of branching passageways available to water flow and thereby reduce k_{rw} from unity to a value in the range of 0.3–0.6.

Between the two extremes is the situation that exists when oil and water flow simultaneously. Both k_{rw} and k_{ro} are substantially less than one, and in fact their sum is also significantly less than unity.

In order to predict the rates at which water and oil will be produced from a reservoir rock, we need to know the relative permeabilities at the existing water saturation as well as the absolute permeability. The effective permeabilities, k_o and k_g, required for Eqs. 8.1–8.6 are

$$k_o = k \cdot k_{ro} \tag{8.7}$$

$$k_g = k \cdot k_{rg} \tag{8.8}$$

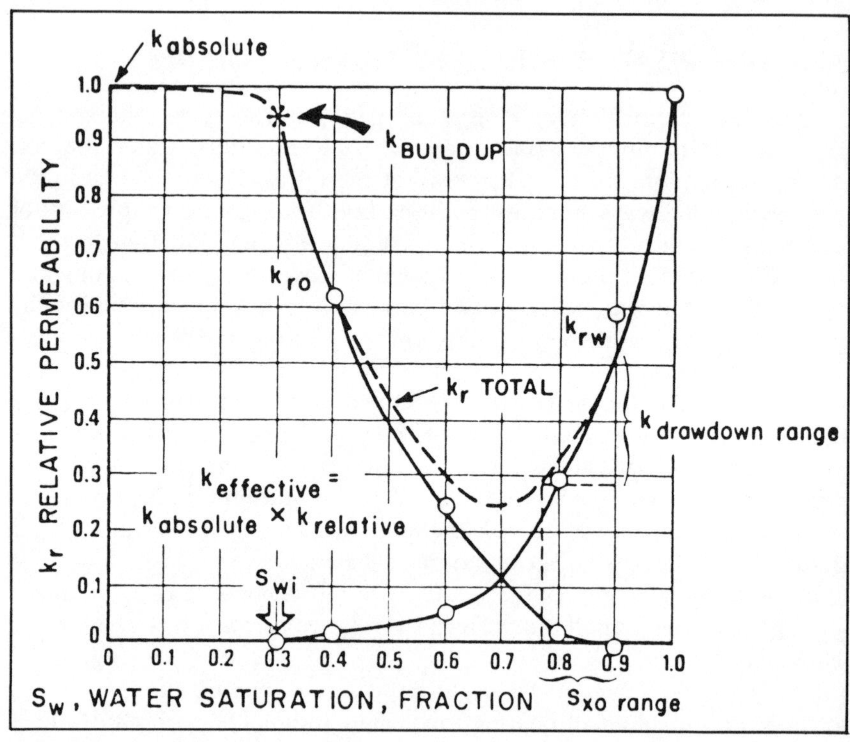

Fig. 8–1 Relative permeability to oil and water in an oil-bearing formation (courtesy Schlumberger, © SPE-AIME)

Empirical relations exist that allow estimation of relative permeabilities if both actual and irreducible water saturations are known. One of the simpler versions for oil and water is[4]

$$k_{ro} = [(0.9 - S_w)/(0.9 - S_{wi})]^2 \quad (8.9)$$

$$k_{rw} = [(S_w - S_{wi})/(1 - S_{wi})]^3 \quad (8.10)$$

With the foregoing equations, the water cut—that fraction of liquid production that is water—can be derived. From Eq. 8.1, the water-oil ratio is derived as

$$WOR = (k_{rw} \cdot \mu_o \cdot B_o)/(k_{ro} \cdot \mu_w \cdot B_w) \quad (8.11)$$

Water cut is

$$WC = WOR/(WOR + 1) \quad (8.12)$$

Fig. 8–2 shows water cuts so calculated for light, medium, and heavy oils.[5] When $S_w = S_{wi}$, no water flows and $WC = 0$. When $S_w > S_{wi}$, some water will be produced. The heavier the oil, the more reluctant it is to move and the greater the water cut. The problem in applying these charts is that the value of S_{wi} is difficult to determine accurately.

IRREDUCIBLE WATER SATURATION

The water saturation profile in a thick, extremely homogeneous, water-wet, hydrocarbon-bearing reservoir that has stabilized over geologic time will be as illustrated in Fig. 8–3a.[6] Consider first curve B. Proceeding upward from the free water level, water saturation remains close to unity for a few feet, drops rapidly over the next 50 ft in the so-called transition zone, and finally levels off at a more-or-less constant value considered the irreducible saturation.

Two opposing forces are at work. Interfacial tension, which may be thought of as an elastic skin between water and hydrocarbon, tends to hold water at grain contacts and in small pores. Its force is essentially proportional to 1/r, where r is the radius of curvature of the water surface, closely related to grain and pore size. The smaller the grains and pores, the more tightly the water is held. Opposing that force is gravity tending to pull the heavier water below the lighter hydrocarbon. This force is proportional to the height above the water table and the density difference between water and hydrocarbons.

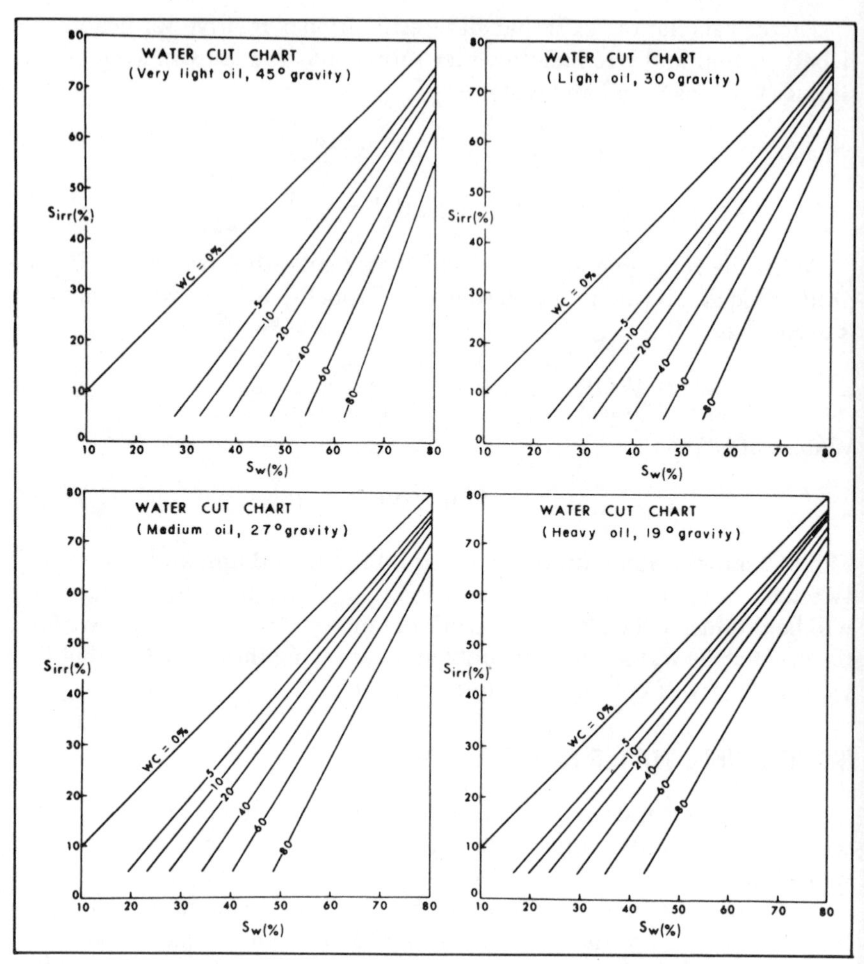

Fig. 8–2 Calculated water cut as a function of S_w and S_{wi} (courtesy SPWLA, reprinted in Schlumberger)

At any given level, water remains in only those interstices small enough that capillary pressure created by interfacial tension balances the gravitational force. The equilibrium equation is[7]

$$\sigma/7r \approx h(\rho_w - \rho_h)/2.3 \tag{8.13}$$

where

σ = interfacial tension between water and hydrocarbon, dynes/cm (25 for water-oil and 72 for water-gas).

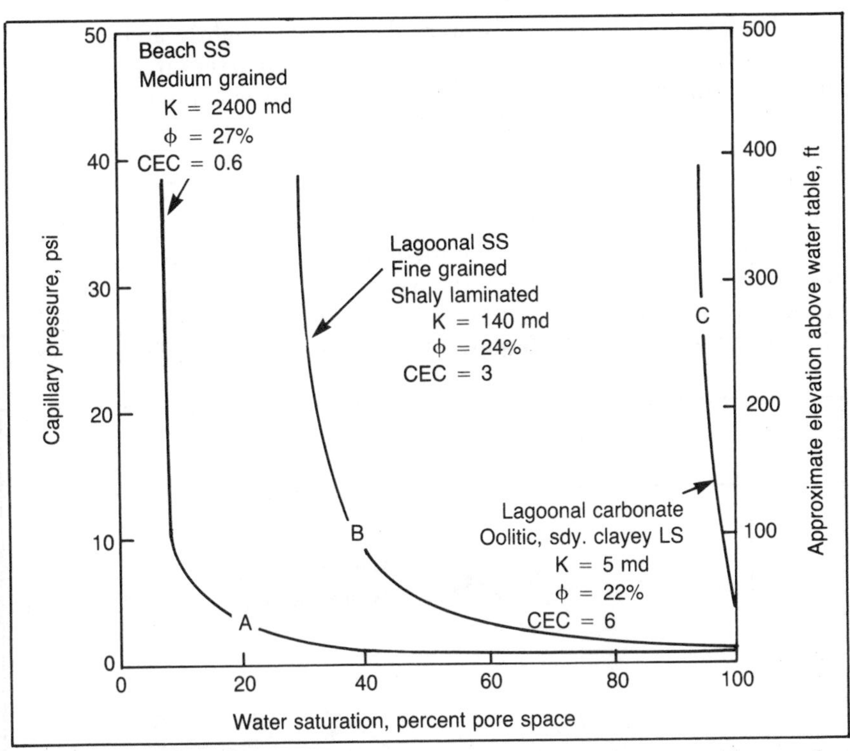

Fig. 8–3 (a) Water saturation vs capillary pressure for several types of formations (courtesy Core Lab, © SPE-AIME)

r = smallest surface radius associated with water held at grain contacts or in pores, μ

h = elevation above the free water table, ft

ρ_w = water density, g/cc

ρ_h = hydrocarbon density, g/cc

The left-hand side is the capillary pressure and the right side is the gravitational force, both in psi.

Ascending from the water table, then, oil displaces no water until the gravitational head exceeds the capillary pressure corresponding to the largest pores. Thus, the effective water-oil contact is slightly above the free water level. Above that point oil displaces water first in the large pores, then in the medium pores, and finally in small pores as elevation increases. The transition zone will be sharp if all pores are about the same size but more gradual if there is a wide distribution of pore sizes (Fig. 8–3b).

At an elevation of about 10 ft, water has been displaced from all pores greater than about 10μ in size. From that point upward, water saturation continues to decrease very slowly until only water electrostatically bound to the grains remains. This water is the true irreducible water saturation, S_{wi}. Its exact value is difficult to pinpoint because it is approached so gradually. However, the S_w value 200 ft above the water table is generally close to S_{wi}.

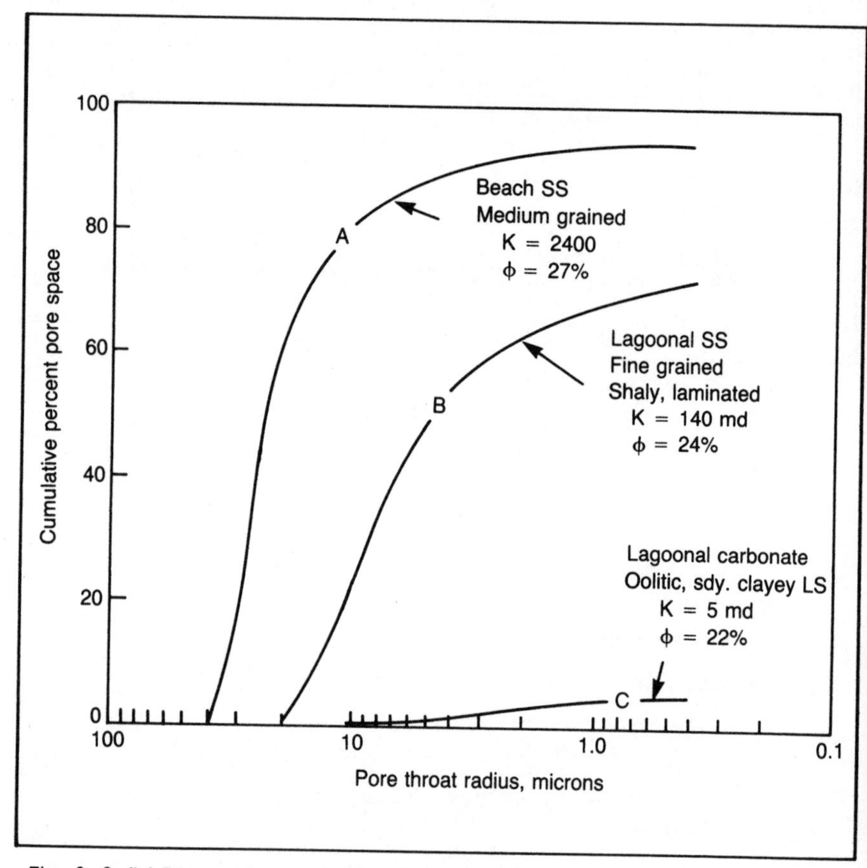

Fig. 8–3 (b) Pore entrance distributions for the same formations (courtesy Core Lab, © SPE-AIME)

The saturation profile of curve B is typical of an average shaly sand with irreducible water saturation of about 30%. Curve A illustrates the one extreme of an almost clean coarse-grained sand. The transition zone is much sharper, and the irreducible water saturation is about 8%. At the other extreme, curve C represents a sandy carbonate containing a large percentage of clay. The transition is extremely gradual, and the irreducible water saturation is over 90%. The correlation between clay content (indicated CEC values) and irreducible saturation is evident.

When a reservoir is produced from some level above the water-oil contact, the oil production is accompanied by pressure drawdown in that phase, which essentially adds to the gravitational pressure difference trying to remove water from grain contacts. This plus friction effects between moving oil and stationary water will cause loosely held water to break away and flow with the oil. Water production will therefore be large if completion is in the transition zone but small if completion is in the irreducible zone.

It is apparent that the relative amounts of hydrocarbons and water produced will be related to the difference between actual and irreducible saturation, $S_w - S_{wi}$. Stated another way, effective permeability is also a function of that difference. The empirical equations 8.9 and 8.10 portray this point.

Estimation of Irreducible Water Saturation

The pertinent question is how to determine S_{wi} from logs. In the ideal case of a long (200-ft), homogeneous, hydrocarbon-bearing zone, there is no problem because the actual water saturation, S_w, at the top of the sand is close to the irreducible value, and this value of S_{wi} can be applied throughout the zone. Such formations, however, are rare. More often the formation is short, is not homogeneous, and does not have a readily identifiable water table.

It is clear that S_{wi} cannot be less that S_b, the fraction of pore water bound by clay, as discussed in chapter 7.

$$S_{wi} \geq S_b \tag{8.14}$$

This means from Eq. 7.26

$$S_{wi} \geq V_{sh} \cdot \phi_{tsh}/\phi_t \tag{8.15}$$

or

$$S_{wi} \cdot \phi_t \geq V_{sh} \cdot \phi_{tsh} \tag{8.16}$$

where

ϕ_t = total porosity of the formation in question
V_{sh} = volumetric fraction of shale
ϕ_{tsh} = total porosity of the adjacent shale

The product $V_{sh}\,\phi_{tsh}$ can range from zero in perfectly clean formations to about 0.12 in very shaly formations.

To the above value of S_{wi} should be added the irreducible water associated with clean matrix grains. Sand grains have a surface charge similar to that of clays, sufficient to attract electrostatically and immobilize about two molecular layers of water. The volumetric fraction of pore water so bound, $S_{wi}\phi$, will be dependent on the surface area of the sand grains. However, calculations indicate that the maximum value, even where a large portion of the clean matrix is composed of very fine particles (1 μ), should not be greater than approximately 0.005. The conclusion is that the irreducible water fraction is essentially determined by clay content and that

$$S_{wi} \approx S_b \qquad (8.17)$$

An exception is the condition where a significant fraction of the pores is cemented so that they are hydraulically isolated. The outstanding example is that of chalks, which may be clay-free but have quite high S_{wi} values.

In principle, then, S_{wi} is determinable from logs by the methods explained in chapter 7 for estimating S_b. Basically these rest on determining V_{sh} from shale indicators, on the assumption that clays in sands are of the same type as those in adjacent shales where the indicators are calibrated. One case study showed good correlation in individual formations between V_{sh} from GR and SP logs and S_{wi} values measured on cores, but the correlations differed between formations, implying the foregoing assumption is invalid.[8] This is to be expected with authigenic clays. Once again the need for a Q log from which S_b can be obtained directly is evidenced.

To predict relative permeability and water cut with any confidence, an accurate value of $(S_w - S_{wi})$ is required. This is difficult to obtain from logs, given the imprecision in determining V_{sh}. In particular if V_{sh} is overestimated, S_{wi} will be overestimated, S_w will be underestimated, and the difference will be magnified accordingly.

Empirical Method

A method of circumventing this difficulty to some extent is as follows. It has long been observed in the field that the product $S_w\phi$ tends to be constant in the irreducible zone of a reservoir. This is predicted by Eq. 8.16 if the shaliness of the interval is essentially constant. Consequently, if the product of *actual* water saturation, S_w, and total porosity, ϕ_t, is calculated for a number of levels over an interval where ϕ_t varies and the product is found to be constant, then it is assumed those levels are in the irreducible zone. If $S_w\phi_t$ shows higher values at lower levels, then those points are considered to be in the transition zone.

Fig. 8–4a is a plot of S_w against ϕ for a Wyoming sand, and Fig. 8–4b is a similar plot for an East Texas carbonate.[9] In both cases most points conform

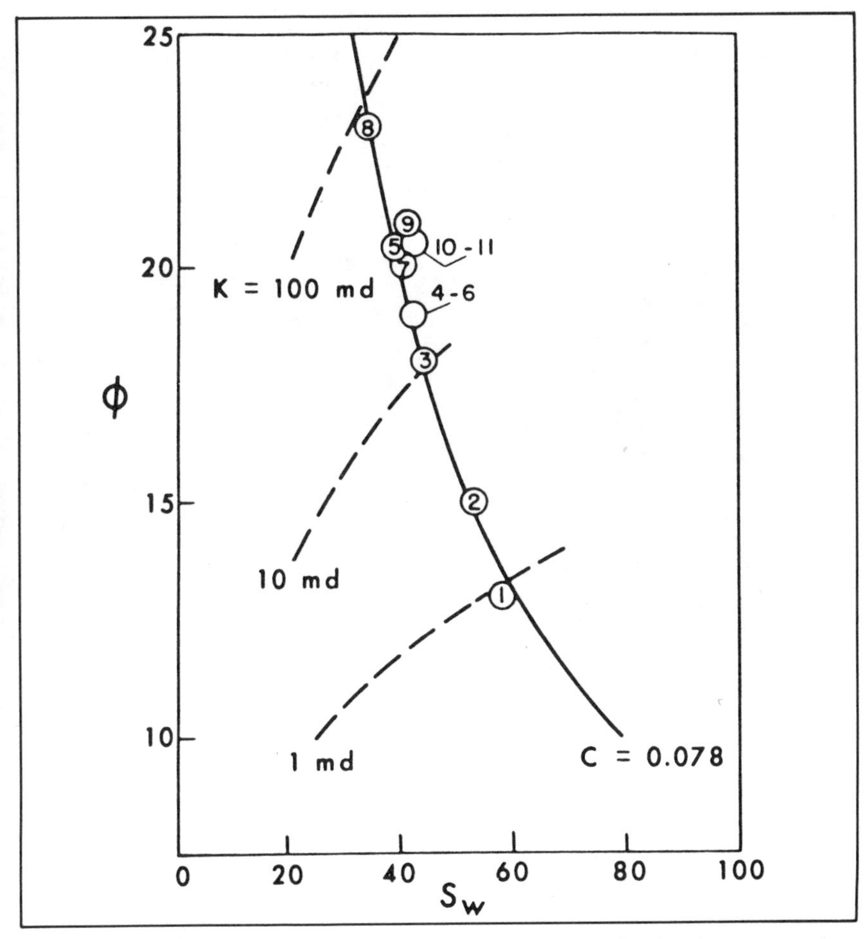

Fig. 8–4 (a) S_w vs ϕ for a Wyoming sand (courtesy Schlumberger and SPWLA)

to a hyperbolic line indicating $S_w\phi$ = constant. Those points are considered to be in the irreducible water zones. The exception is the open circles on the carbonate plot, which correspond to the bottom 10 ft of that formation. These show higher $S_w\phi$ values and are interpreted to be in the transition zone. Note that the constant is 0.078 for the sand and 0.008 for the carbonate, indicating the former is much more shaly than the latter.

If the Gamma Ray log or other shale indicators show that V_{sh} is not constant over the interval of interest, then the quantity $(S_w \cdot \phi_t)/V_{sh}$ should be tested for constancy to determine if in fact S_w represents S_{wi}. However,

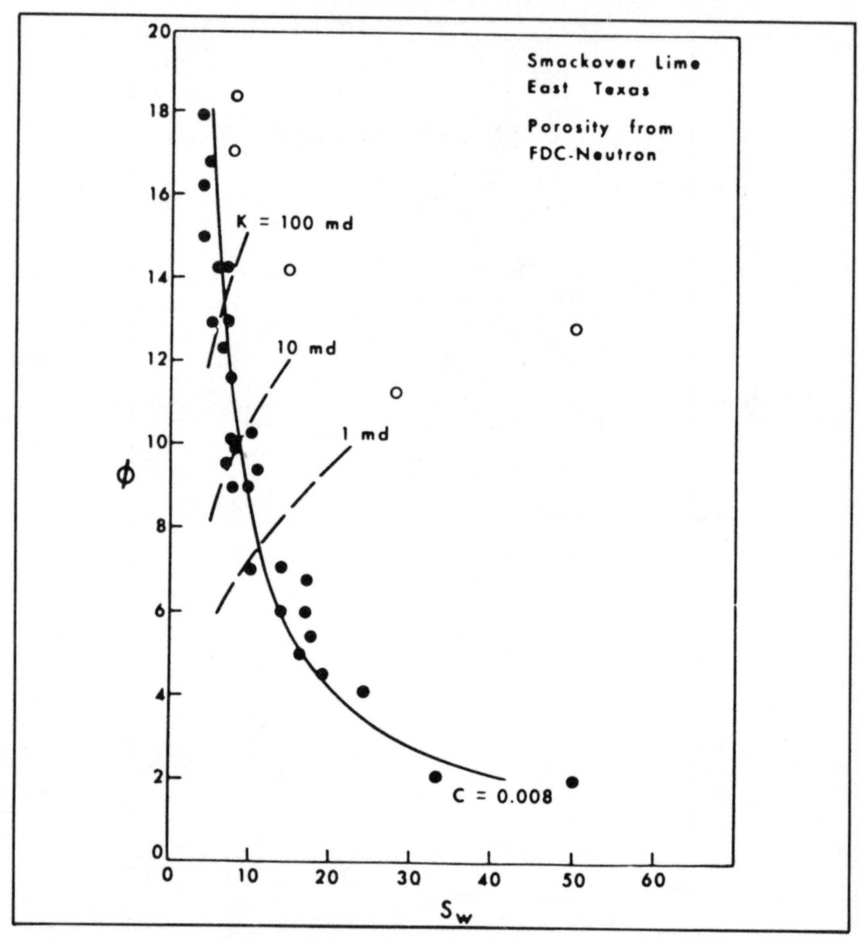

Fig. 8–4 (b) S_W vs ϕ for an East Texas carbonate (courtesy Schlumberger and SPWLA)

this introduces the uncertainty of V_{sh} determination and reduces the reliability of the method.

In short, irreducible water saturation is a very important parameter in production estimation, but it is an elusive quantity that is difficult to obtain from logs. The one log that comes close to defining S_{wi} is the Nuclear Magnetism (NML) log. In sandstone it measures the amount of fluid (of normal viscosity) free to move in the pore space. Therefore

$$S_{wi} = (\phi_t - \phi_{NML})/\phi_t$$

where

ϕ_t = total porosity measured by Neutron-Density
ϕ_{NML} = porosity indicated by NML

In carbonates, however, the NML measures total porosity. The NML tool is not discussed in this book because it is a very specialized device not yet commonly run.

ESTIMATION OF PERMEABILITY FROM LOGS

For intergranular-type formations, permeability increases rapidly with porosity. However, permeability is also quite dependent on the surface area presented by the grains. A clayey, fine-grained formation will have a lower permeability than a clean, coarse-grained formation of the same porosity. Flow paths are finer and more tortuous in the former case.

Empirical Relations

Irreducible water saturation reflects surface area. Therefore, several empirical correlations have come into use relating absolute permeability to porosity and irreducible water saturation for intergranular formations. A well-documented correlation is that of Timur, who made careful laboratory measurements on 155 sandstone cores from the Gulf Coast, Colorado, and California.[10] His correlations between porosity, irreducible water saturation, and permeability are shown in Fig. 8–5. With this data he derived the following relation

$$k = (93\phi^{2.2}/S_{wi})^2 \qquad (8.18)$$

where k is in millidarcies, and ϕ and S_{wi} are fractional. This equation

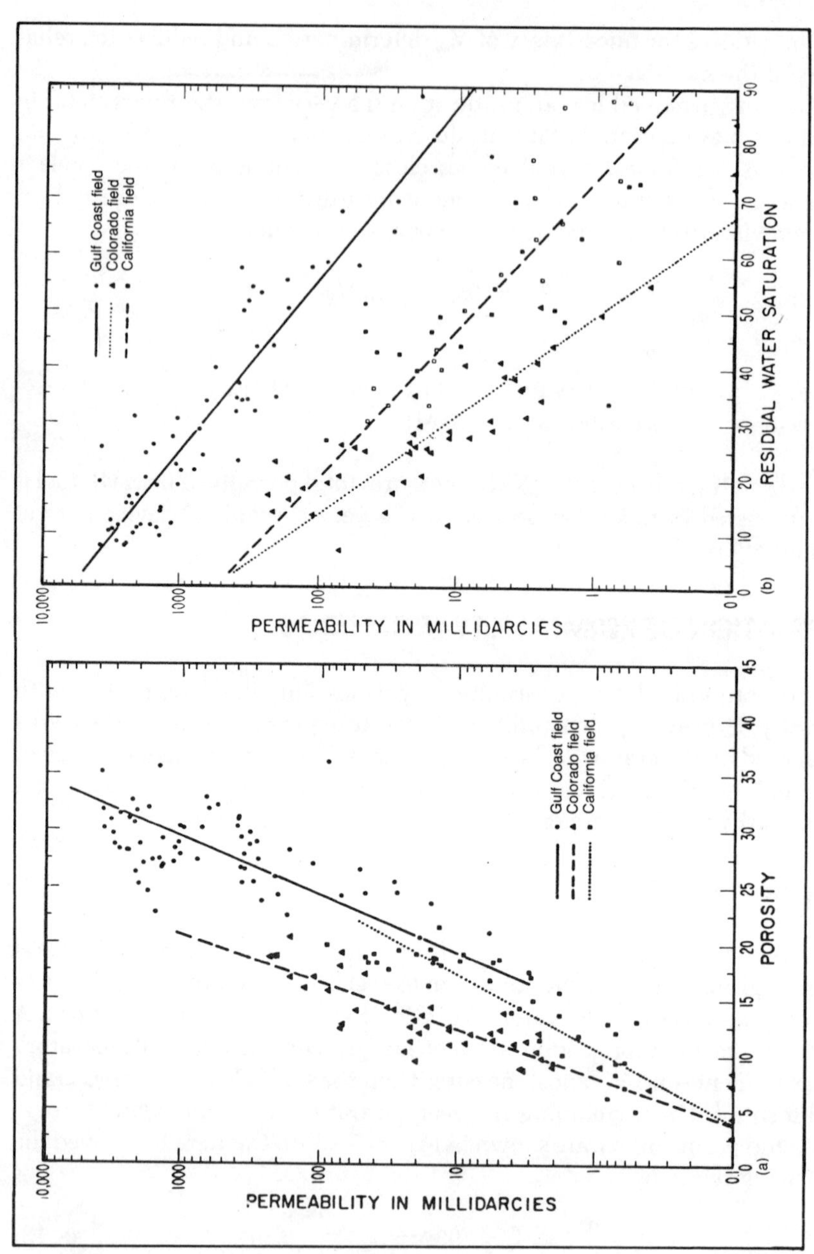

Fig. 8–5 Permeability vs porosity and irreducible water saturation for various sands (after Timur, courtesy SPWLA)

predicted the measured core permeabilities to within a factor of two (standard deviation), which is quite good for permeability estimation. It should be noted that the porosities measured were total porosities, including bound water, if any. Also, irreducible water saturations were measured at an effective capillary pressure of 50 psi, which should be high enough to eliminate all but clay-bound water.

An alternate empirical expression by Tixier is often used[11]

$$k = (250\ \phi^3/S_{wi})^2 \tag{8.19}$$

The two relations are plotted in Fig. 8–6a and 8–6b. For the same values of ϕ and S_{wi}, they give about the same answers except at extremes of permeability. A more elaborate relation gives similar answers.[12]

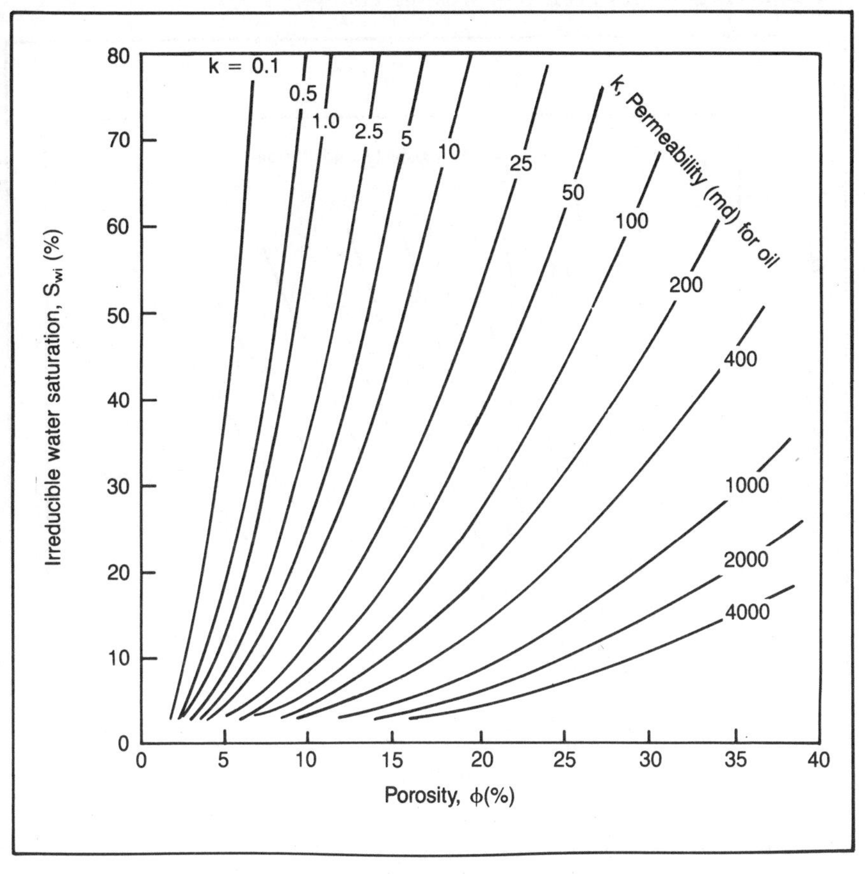

Fig. 8–6 (a) Permeability from the Timur relation (courtesy Schlumberger)

In principal, these equations should apply equally well to oil or gas flow. Experience has shown, however, that k values obtained from the equations must be reduced by a factor of 3 to 10 when the expected production is gas. The reason for this is not clear, although partial explanations have been advanced.[13] One important factor may be that gas flow near the borehole is likely to be turbulent, whereas the equation applies to laminar flow. In any case the recommended expression for absolute permeability for gas flow is:

$$k = \rho_g\,(93\,\phi^{2.2}/S_{wi})^2 \qquad (8.20)$$

where ρ_g is the gas density in g/cc and the bracketed term is obtainable from Fig. 8–6a.

Combining Eqs. 8.5, 7, 9, and 18, the specific productivity index (SPI) for an average oil reservoir is approximately

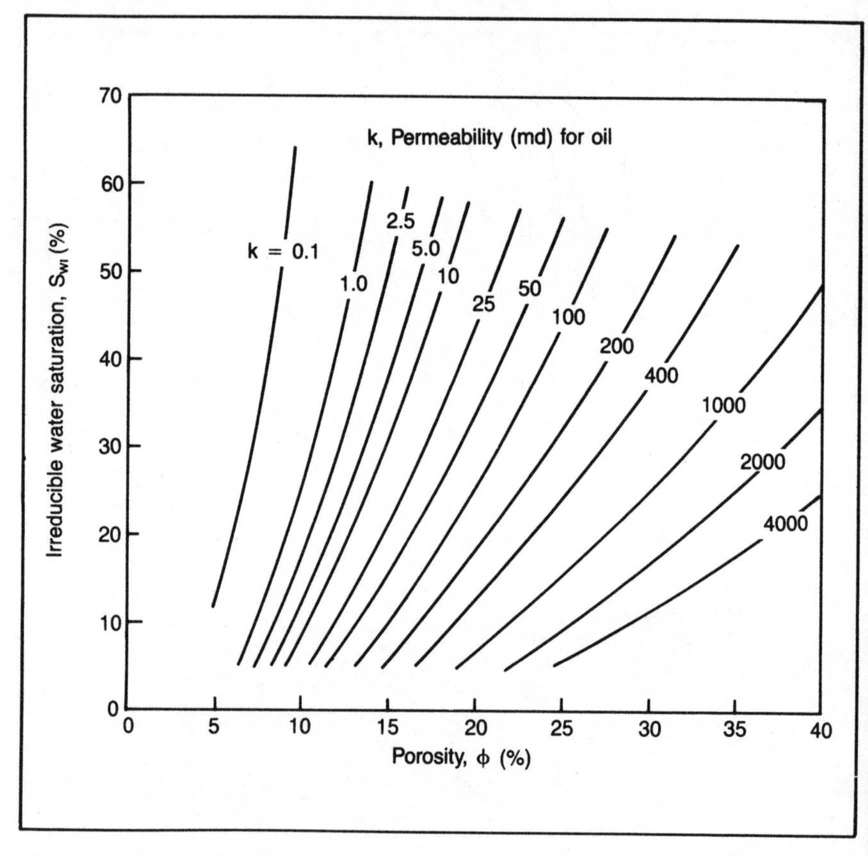

Fig. 8–6 (b) Permeability from the Tixier relation (courtesy Schlumberger)

$$SPI \approx 8.6\,(\phi^{2.2}/S_{wi})^2 \cdot [(0.9 - S_w)/(0.9 - S_{wi})]^2 \qquad (8.21)$$

This is an appropriate relation if S_w and if S_{wi} are known.

Using S_w for S_{wi}

In the more usual case where S_{wi} cannot be reliably determined, the recommended approach is to use the actual S_w in Eq. 8.18 and consider the permeability so obtained as the effective rather than the absolute value. This leads to the simple expression

$$SPI \approx 8.6\,(\phi^{2.2}/S_w)^2 \text{ stb/d-psi-ft} \qquad (8.22)$$

There is justification for this procedure. At one extreme, when $S_w = S_{wi}$, Eqs. 8.21 and 8.22 are identical. At the other extreme, when $S_w > S_{wi}$, they can give similar values, depending on the numbers. For example, if $S_w = 0.6$ and $S_{wi} = 0.3$, answers are identical. This simply means that permeability calculated with S_w instead of S_{wi} can be a good approximation to effective permeability.

A similar expression for the specific absolute open-flow potential of gas is

$$SAOFP \approx 170\,\rho_g \cdot (\phi^{2.2}/S_w)^2 \cdot P_r^2 \text{ scfd/ft} \qquad (8.23)$$

To summarize, specific flow potentials for intergranular formations, typically sandstones, can be calculated from log data to within a factor of about 3 if S_{wi} is reasonably well known. With the usual uncertainty in S_{wi}, however, accuracy is probably no better than a factor of about 5. And of course there is no way to take into account skin damage, if it exists.

For nonintergranular rock the above approach to productivity estimation is not applicable. Formations containing solution channels, vugs, or fractures may have quite high permeability even though overall porosity is low, as indicated in Fig. 8–7. At the other extreme, formations with isolated pores such as chalks may have high porosity but quite low permeability. Both conditions are more prevalent in carbonates than in sandstones. For these situations one must resort to movable oil calculation or formation testing or localized log-core correlations.[14]

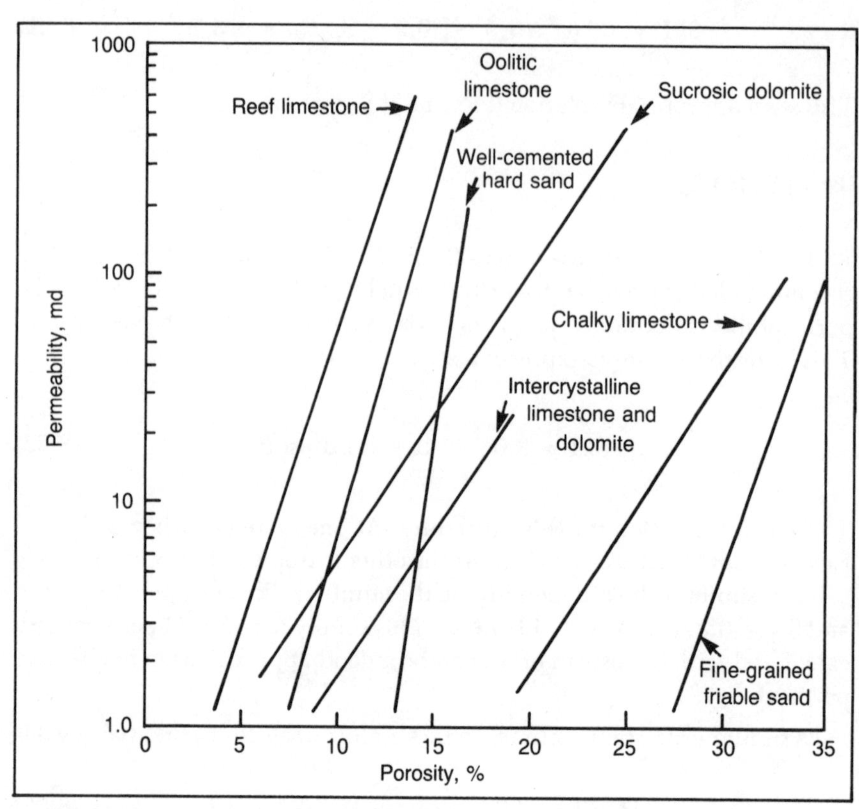

Fig. 8–7 Variation of permeability with rock type (courtesy Core Lab)

THE MULTIPLE FORMATION TESTER

The Multiple Formation Tester is a tool that can be run in open hole, positioned at a desired level, actuated to measure reservoir pressure and flow capability, and, if desired, further activated to take a 1–12-gal sample of reservoir fluid.*

*Designated Repeat Formation Tester (RFT) by Schlumberger, Formation Multi-Tester (FMT) by Dresser Atlas, Multiset Tester by Welex, and Selective Formation Tester (SFT) by Gearhart.

Any number of pressure measurements can be made at various levels.[15,16] From these, reservoir pressures can be accurately determined and formation permeabilities calculated. However, only two fluid samples can be retrieved, taken at the same or different levels. Analysis of these samples allows prediction of gas-oil ratios and water-oil ratios on production.

Operation

The Formation Tester has a hydraulically activated pad that goes downhole in the retracted position. The tool is positioned at the desired level using a simultaneously run GR or SP log, and the pad is extended. This causes a small circular packer to seal against the mud cake on one side of the hole, as shown in Fig. 8–8. A metal snorkel tube about 0.5 in. in diameter,

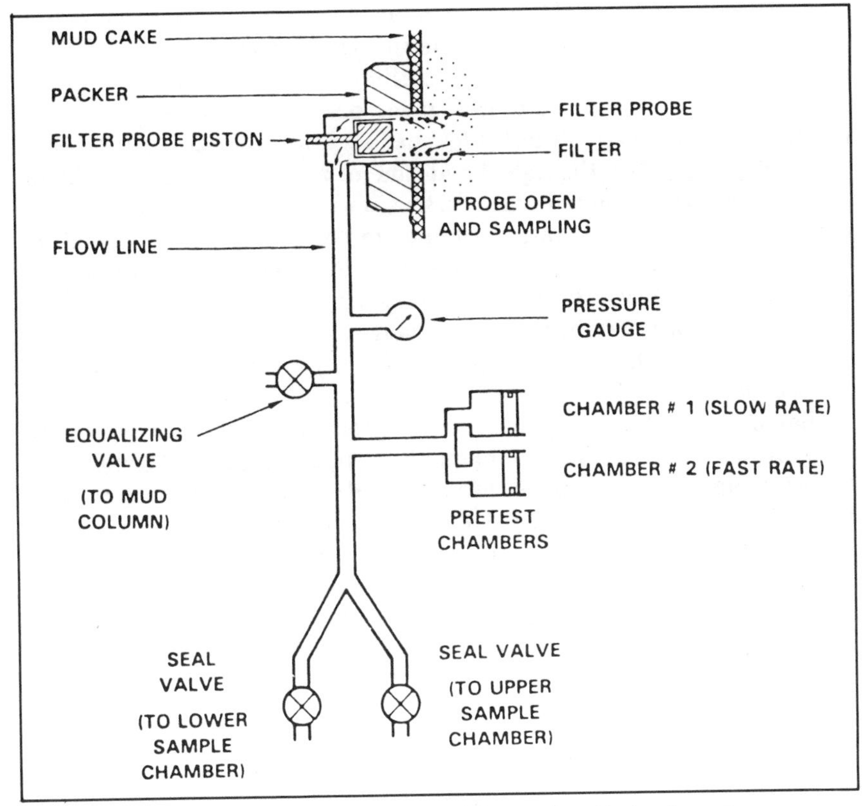

Fig. 8–8 Flow path in the Repeat Formation Tester (courtesy Schlumberger, © SPE-AIME)

positioned in the center of the packer, is then forced through the mud cake into the formation. Simultaneously, a piston inside the tube retracts, uncovering filter holes and allowing reservoir fluid to flow into the tool.

Reservoir fluid successively enters two small pretest chambers, each having 10 cc of capacity (in the RFT). Flow is first into the upper chamber at a rate of approximately 0.7 cc/sec, determined by the speed at which a piston in that chamber is hydraulically retracted. When that chamber is full, the lower one fills. Its piston is retracted twice as fast so the inflow rate is approximately 1.5 cc/sec. Typically, both chambers fill in 15–30 sec, depending on the particular tool.

Pressure Recording

During tool setting and filling, pressure at the probe entrance is continuously monitored and recorded at the surface by means of the indicated pressure gauge. A typical pressure recording is shown in Fig. 8–9. Time is proceeding from top to bottom at 6 sec per chart division. Pressure is recorded on an analog scale on the left-hand side for qualitative reading and in a digital format on the right-hand side for precise reading. The digital recording has a thousands, hundreds, tens, and units track. In each track the recording can be on only one of 10 discrete positions, either on one of the vertical chart divisions or halfway between. At level A, for example, pressure would be read as 4,351 psi. Thus, pressure can be read to 1 psi, which is the resolution of the normal pressure gauge. More recent recordings have pressure values to the nearest psi printed directly in the depth column every few seconds.

Beginning at the top of Fig. 8–9, the pressure gauge reads hydrostatic pressure of the mud (point A). At point B the pad sets against the mud cake. Flow into the first test chamber starts at C and ends at E. Flowing pressure is read at D as 1,850 psi. The second test chamber begins to fill at point E and is full at point G. Flowing pressure is read at F as approximately 50 psi. At point G, the test chambers and flow lines in the tool are full and pressure builds back up to reservoir pressure. It is read at point H as 3,848 psi. This pressure is normally several hundred psi less than the hydrostatic pressure of the mud.

Pad Sealing

The pretest phase indicates quickly whether a good seal to the formation has been effected, whether the probe is tending to plug, or whether the

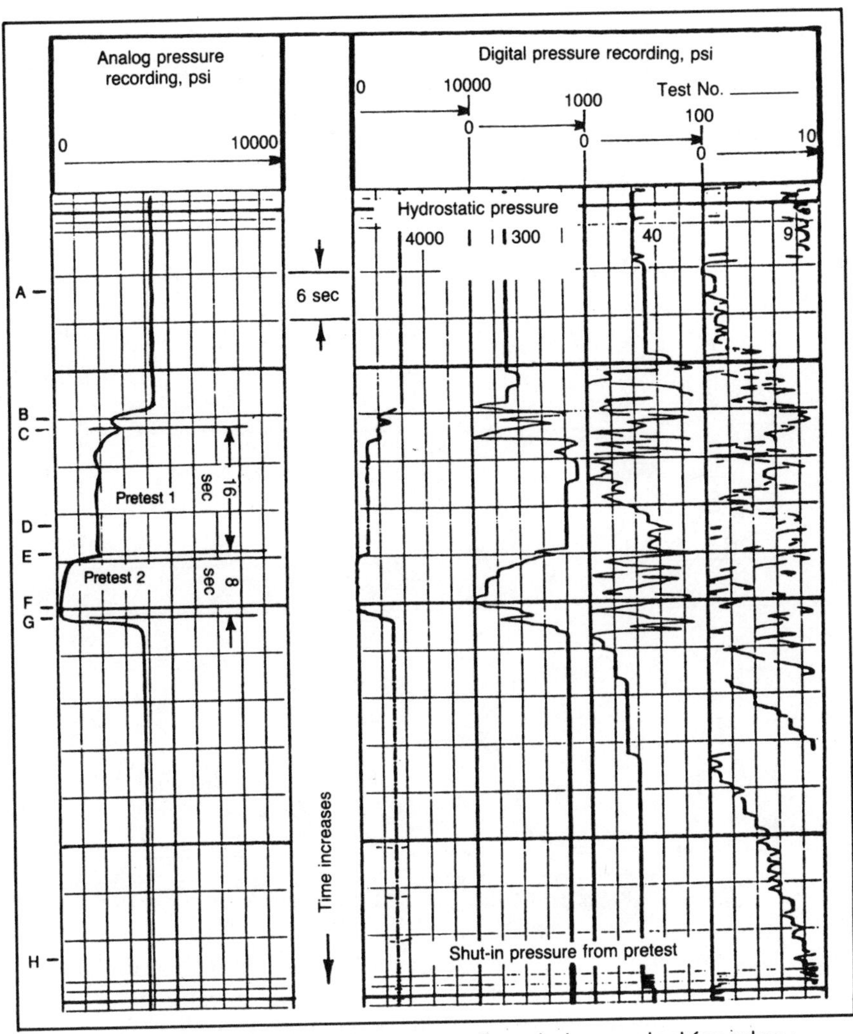

Fig. 8–9 Typical RFT pressure recording during pretest (courtesy Schlumberger, © SPE-AIME)

formation is completely tight. Figs. 8–10, 8–11 and 8–12 are examples of these problems. Poor seal is characterized by quick return to mud hydrostatic pressure. Plugging appears as erratic flowing pressure. A tight formation is characterized during the first pretest by the flowing pressure dropping essentially to zero and the chamber not filling in the usual 12–18

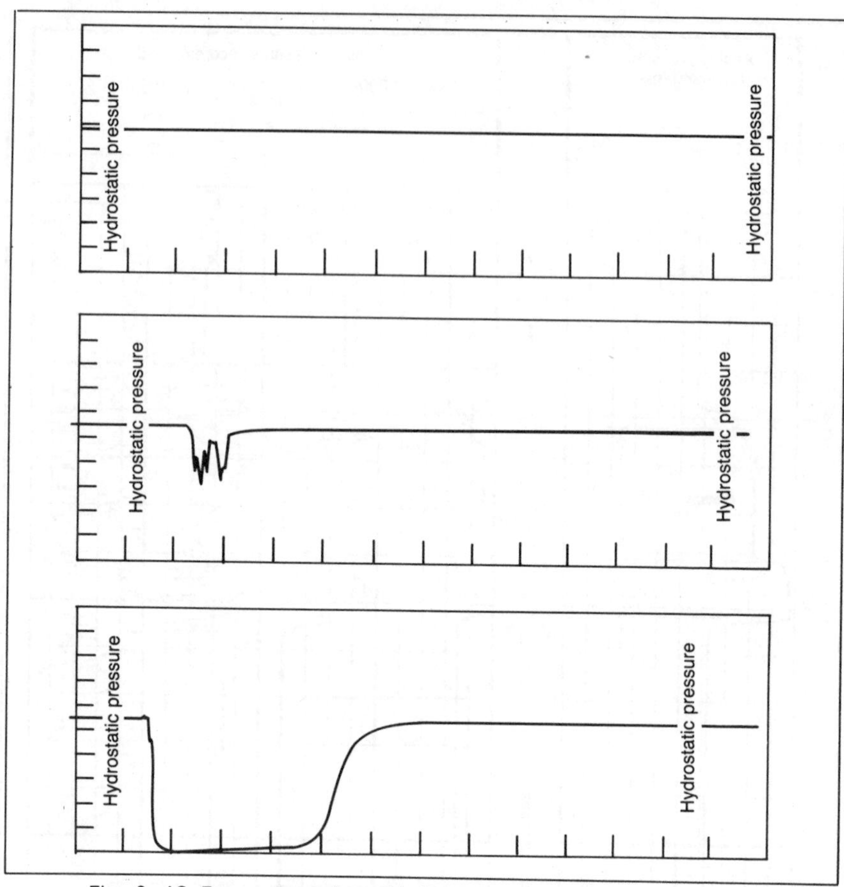

Fig. 8–10 Examples of seal failures (courtesy Schlumberger)

sec. When this happens, the recording should be continued for about 3 min before condemning the formation as essentially impermeable (or assuming the probe is completely plugged). The formation may simply have such a low permeability that it cannot supply fluid at the rate of 0.7 cc/sec. Nevertheless, if the pretest chambers eventually fill, permeability can be calculated, which is desirable.

When problems occur, the normal procedure is to unseat the tool, reposition it 6 in. or more away, and try again. A good test in one out of two attempts is not uncommon.

If the pretest phase indicates a good seat along with reasonable reservoir flowability, then one of the seal valves (Fig. 8–8) can be opened and a 1–12-

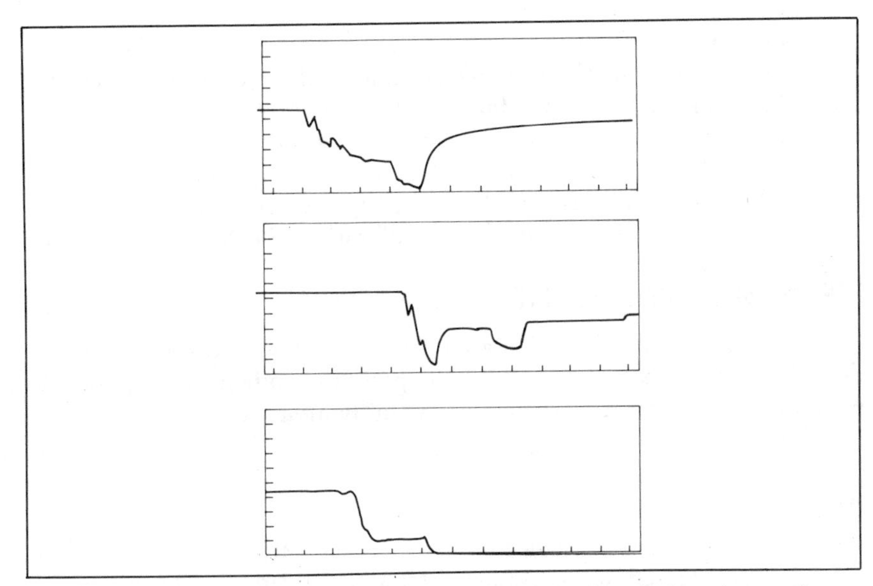

Fig. 8–11 Examples of probe plugging (courtesy Schlumberger)

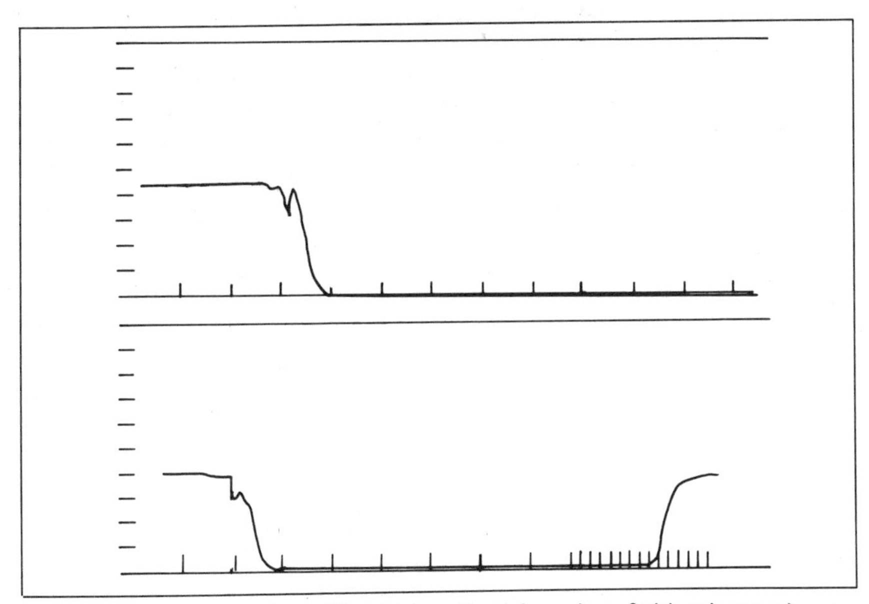

Fig. 8–12 Examples of tight formations (courtesy Schlumberger)

gal sample of formation fluid withdrawn if desired. When the sample chamber is filled or sampling is to be terminated, the seal valve is closed and the tool is unseated in readiness for repositioning or retrieving. Alternately, the other seal valve can be opened and the second chamber filled with reservoir fluid from the same level. The second sample should contain a smaller fraction of mud filtrate than the first, which should allow more accurate prediction of water-oil and gas-oil ratios expected on production.

PERMEABILITY FROM PRETEST DRAWDOWN

Permeability may be calculated from the flow rates and pressure drawdowns that occur while filling the pretest chambers. The applicable equation for drawdown permeability, in millidarcies, is

$$k = 5{,}650\ q\mu/\Delta p \tag{8.24}$$

where

- 5,650 = constant that takes into account the entrance diameter of the probe and the quasispherical flow of fluid into it. This constant is for the RFT; an earlier constant was 3,300. For the FMT it is 2,760.
- q = average flow rate during chamber filling, cc/sec. Chamber volume is known and filling time is taken from the pressure recording
- μ = viscosity of fluid at reservoir conditions, cp. Intaken fluid is considered to be mud filtrate, so μ is normally taken as 0.5 cp
- Δp = pressure drawdown, the difference between reservoir pressure (not mud pressure) and flowing pressure measured near the end of the filling period, psi

This calculation must be made separately for each pretest chamber. For the recording of Fig. 8–9

Pretest chamber #1: $q = 10/16 = 0.625$ cc/sec
$\Delta p = 3{,}848 - 1{,}850 = 1{,}998$ psi
$\mu = 0.5$ cp
hence: $k = 0.88$ md
Pretest chamber #2: $q = 10/8 = 1.25$ cc/sec
$\Delta p = 3{,}848 - 50 = 3{,}798$ psi
hence: $k = 0.93$ md

Average drawdown permeability is therefore 0.9 md.

Drawdown permeability represents the effective permeability to water in the invaded zone. This may represent only 30–50% of the absolute permeability when residual oil or gas is present in that zone, as illustrated in Fig. 8–1. Further, the measurement has very shallow depth of investigation — only an inch or two — since practically all of the pressure drawdown occurs very close to the probe entrance. The calculated value can therefore be too low as a result of formation damage close to the borehole, such as clay swelling. Sometimes formation cleanup during chamber filling can actually be observed, as illustrated in Fig. 8–13.

Because of these two factors, drawdown permeability is often considered a lower limit on absolute formation permeability. On the other hand, in consolidated formations the forced insertion of the probe can cause microcracks that stimulate rather than reduce permeability. This can lead to abnormally high permeability values.

Fig. 8–14 is a plot of Eq. 8–24 for various pretest flow rates. This can be used to make a quick estimate of permeability as soon as the pretest drawdown pressure is seen. The range of permeabilities measurable is from approximately 0.1–200 md. The upper limit is determined by the need to have at least 10 psi drawdown to obtain a reasonably accurate reading with a pressure recording of 1 psi resolution. Newer systems with resolution of 0.1 psi increase the range to 2,000 md. The lower limit is achieved by waiting for about 3 min for the pretest chambers to fill if flowing pressure drops to essentially zero. In this case both chambers fill at the same formation-determined rate, calculated as total pretest chamber volume divided by total filling time, and only one permeability calculation is made.

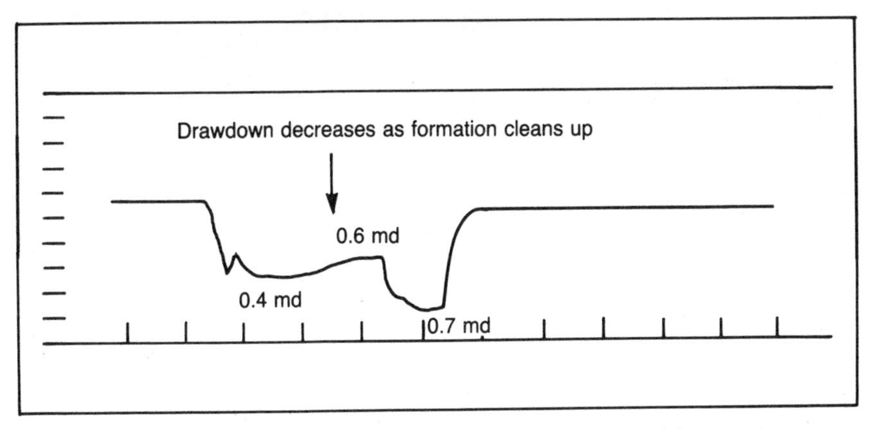

Fig. 8–13 Example of formation cleanup during pretest (courtesy Schlumberger)

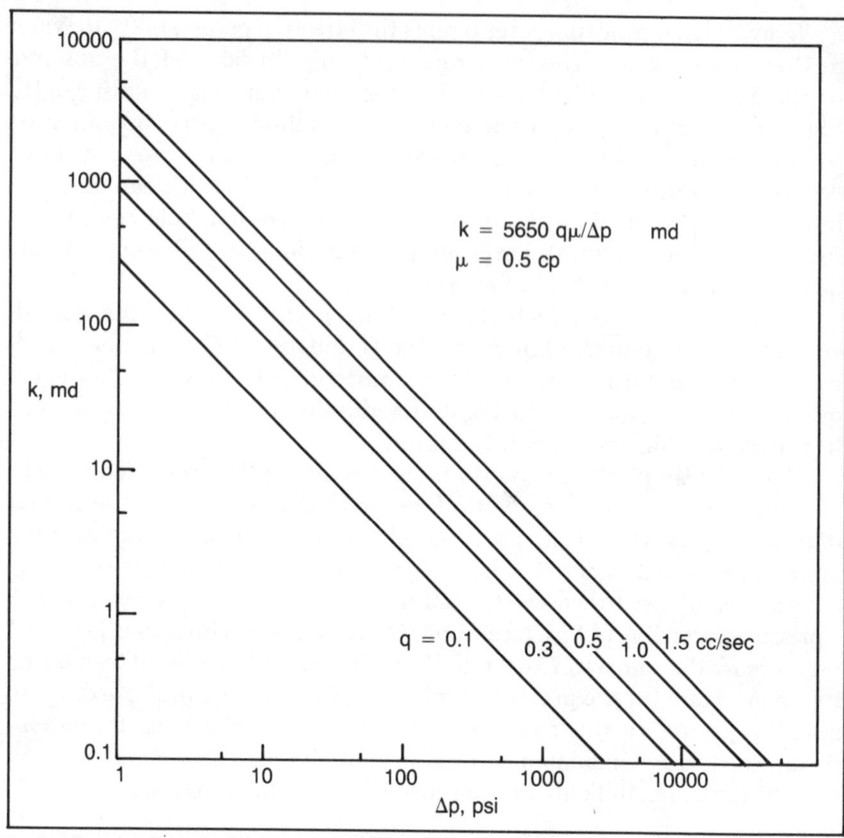

Fig. 8–14 Permeability from RFT pretest drawdown

PERMEABILITY FROM PRESSURE BUILDUP

Following cessation of flow into the test chambers, the pore space in the vicinity of the probe from which reservoir fluid has been withdrawn is repressurized by fluid flow from farther back in the reservoir. During this time, pressure at the probe builds back up to reservoir pressure, as illustrated by the G-H portion of Fig. 8–9. The greater the permeability, the faster the buildup.[17,18,19]

Permeability can be determined from the rate of buildup. It is more difficult to calculate than drawdown permeability, but the value derived is not influenced by formation alteration near the borehole since the depth of investigation is a few feet rather than a few inches.

Importance of Extended Measurement

After the second pretest chamber fills, following point G of Fig. 10–9, a small amount of flow continues into the RFT tool as the fluid in the flow lines and pretest chambers compresses under the increasing pressure. This effect is called *afterflow*. The duration of afterflow, T_a, depends very strongly on the compressibility of the fluid in the tool. It is given approximately, in seconds, by

$$T_a = 10^6\ C/k \tag{8.25}$$

where

C = compressibility of the fluid, psi^{-1}
k = formation permeability, md

Usually the fluid is mud filtrate that has low compressibility, $C \approx 3 \times 10^{-6}$, in which case the afterflow lasts for a period of approximately 3/k sec. The duration is negligible if k is greater than 1 md.

However, if gas is drawn into the tool, the compressibility can increase drastically, in which case the afterflow will be much longer. Gas is indicated if drawdown pressure increases significantly during pretest chamber filling rather than holding constant, as illustrated in Fig. 8–15. The effect of afterflow is to slow the pressure buildup and render the affected part of the buildup unusable for permeability determination.

Near-wellbore damage or stimulation also makes the early part of the pressure buildup unreliable for determining formation permeability. Consequently, the late part of the buildup, essentially the last 10 or 15 psi extending from point H of Fig. 8–9 and beyond, is critical. **Recording must be continued until pressure returns to within a few psi of the apparent reservoir pressure**, as guessed by extrapolating the units scale of the digital recording. Otherwise, valuable information will be lost.

Permeability determined from the late part of the pressure buildup reflects primarily the flow taking place several feet out in the formation beyond normal invasion. The value so obtained should be the absolute permeability to hydrocarbon, as indicated in Fig. 8–1, if the undisturbed formation is at irreducible water saturation. In this case the buildup permeability should be higher than the drawdown value if there is no local stimulation caused by the probe.

The method of calculating buildup permeability depends on whether the repressurizing flow is considered to be spherical or radial. The flow starts essentially spherical close to the probe. But as it moves outward it may encounter upper and lower impermeable beds that cause it to become radial

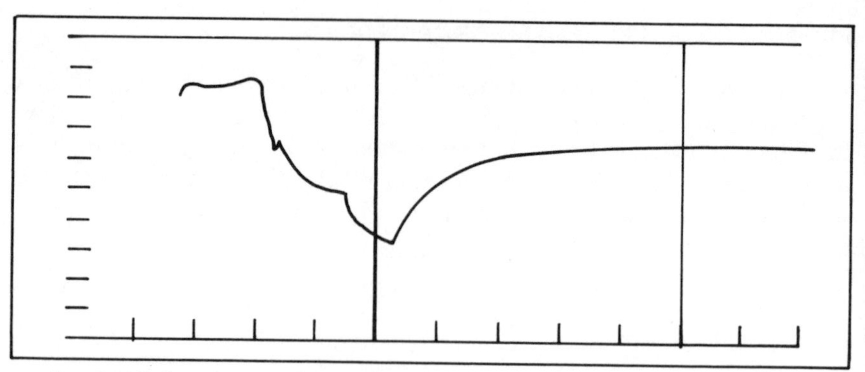

Fig. 8–15 Example of gas in flow line or pretest chambers (courtesy Schlumberger)

in the late stages. By comparing radial and spherical flow analyses, the nature of the late flow can be deduced.

Radial Flow

Assuming radial-cylindrical flow, radial permeability, in millidarcies, is given by

$$k_r = -88.4\, q \cdot \mu/(m \cdot h) \tag{8.26}$$

where

- q = average flow rate during drawdown, i.e., total chamber volume divided by total flow time, cc/sec
- μ = viscosity of the repressurizing fluid, cp. It is the value for water at the temperature of interest if the formation is water-bearing or the value for hydrocarbon in a hydrocarbon-bearing formation with shallow invasion. If invasion is deep, it should be an intermediate value
- m = slope of the Horner buildup plot, psi/cycle, where pressure on a linear scale is plotted against the time function, $(T + \Delta t)/\Delta t$, on a logarithmic scale. T is the total flowing time (24 sec on Fig. 8–9) and Δt is time elapsed since cessation of flow, i.e., time beginning at point G on Fig. 8–9. Pressure is read at Δt increments and $(T + \Delta t)/\Delta t$ vs pressure is tabulated in preparation for making the Horner plot
- h = vertical thickness of formation sampled, ft. This is a variable figure since it is not defined by the tool but by the separation between upper and lower impermeable beds straddling the test point. It could be as much as several feet in a homogenous forma-

tion and as little as an inch in a thin sand bed confined by shale stringers. In lieu of other information, it is generally taken as 0.5 ft

Table 8–1 tabulates pressure vs $(T + \Delta t)/\Delta t$ for the example in Fig. 8–9 in Δt increments of 6 sec. Values for Δt greater than 48 sec, which are critical, are from the very late portion of the buildup not shown in Fig. 8–9.

The corresponding Horner plot is shown in Fig. 8–16 with time after flow progressing from left to right. A straight line is drawn through the late-

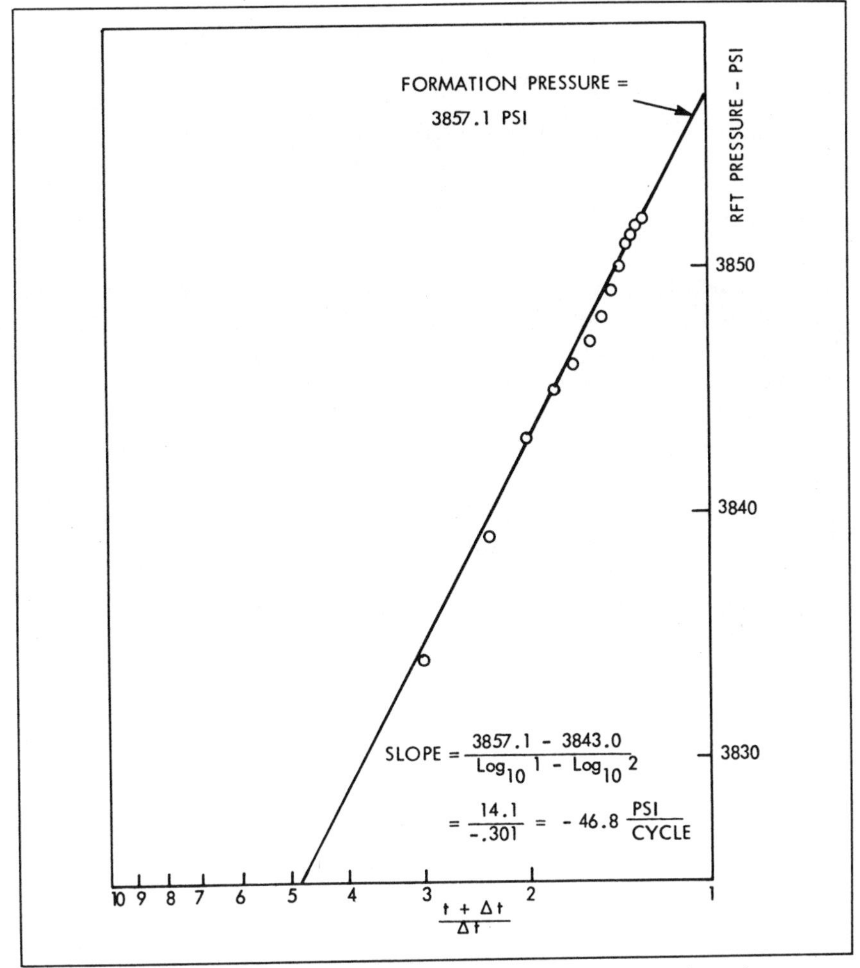

Fig. 8–16 Radial flow pressure buildup plot of Horner (courtesy Schlumberger)

TABLE 8-1 TABULATION OF PRESSURES AND TIME FUNCTIONS FOR RADIAL AND SPHERICAL FLOW PLOTS

	Radial (Horner)			Spherical					
P (psi)	Δt	T + Δt	$\frac{T + \Delta t}{\Delta t}$	8 + Δt	24 + Δt	(a) $\frac{2}{\sqrt{\Delta t}}$	(b) $\frac{1}{\sqrt{8 + \Delta t}}$	(c) $\frac{1}{\sqrt{24 + \Delta t}}$	(a) − (b) − (c)
3,824	6	30	5.00	14	30	0.8165	0.2673	0.1826	0.366
3,834	12	36	3.00	20	36	0.5773	0.2236	0.1667	0.186
3,839	18	42	2.33	26	42	0.4714	0.1961	0.1543	0.121
3,843	24	48	2.00	32	48	0.4082	0.1768	0.1443	0.0871
3,845	30	54	1.8	38	54	0.3651	0.1622	0.1361	0.0668
3,846	36	60	1.67	44	60	0.3333	0.1507	0.1291	0.0535
3,847	42	66	1.57	50	66	0.3086	0.1414	0.1231	0.0441
3,848	48	72	1.50	56	72	0.2887	0.1336	0.1178	0.0373
3,849	54	78	1.44	62	78	0.2722	0.1270	0.1132	0.0320
3,850	60	84	1.40	68	84	0.2582	0.1213	0.1091	0.0278
3,851	66	90	1.36	74	90	0.2462	0.1162	0.1054	0.0246
3,851.3	72	96	1.33	80	96	0.2357	0.1118	0.1021	0.0218
3,851.7	78	102	1.31	86	102	0.2264	0.1078	0.0990	0.0196
3,852	84	108	1.29	92	108	0.2182	0.1043	0.0962	0.0177

time points on the plot, ignoring early-time points that may fall below or above it because of afterflow and near-wellbore damage or stimulation. In this case the drawdown pressure shows no sign of gas flow into the pretest chambers and permeability is about 1 md, so afterflow should disappear in a few seconds. Extrapolating the line to $(T + \Delta t)/\Delta t = 1$, that is, to essentially infinite recovery time, gives the shutin reservoir pressure. In this example it is 3,857 psi.

Determination of the slope of the straight line in psi/cycle, where cycle means a factor of 10 on the logarithmic time scale, is a little tricky if the line does not extend over one complete cycle. In that case the slope is best obtained by taking the difference in pressure between the points at $(T+\Delta t)/\Delta t$ equal to 1 and 2 and dividing by (log 1 − log 2), which is $0-0.30$ as illustrated on Fig. 8–16. In the example, m is obtained as −46.8 psi/cycle. Consequently for Fig. 8–9

$q = 20$ cc/24 sec $= 0.83$ cc/sec
$\mu = 1.0$ cp, assumed
$h = 0.5$ ft, assumed temporarily

Therefore, a first calculation of radial permeability from Eq. 8.26 gives

$$k_r = (-88.4)(0.83)(1.0)/(-46.8)(0.5) = 3.2 \text{ md}$$

This will be recalculated after a value of h is derived from the combined radial and spherical plots.

Spherical Flow

Assuming spherical flow during the late stages of repressurizing, spherical permeability is given by the relation

$$k_s = 1{,}856\,\mu(q/-m)^{2/3}(\phi \cdot C)^{1/3} \tag{8.27}$$

where

q, μ = as defined for radial flow

m = slope of a buildup plot, psi-$\sqrt{sec}$, wherein pressure is plotted linearly against a time function, f(t), given by

$$f(t) = (q_2/q_1)/\sqrt{\Delta t} - (q_2/q_1 - 1)/\sqrt{T_2 + \Delta t} - 1/\sqrt{T_1 + T_2 + \Delta t} \tag{8.28}$$

in which q_1 and q_2 are the first and second flow rates, T_1 and T_2 are the first and second flow times, and Δt is the time after cessation of flow*

ϕ = porosity, fraction

C = compressibility of the overall fluid (not just that which moves) in the first few feet from the borehole. For a water-bearing formation the applicable compressibility is that of water, $3 \times 10^{-6}\,psi^{-1}$. For a shallowly invaded formation containing light oil, an appropriate value would be that of the oil, $3 \times 10^{-5}\,psi^{-1}$; if deeply invaded, perhaps $10^{-5}\,psi^{-1}$.

For the example of Fig. 8–9, $q_2/q_1 = 2$, $T_1 = 16$, and $T_2 = 8$ so that

$$f(t) = 2/\sqrt{\Delta t} - 1/\sqrt{8 + \Delta t} - 1/\sqrt{24 + \Delta t} \tag{8.29}$$

Values of P vs f(t) are tabulated in Table 10–1 and are plotted in Fig. 8–17.

The early points define a straight line extrapolating to a lower shutin pressure (3,851.0 psi) than indicated on the Horner plot, and the last points define a concave-upward curve. This behavior, combined with the late points defining a straight line on the Horner plot, signifies that the late flow is radial rather than spherical. If the late flow were spherical, the final points on the spherical plot would define a straight line, extrapolating to the proper shutin pressure, and the Horner plot would be a concave-downward curve extrapolating to a lower pressure.

The slope of the straight-line portion of the spherical plot is the appropriate value for m in Eq. 8.27. For this case $m = -104$ psi-$\sqrt{sec}$, as indicated on Fig. 8–17. With the following values of other parameters

---

*For a single rate test, $T_2 = 0$ and the equation simplifies to

$$f(t) = 1/\sqrt{\Delta t} - 1/\sqrt{T + \Delta t}$$

where T is the flowing time and Δt the time after flow.

$\phi = 0.10$
$\mu = 1.0$ cp
$C = 10^{-5}$ psi^{-1} for a mixture of water and oil

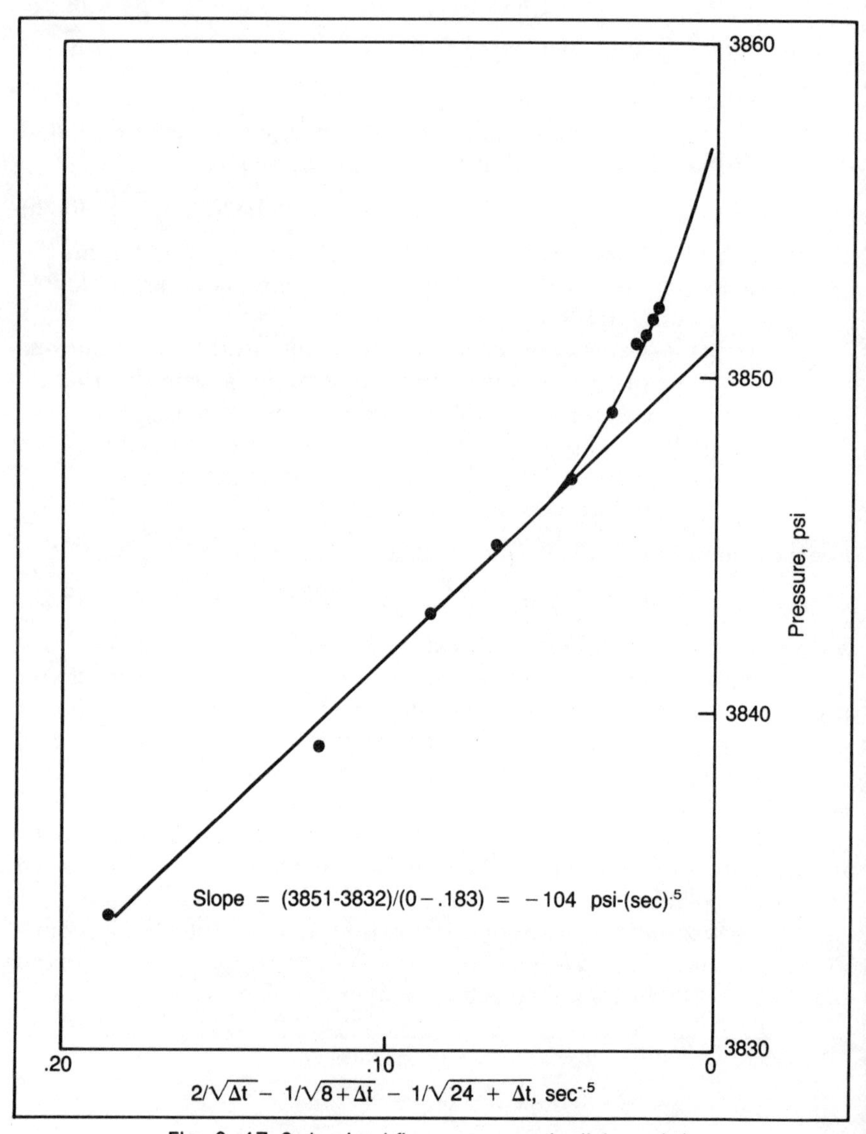

Fig. 8–17 Spherical flow pressure buildup plot

Eq. 8.27 gives

$$k_s = (1{,}856)(1.0)(0.83/104)^{2/3}(0.10 \times 10^{-5})^{1/3} = 0.74 \text{ md}$$

The spherical permeability, k_s, is actually a combination of the radial permeability, k_r, and vertical permeability, k_z, in accordance with the relation

$$k_s = (k_r^{\,2} \cdot k_z)^{1/3} \tag{8.30}$$

Typically, k_z is significantly lower than k_r by a factor from 2 to 10 due to grain orienting on deposition. We shall assume an anisotropy factor, k_z/k_r, equal to 0.5.

Estimation of Formation Thickness

An estimate can now be made of the thickness, h, of the formation sampled. It is based on the difference in pressures obtained by extrapolating the straight lines on the Horner and spherical plots to $\Delta t = \infty$ on the time axis. Assuming the probe is in the center of the bed, its thickness is given by

$$h = 0.039\,\{VA/[4\pi \cdot (P_h - P_s) \cdot \phi \cdot C]\}^{1/3} \tag{8.31}$$

where

V = volume of fluid taken into the tool, cc
A = anisotropy ratio, k_z/k_r
P_h = extrapolated pressure on the Horner plot, psi
P_s = extrapolated pressure on the spherical plot, psi
ϕ = porosity
C = compressibility of reservoir fluid, psi^{-1}

For the example, $V = 20$, $A = 0.5$, $P_h = 3{,}857$, $P_s = 3{,}851.0$, $\phi = 0.10$, and $C = 10^{-5}$, which gives

$$h = 1.9 \text{ ft}$$

Recalculating the radial permeability with this value of h leads to

$$k_r = 0.84 \text{ md}$$

For comparison

$$k_s = 0.75 \text{ md}$$
$$k \text{ (drawdown)} = 0.9 \text{ md}$$

All of these values are consistent, which gives some confidence that they are close to the true value.

With a gauge of 1 psi resolution, the maximum permeability determinable from pretest buildup is approximately 10 md (corresponding to a minimum measurable slope of 10 psi/cycle on the Horner plot). The range is extended to 100 md with a gauge of 0.1 psi resolution. To reach beyond that limit requires that permeability be measured on sample buildup.

Advantages of Measuring Permeability After Sampling

It is highly desirable to measure buildup permeability following fluid sampling for several reasons.

1) Flow rates (q) can be 10–100 times greater into the sample chamber than into the pretest chambers, which extends the permeability range in direct proportion, as indicated by Eqs. 8.26 and 8.27.
2) Viscosities and compressibilities to be used in the permeability equations can be determined from analysis of the fluid sample at the surface. The viscosity of a mixed sample is

$$\mu_m = (V_g \cdot \mu_g) + (V_o \cdot \mu_o) + (V_w \cdot \mu_w) \quad (8.32)$$

where V_g, V_o, and V_w are the volumetric fractions of gas, oil, and water reconverted to reservoir temperature and pressure; μ_g, μ_o, and μ_w are their viscosities at reservoir conditions.

Similarly, mixture compressibility is

$$C_m = (V_g \cdot C_g + V_o \cdot C_o) + (V_w \cdot C_w) \quad (8.33)$$

where C_g, C_o, and C_w are gas, oil, and water compressibilities at reservoir conditions. Charts are available giving viscosities and compressibilities of gas, oil, and water at reservoir temperatures and pressures.[20]

The quantity of gas in the sample can have a large effect on the parameters, decreasing mixture viscosity and increasing mixture compressibility. The sample may not be fully representative of the fluid investigated in the buildup, but it is the best estimate that can be made.
3) The volume of rock investigated by the buildup is proportional to the quantity of fluid withdrawn, V(cc), which is far greater for sampling than for pretesting. The radius of investigation is given by

$$r_i = 0.020\,[V/(4\pi \cdot \delta p \cdot \phi \cdot C]^{1/3} \quad (8.34)$$

where δp is the gauge resolution in psi. With $\delta p = 1$, the radius investigated in a formation of average porosity and compressibility ($\phi = 0.20$, $C = 10^{-5}$ psi^{-1}) is 2 ft for pretest buildup ($V = 20$ cc) and 15 ft for buildup following 2¾-gal sampling (10,000 cc). The latter radius extends well beyond any invasion effects.

It is important when sampling to close the seal valve to the sample chamber purposely either before or at the time the pressure starts to build up from flowing pressure and to mark this time on the log as cessation of flow. Otherwise, afterflow obscures the start of buildup and may make much of the buildup plot unreliable, especially if the chamber contains gas. The flow rate, q, to be used in the buildup equations is the total volume of fluid recovered, in cc, converted to reservoir conditions and divided by the total sampling time.

General Comments

There has been a good deal of skepticism concerning the reliability of pretest permeabilities because of the small amount of fluid withdrawn, the uncertainty in formation thickness sampled, and the inaccuracy in buildup measurement. However, matters are improving. Tools are being introduced with pressure resolution increased by a factor of 10 (0.1 psi vs 1 psi) and with higher pretest flow rates and volumes. Horner and spherical plots can now be computer generated at the wellsite. Thickness can be estimated from the dual plots. All of these improvements should facilitate more accurate and rapid evaluation.

Bear in mind that a single measurement is insufficient to characterize the average permeability of a zone because permeability can vary drastically foot by foot. At least one measurement every 2 ft in a hydrocarbon-bearing zone of interest is desirable.

A technique that has been used for productivity prediction is to compute permeability from the Formation Tester and insert it, with log-derived porosity, into Fig. 8–6a (or an equivalent) to derive the irreducible water saturation, S_{wi}.[21] This, along with actual S_w from logs, is then entered into charts such as Fig. 8–2 to predict water cut. The method is appealing, but the inaccuracies are such that it can only be successful if carefully calibrated against prior production in a given field.

SAMPLING AND SAMPLE ANALYSIS

Fluid sampling with the RFT is accomplished by opening one of the seal valves (Fig. 8–8) and allowing reservoir fluid to flow into the selected chamber.[22] As the fluid enters the upper part of the chamber, it forces a floating piston to expel water from the lower part, through a choke or several chokes in series, and into a ballast chamber below that is initially filled with air at atmospheric pressure. The size of the choke is chosen to restrict formation flow rate, if necessary, to prevent breakdown of soft formations with high permeabilities.

Prior to sampling, an estimate of the flow rate and flowing pressure to be expected can be obtained from Fig. 8–18. Shutin reservoir pressure, P_r, is marked on the vertical scale, and a value q_{max} is marked on the horizontal scale where

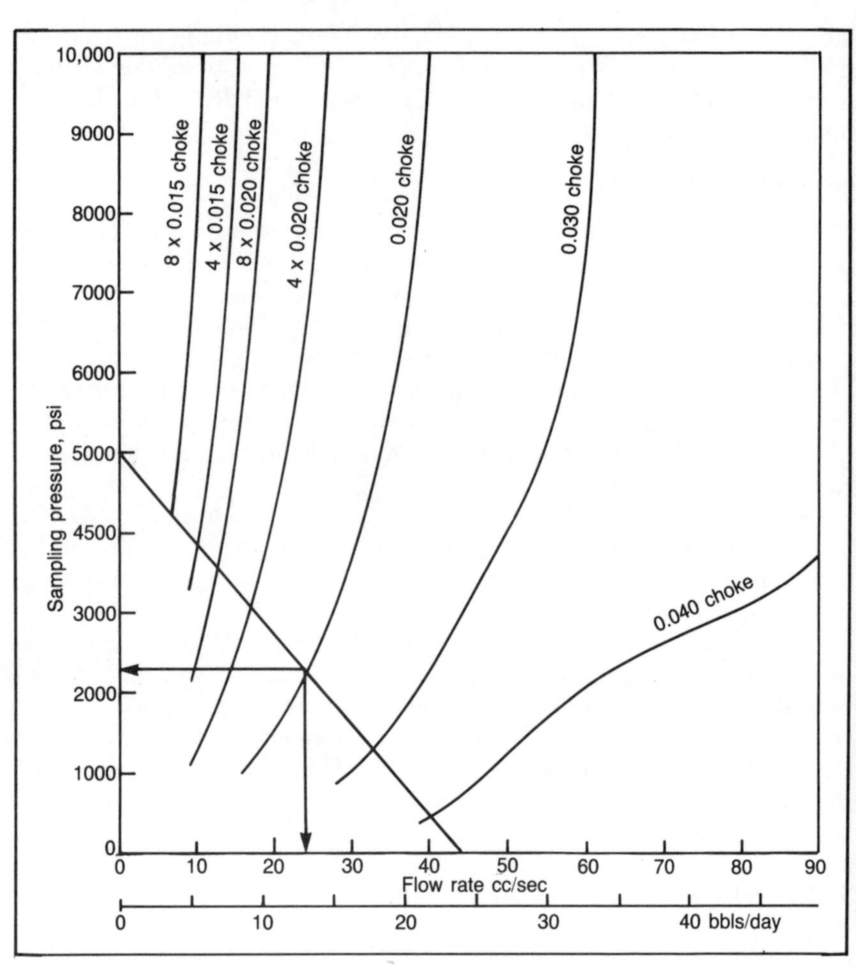

Fig. 8–18 Estimation of flow rate during sampling (courtesy Schlumberger)

$$q_{max} = k \cdot P_r/2{,}820 \text{ cc/sec} \tag{8.35}$$

where k is the drawdown permeability in md. The two points are joined by a straight line, and the intersection of this line with the appropriate choke line gives the anticipated flow rate and flowing pressure. The example shown for P_r = 5,000 psi, k = 25 md, and 0.020 in. choke indicates a flow rate of 24 cc/sec and a flowing pressure of 2,300 psi. Knowing the sample chamber volume (1 gal = 3,785 cc), the fill-up time of the chamber can then be estimated.

If maximum flow rate from the formation is desired, then an air cushion instead of a water cushion can be used below the sample chamber and the flow will not be restricted. Sampling pressure will drop essentially to zero and flow rate will be approximately q_{max}, given by Eq. 8.35.

When the tester tool is brought back to the surface, it is stopped at the rotary table and a portable separator of the type shown in Fig. 8–19 is

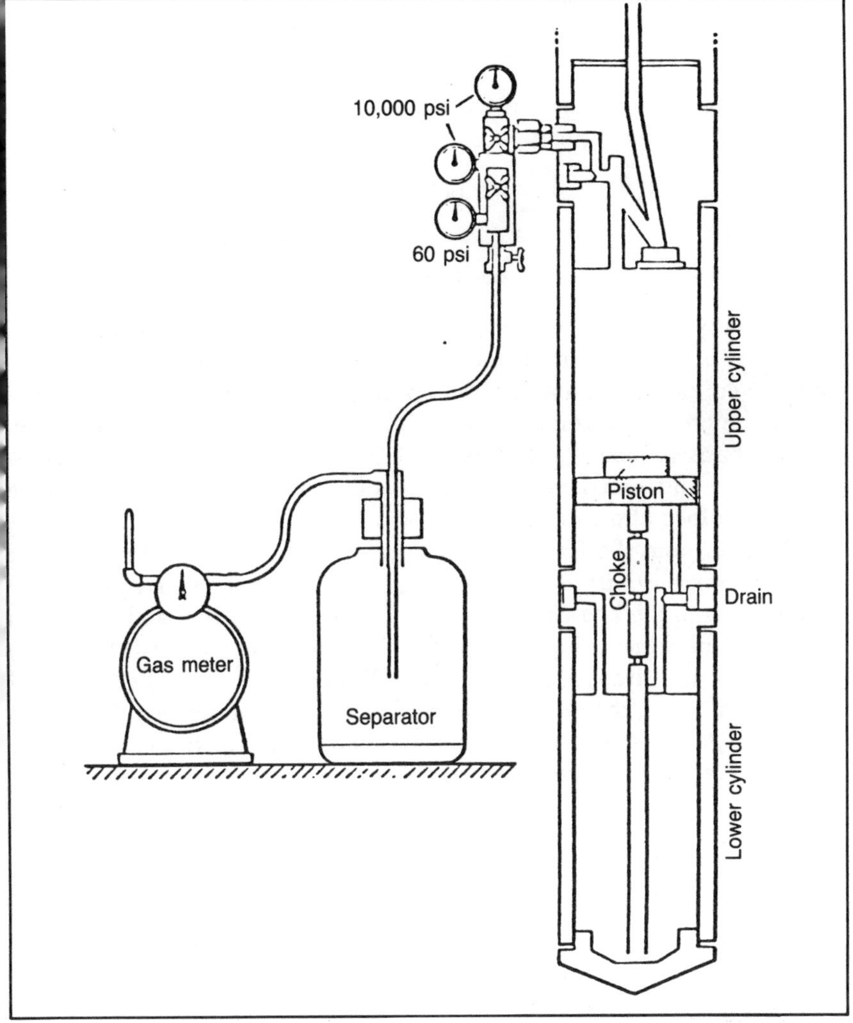

Fig. 8–19 Portable separator attached to sample chamber (courtesy Schlumberger)

attached. This consists of a pressure regulator, a calibrated plastic bottle, and a gas meter. When the valve is opened, the fluid sample is forced out of the chamber through the separator. Liquid drops out in the bottle and the gas passes on through the gas meter, which registers the amount in cubic feet at surface conditions. The liquid in the bottle is allowed to segregate into water on bottom and oil on top, and the amount of each is read from a graduated scale on the side of the bottle in cc.

Calculating GOR and Water Cut

From the recovered amounts the gas-oil ratio (GOR) in cu ft/bbl and the water cut (WC) can be computed. These are given by

$$\text{GOR} = 159{,}000 \times (\text{cu ft of gas})/(\text{cc of oil}) \tag{8.36}$$

$$\text{WC} = \frac{\text{cc of formation water}}{\text{cc of formation water} + \text{cc of oil}} \tag{8.37}$$

Normally the recovered water will contain a substantial amount of mud filtrate. It is assumed that if many samples were withdrawn at the same point, eventually the quantity of mud filtrate would drop to zero but the *relative* amounts of formation water, oil, and gas would remain the same. On this basis (which is debatable) the amount of mud filtrate must be determined and subtracted from the total water recovery to obtain the quantity of formation water.

To determine the fraction of formation water, the logging engineer measures the resistivity, R_{rf}, of the recovered water in a resistivity cell. This is compared to the known resistivities of mud filtrate, R_{mf}, and formation water, R_w, all values being converted to the same temperature. If $R_{rf} = R_{mf}$, the water is all filtrate; if $R_{rf} = R_w$, it is all formation water. For the usual intermediate case, the ratios R_{mf}/R_{rf} and R_{mf}/R_w are entered on the two axes of Fig. 8–20 and the percentage of formation water is read from the numbers on the curved lines. As an example, assume the following recovery

Gas: 0.51 cu ft
Oil: 350 cc
Water: 3,200 cc

Resistivities

R_{rf} = 0.54 ohm-m at 75°F
R_{mf} = 0.76 ohm-m at 75°F
R_w = 0.080 ohm-m at 200°F (formation temperature)

The gas-oil ratio is directly

$$\text{GOR} = \frac{0.51}{350} \times 159{,}000 = 230 \text{ cu ft/bbl}$$

Converting R_w to 75° F

$$R_w = 0.080\,(200 + 6.7)/(75 + 6.7) = 0.20$$

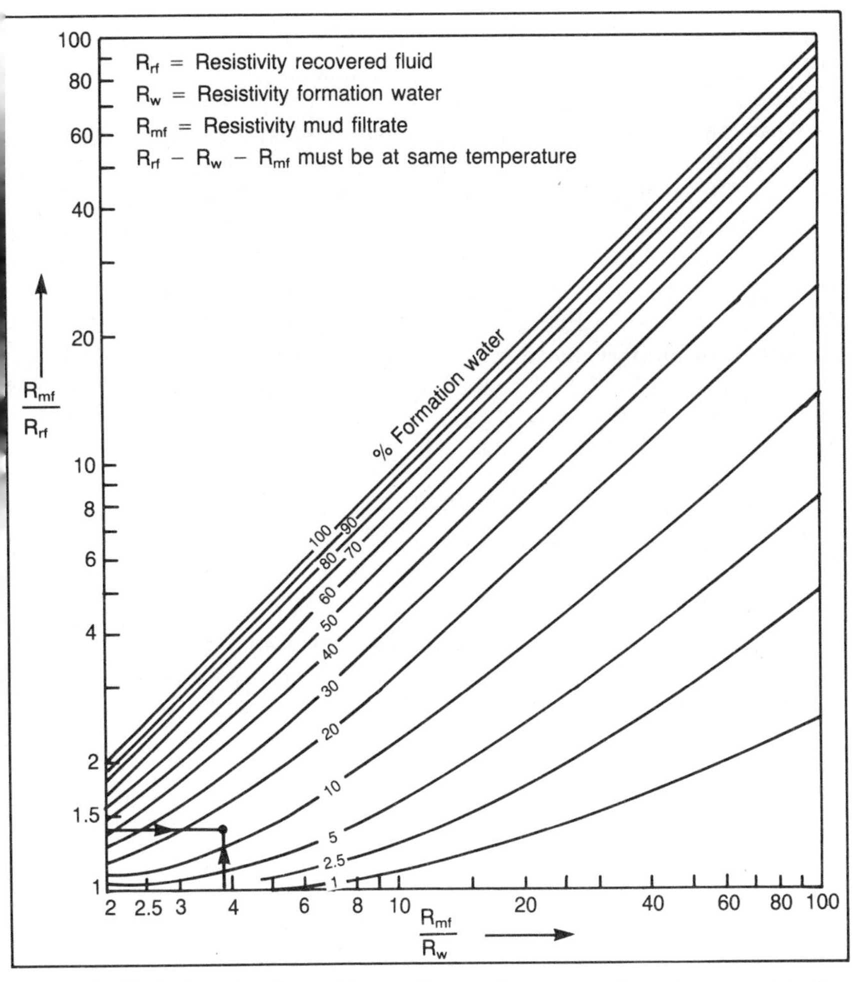

Fig. 8–20 Determination of formation water percentage in recovered sample (courtesy Schlumberger)

Ratios are

$$R_{mf}/R_{rf} = 0.76/0.54 = 1.41$$
$$R_{mf}/R_{w} = 0.76/0.20 = 3.8$$

Entering Fig. 8–20 with these values, as illustrated, shows 15% formation water. Thus, the quantity of formation water is (0.15 × 3,200) or 480 cc. Hence

$$WC = 480/(480 + 350) = 0.58$$

Fig. 8–20 is appropriate for fresh muds where R_{mf} is at least $2R_w$. The greater the R_{mf}/R_w ratio, the better the accuracy. It is not applicable in salt muds where $R_{mf} \leq R_w$.

Cases have arisen where the recovered water resistivity, R_{rf}, has been higher than the R_{mf} of the circulated mud sample, which should not be possible if $R_{mf} > R_w$. The only conclusion is that the circulated sample is not representative of the filtrate in the formation at the tested level. Perhaps the mud characteristics were changed significantly after the tested interval was drilled. In any case, when this happens predictions will be unduly optimistic, i.e., zero or low values of water cut. Therefore, it is advisable when possible to use an R_{mf} value obtained from the logs. That is, determine R_w by applying Archie's Eq. 2.8 (with $S_w = 1$) in a clean water sand near the tested zone, read the SSP in the same sand, and work backward through the SP charts, Figs. 3–7 and 3–8, to obtain R_{mf}.

Rules of Thumb on Sample Interpretation

The foregoing is the general interpretation procedure. A few rules of thumb are the following:

Large recoveries: In high-porosity, high-permeability intervals where invasion is normally shallow, the tested interval will produce oil and/or gas if more than 10 cu ft of gas or more than 1,000 cc of oil are recovered using a 2¾-gal chamber. If less than 600 cc of formation water are recovered, production is usually water free.

Small recoveries: With deep invasion the recovered water may be all mud filtrate. In this case even a small show of oil or gas, a fraction of a cubic foot of gas, or a fluorescent film of oil may indicate a productive zone. If some oil or gas is recovered and if the fraction of formation water is less than 10% of the total water recovered, the tested interval will produce gas or oil with little water cut. If the fraction is 10–25%, there will be some water cut; if greater than 25%, water cut will be excessive. By and large, however, prediction of hydrocarbon production with small hydrocarbon recoveries is reliable only with a good deal of local experience.

APPLICATIONS OF PRESSURE MEASUREMENTS

Perhaps the most important feature of the formation tester is its ability to measure reservoir pressures at multiple levels much more quickly and less expensively than by any other method. Pressure measurements are the best means of determining communication between zones in the same well or in different wells. In essence they define reservoir continuity.[23,24,25]

It is virtually certain that permeable zones which correlate between wells communicate hydraulically if their pressures are exactly the same when adjustment is made for any difference in elevation between the measuring points and there is no flow in the zones. If adjusted pressures are significantly different, then a pinchout, fault, or other type of permeability block exists between the wells. Such information is needed if production is to be optimized.

Pressure Gauge Characteristics

The usefulness of pressure measurements is dependent to a large extent on gauge resolution, repeatability, and absolute accuracy. Resolution is the smallest pressure increment that can be reliably recorded. Repeatability is the degree to which a pressure gauge will repeat a measurement after being exposed to interim changes in temperatures and pressure. It is primarily a function of the temperature stability and cycling behavior of the gauge and associated circuits. Generally, it is considerably poorer than the resolution. Absolute accuracy is, as implied, the nearness of a pressure reading to the true pressure. It can be no better than the repeatability and may be much worse.

In measuring pressure drawdown and buildup at a given level, the resolution is important. The recording is made in seconds or minutes, and the gauge hardly has time to drift. For measurement of pressure differences at a number of closely spaced levels in a given well to establish pressure gradients, repeatability is the key factor. The measurements may take a few hours, and each one subjects the gauge to the pretest pressure transients so that cycling stability is important. For a comparison of pressure measurements in one well with those in another, the absolute accuracy is paramount inasmuch as different gauges may be used in different wells.

The standard gauge in Formation Testers is a strain gauge. Typical values for a 10,000-psi unit along with its associated circuits are

Resolution: ± 1 psi in most existing tools
± 0.1 psi in recent RFT units

Repeatability: ± 5 psi

Absolute accuracy: ±25 psi with normal field calibration
±15 psi with special field calibration

As already illustrated, a 1-psi resolution is barely adequate for drawdown and buildup measurements. The 0.1-psi resolution of newer tools is highly desirable and should lead to significant improvement in permeability determination.

Pressure gradients in permeable zones of a given well vary from approximately 0.46 psi/ft for salt water to 0.40 psi/ft for light oil to 0.10 psi/ft for gas. Over a 20-ft interval the pressure differential from top to bottom would be 9.2 psi for water, 8 psi for oil, and 2 psi for gas. Distinguishing between water and oil is therefore difficult with the standard strain gauge of 1-psi resolution. It really demands a gauge of 0.1-psi resolution and also demands that the stability of the gauge under these circumstances be far better than 5 psi. Fortunately, one can monitor repeatability by observing the hydrostatic pressure of the mud between each measurement. Correction can be made for slow drift in response, provided the mud level in the well remains absolutely constant and solid particles in the mud are not settling out. In any case the present strain gauge is pushed to its limit in gradient measurements.

Strain gauge measurements of pressures in adjacent wells are adequate to conclude that communication does not exist between two zones whose pressures differ by more than about 50 psi but are inadequate to conclude that communication does exist if pressures differ by less than 50 psi. For producing reservoirs this may cause no problem since pressure differentials may be hundreds or thousands of psi apart in correlatable intervals that do not communicate. In virgin reservoirs, however, much better absolute accuracy is required.

The High-Accuracy Gauge

For high-accuracy applications a quartz-crystal gauge (also called the Hewlett-Packard or HP gauge) can be run with the RFT. Its characteristics are resolution of ±0.01 psi, repeatability of ±0.4 psi, and absolute accuracy of ±0.5 psi.

This gauge has amazing precision—10 to 100 times better than that of a strain gauge. It would be in universal use except for two factors. One is that it is expensive and fragile. The other is that it has a settling time of about 10 min. That is, each time a pressure change occurs, the gauge requires 10 min or so to stabilize. This makes it useless for short-term pressure buildup measurements. However, it is ideal for intrawell pressure gradient or interwell absolute pressure recording. The prime use of this gauge, however (not necessarily in the RFT), is for interference tests where a flow rate change is made in a given well and the corresponding pressure change is observed in an adjacent well. Such tests, which require extreme resolution and stability in pressure measurement, are very valuable in establishing transmissibility and storage capacity between the wells of a reservoir.

Single Well Pressure Applications

Overpressuring: Formations may be highly overpressured to the extent that pore pressure may be twice the hydrostatic head of salt water at that depth. While this must be anticipated prior to drilling, once a wildcat is drilled it is desirable to obtain a pressure profile in the well to optimize both production and offset drilling. For this purpose great accuracy is not required. The strain gauge is adequate.

Gas-oil-water-contacts: Pressure measurements can assist logs in defining such contacts. Fig. 8–21 shows RFT pressure measurements vs depth in a Middle East well. Three separate pressure gradients are apparent: an upper one of 0.17 psi/ft corresponding to a fluid density of 0.39 g/cc, a middle one of 0.29 psi/ft corresponding to 0.67 g/cc, and a lower one of 0.47 psi/ft corresponding to 1.09g/cc. Thus, a water-oil contact is fixed at point A and an oil-condensate contact is fixed at point B. In this case, which is unusual, the three zones are long enough (about 100 ft each) that pressure gradients can be determined to sufficient accuracy with the standard strain gauge.

In measuring pressure gradients, several points must be kept in mind. First, the gradient of the mobile fluid phase behind the invaded zone is being measured. The irreducible water does not influence the reading because it is held in place by surface tension, nor does the presence of mud filtrate influence the pressure—the invaded zone is too thin. Second, in a low-permeability zone reservoir pressure must be obtained by extrapolation of the Horner or spherical plot to obtain an accurate value. Third, in extremely low permeabilities the extrapolated pressure can be abnormally high because of supercharging.

Supercharging is temporary overpressuring caused by invasion. The differential pressure between mud and reservoir at a given level, which is responsible for invasion, is divided between the mud cake and the formation according to their respective flow resistances. Normally the mud cake has much higher resistance than the formation so that the excess pressure created in the latter is negligible. However, once formation permeability drops below about 1 md, this condition no longer holds and the formation next to the borehole may become supercharged. The magnitude is strongly dependent on downhole water-loss characterisitics of the mud and is not predictable. However, it is believed to be on the order of l/k psi where k is the formation permeability in md.

Point 14 of Fig. 8–21 is a case in point. Permeability at that level was found to be 0.024 md, which would lead to an estimate of 40 psi of supercharging. Points such as these must be ignored on a gradient plot.

Pressure depletion: In a reservoir producing from several zones, pressure measurements in infill or observation wells provide an opportunity to monitor the depletion of the individual intervals. Fig. 8–22 shows a case where zones depleted at quite different rates. Such information is needed to plan stimulation of nonproducing zones or to prevent depleted intervals from becoming thief zones.

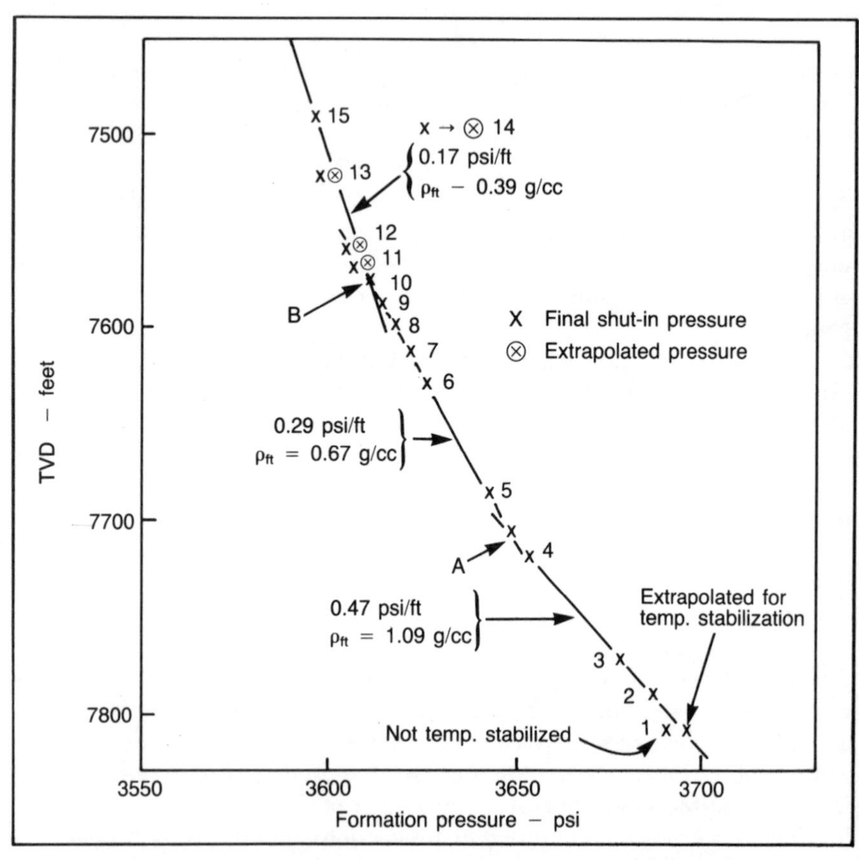

Fig. 8–21 Pressure gradients in a Middle East reservoir (courtesy Schlumberger, © SPE-AIME)

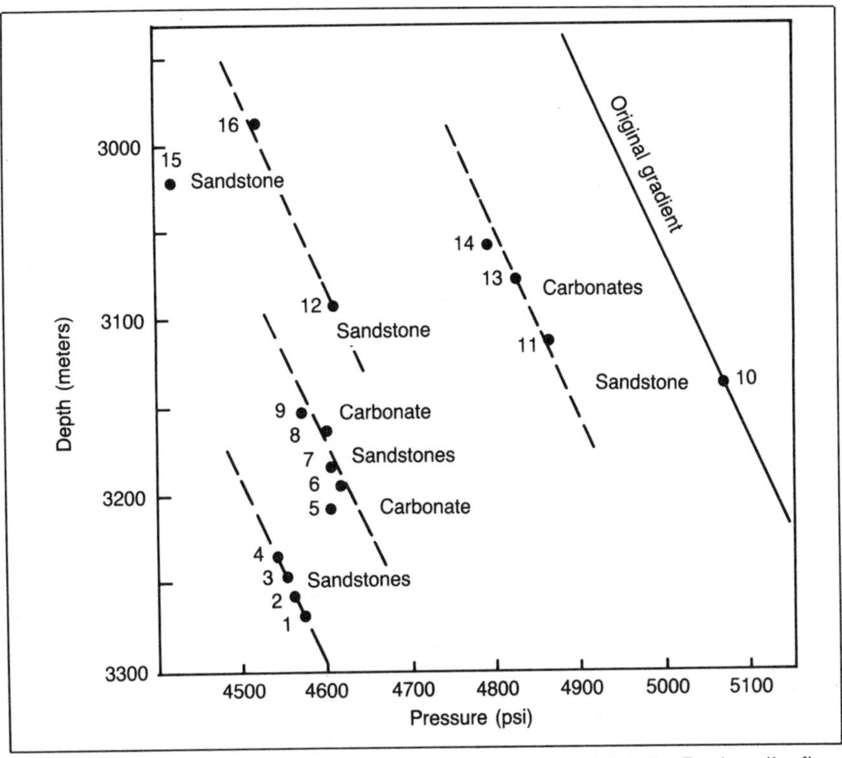

Fig. 8–22 Pressures in various producing zones of a Middle East well after several years of production (courtesy Schlumberger, © SPE-AIME)

Multiple Well Pressure Applications

Ideally when exploration wells are drilled to define reservoir limits, pressure measurements should be made to determine whether the productive horizons defined by the logs actually communicate between wells. Accurate measurements of static pressures are good; interference tests are even better. This practice is slowly advancing.

It becomes imperative to know interwell communication when secondary recovery, usually waterflooding, is undertaken. Many cases are documented where unsuspected permeability blocks caused poor performance of the flood. Fig. 8–23 illustrates pressure recordings over a 600-ft producing interval in 12 infill wells drilled into an old reservoir in Colorado preparatory to additional water flooding. The separation between vertical lines represents 4,000 psi. Great lateral as well as vertical variations in

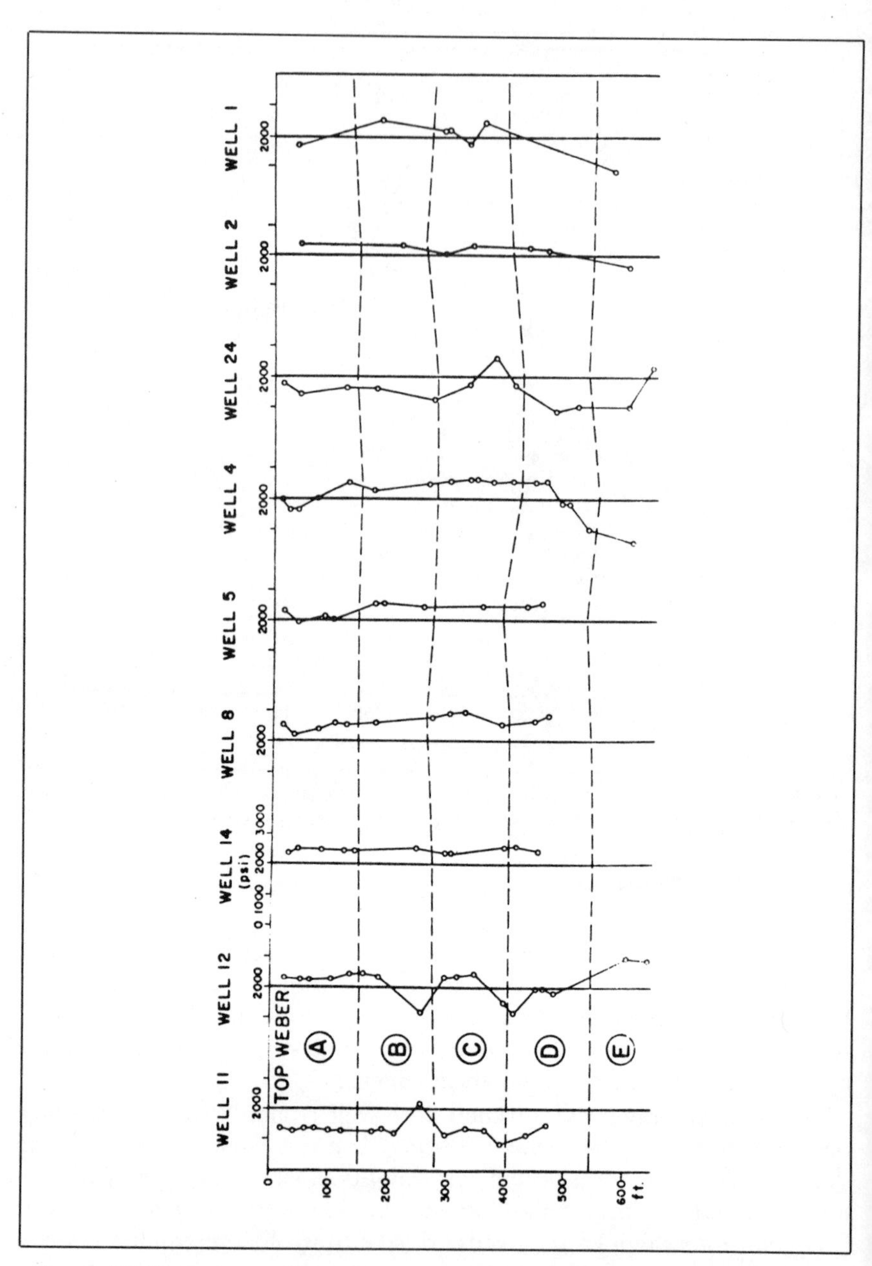

Fig. 8–23 Pressure profiles across a producing field, Colorado (courtesy Schlumberger, © SPE-AIME)

pressure are observed, indicative of many permeability barriers. For example, the lowermost sand in Zone B differs in pressure by about 2,000 psi between adjacent wells 11 and 12, even though they are at the same stratigraphic level. These zones are certainly isolated from each other and even from zones above and below in the same well.

Forthcoming Tool Improvements

Other applications of formation tester pressure measurements are continually coming to light.[26] Each advancement in tool accuracy and versatility opens new avenues. Major improvements in the RFT just being introduced are

1. Pressure resolution of 0.1 psi and absolute accuracy of ± 10 psi (with monthly deadweight tester calibration).
2. Fast flow option with a probe approximately five times larger in area than the previous one. This will increase flow rates accordingly during pretest and sampling and will extend the measurable permeability range to higher values. The appropriate drawdown constant (Eq. 8.24) for this probe is 2,395 instead of 5,650.
3. Large pretest volume, 1,000 cc instead of 10 or 20 cc. This is necessary with the fast-flow option but will also increase the radius of investigation on pretest buildup from about 2 to 7 ft (Eq. 8.34). However, meaningful fluid samples cannot be retrieved as the pretest chamber is exhausted into the sample chamber after each pretest.

These and other mechanical improvements should significantly improve the usefulness of the RFT and similar tools.

SUMMARY

The Multiple Formation Tester is a tool that can be run in open hole, positioned at a desired level, and actuated to take a sample of reservoir fluid (2–12 gal) and to measure reservoir fluid pressure.

Only two fluid samples can be brought back to the surface for analysis. However, any number of pressure measurements can be made at any level desired. Information obtainable is

1) From analysis of fluid samples at the surface
   - An estimate of the gas-oil ratio expected on production
   - An estimate of the water cut expected on production

The sample normally contains a substantial amount of mud filtrate. The fraction of recovered water that is virgin water is obtained by comparing the resistivity of the recovered sample, R_{rf}, with that of the mud filtrate, R_{mf}, and the formation water, R_w.

2) From downhole pressure measurements
   - Reservoir permeability
   - Reservoir pressure
   - Location of gas-oil-water contacts
   - Communication between zones

The MFT is the best wire-line method of obtaining permeability. It also gives valuable information on reservoir connectivity for secondary recovery.

REFERENCES

[1]T.C. Frick, *Petroleum Production Handbook* (New York: McGraw-Hill, 1962), chapter 23.

[2]Frick, chapter 25, Ibid.

[3]J. Pelissier-Combescure, D. Pollock, and M. Wittman, "Application of Repeat Formation Tester Pressure Measurements in the Middle East," *SPE 7775*, Manama Bahrain (March 1979).

[4]J.P. Jones, "Production Engineering and Reservoir Mechanics," *OGJ* (1945), pp. 45–46.

[5]R.L. Morris and W.P. Biggs, "Using Log-Derived Values of Water Saturation and Porosity," *SPWLA Logging Symposium Transactions* (1967).

[6]D.K. Keelan, "Core Analysis for Aid in Reservoir Description," *J. Pet. Tech.* (November 1982), pp. 2483–2491.

[7]Frick, chapter 23, Ibid.

[8]R.P. Murphy and W.W. Owens, "A New Approach for Low Resistivity Sand Log Analysis," *SPE 3569*, New Orleans (October 1971).

[9]Morris and Biggs, Ibid.

[10]A. Timur, "An Investigation of Permeability, Porosity and Residual Water Saturation Relationships," *SPWLA Logging Symposium Transactions* (June 1968).

[11]Schlumberger, *Log Interpretation Charts* (1979), p. 83.

[12]G.R. Coates and J.L. Dumanoir, "A New Approach to Improved Log-Derived Permeability," *SPWLA Logging Symposium Transactions* (May 1973).

[13]L.L. Raymer, "Elevation and Hydrocarbon Density Correction for Log-Derived Permeability Relationships," *The Log Analyst* (May–June 1981), pp.3–7.

[14]A. Brown and S. Husseini, "Permeability from Well Logs, Shaybah Field, Saudi Arabia," *SPWLA Logging Symposium Transactions* (June 1977).

[15]J.J. Smolen and L.R. Litsey, "Formation Evaluation Using Wireline Formation Tester Pressure Data," *SPE 6822* (Denver: October 1977).

[16]D.K. Sethi, W.C. Vercellino, and W.H. Fertl, "The Formation Multi-Tester — Its Basic Principles and Practical Field Applications," *SPWLA Logging Symposium Transactions* (July 1980).

[17]J.H. Moran and E.E. Finklea, "Theoretical Analysis of Pressure Phenomena Associated with the Wireline Formation Tester," *J. Pet. Tech.* (August 1962), pp. 899–908.

[18]Pelissier-Combescure et al., Ibid.

[19]G. Stewart and M. Wittmann, "Interpretation of the Pressure Response of the Repeat Formation Tester," *SPE 8362* (Las Vegas: September 1979).

[20]Frick, Ibid.

[21]G.W. Lockwood and D.E. Cannon, "Production Forecasting," *SPE 9946*, Bakersfield (March 1981).

[22]Schlumberger, "Formation Tester Interpretation Methods and Charts," paper C-11721.

[23]Smolen and Litsey, Ibid.

[24]Pelissier-Combescure et al., Ibid.

[25]G. Stewart and L. Ayestaran, "The Interpretation of Vertical Pressure Gradients Measured at Observation Wells in Developed Reservoirs," *SPE 11132*, New Orleans (September 1982).

[26]G. Stewart, J. Wittmann, and T. Van Golf-Racht, "The Application of the Repeat Formation Tester to the Analysis of Naturally Fractured Reservoirs," *SPE 10181* (San Antonio: October 1981).

Chapter 9

WELLSITE COMPUTED LOGS

Wellsite computed logs are often called "quick-look" logs. They are curves designed to show directly the presence of hydrocarbons. Their principal use is to pick intervals for testing or sidewall sampling and to facilitate a quick decision on whether to set pipe and complete the well or to plug and abandon it.

Before wellsite computers there were four types of quick-look logs in use

- the apparent water resistivity, R_{wa}, log
- the porosity overlay
- the resistivity or F overlay
- the SP or R_{xo}/R_t overlay

These curves were recorded using simple analog computing circuits. No effort was made to correct logs for borehole, invasion, or shale effects.

The introduction of digital computers to surface logging units has brought much greater versatility to the generation of wellsite-computed curves. The basic logs are recorded digitally on tape as they are being run. Following the logging runs, they can be played back, edited, corrected, merged, and computed, and outputs can be presented in a variety of formats. Consequently, there has been a veritable explosion in the variety of wellsite-computed curves in recent years. Programs are available for shaly formation interpretation, true vertical depth adjustment, matrix identification, interpretation of formation tests and dip conputation, to name only the more important categories.[1,2] A fully interpreted log can be available to the well operator two hours after logging is completed.

The four quick-look logs listed above are described in this section, followed by an outline of the general-purpose wellsite interpretation programs. The Schlumberger version of the latter is called Cyberlook and the Dresser-Atlas version is called Prolog.

THE R_{wa} LOG

The R_{wa} log is a quick-scan curve used for direct indication of hydrocarbon-bearing zones, determination of water resistivity, R_w, and estimation of water saturation, S_w.[3] R_{wa} stands for apparent water resistivity. The R_{wa} curve is obtained by continuously calculating R_w on the assumption that $S_w = 1$ everywhere. The Archie equation is

$$S_w = c\sqrt{R_w/R_t}/\phi \quad (9.1)$$

Placing $S_w = 1$, replacing R_w by R_{wa}, and solving for the latter gives

$$R_{wa} = (\phi^2 \cdot R_t)/c^2 \quad (9.2)$$

Typically, the R_{wa} curve is recorded when either Sonic or Density-Neutron logs are run simultaneously with the Induction log. Porosity is obtained from the porosity logs and R_t is obtained from the deep Induction log. R_{wa} is continuously calculated using Eq. 9.2. It is usually presented in Track 1.

If all permeable formations are clean and water bearing, the R_{wa} log is a slowly varying curve whose values are the actual R_w values in those formations. Thus, R_w in a given interval can usually be read directly from the R_{wa} curve.

If, however, a hydrocarbon-bearing zone is penetrated, R_t will be greater than it would be if the zone were water bearing and the R_{wa} value derived from Eq. 9.2 will be higher than R_w. Water saturation can be directly estimated from the R_{wa}/R_w ratio. Eq. 9.1 may be rewritten

$$S_w = c\sqrt{R_w/(\phi^2 R_t)} \quad (9.3)$$

Replacing $(\phi^2 R_t)$ by its value as given by Eq. 9.2, namely $c^2 R_{wa}$, gives

$$S_w = \sqrt{R_w/R_{wa}} \quad (9.4)$$

Therefore

If R_{wa}		then S_w	
	$= R_w$		$= 100\%$
	$= 2\,R_w$		$= 70\%$
	$= 3\,R_w$		$= 58\%$
	$= 4\,R_w$		$= 50\%$

Consequently, the R_{wa} curve can be scaled directly in water saturation, at least over short intervals (a few hundred feet), where R_w is constant. Alternatively, a rule of thumb is that if $R_{wa} > 3R_w$, commercial production is possible and a thorough analysis of the zone should be made.

Fig. 9–1 shows an R_{wa} curve recorded with the ISF-Sonic combination. R_w would be read as 0.05 ohm-m in the water-bearing zone (2). The critical R_{wa} value is therefore 0.15 ohm-m. The hydrocarbon-bearing section clearly shows at levels 7 and 6. At level 7, $R_{wa} = 0.5$ so that $S_w = \sqrt{0.05/0.5}$ or 0.32. The thin zones above and below those levels, with $R_{wa} > 0.15$, are not hydrocarbon bearing. They result from the fact that the Sonic responds to thinner zones than the Induction (2.5 ft vs 5 ft). Averaging the Sonic over 5 ft would eliminate these spurious indications.

When the Sonic is the porosity tool, Schlumberger uses the relation given by Eq. 5.18 to convert travel time to porosity. This avoids the need for a compaction correction in unconsolidated sands. When the Density-Neutron combination is used for porosity, the arithmetic average of the ϕ_D and ϕ_N values is used. In both cases a certain amount of shaliness is tolerable since both Sonic and Density-Neutron tools will give porosities somewhat high but the shaliness will generally cause R_t to be too low. The effects tend to compensate, and R_w, R_{wa}, and S_w values are reasonably good. This is not true if Density only is used for porosity. S_w values will be abnormally high.

The R_{wa} curve works well in medium to high porosity zones where matrix type is fairly constant (Gulf Coast). It is not appropriate for low-porosity regions where matrix is quite variable because porosity values will be too matrix dependent. This is especially true if Sonic or Density is used for porosity, less so if Density-Neutron is used.

POROSITY OVERLAYS

A porosity overlay is a presentation of two porosity curves on the same track with the same scale. Presentation will be in Track 3 if the curves are run simultaneously with resistivity or in Tracks 2 and 3 if they are run separately.

Porosity overlays provide a quick visual indication of matrix, shale, and hydrocarbon effects as well as porosity. The three possible porosity combinations are Density-Neutron, Density-Sonic, and Sonic-Neutron. Of these the Density-Neutron overlay is the most useful for several reasons. First, the true porosity is very close to the average of Density and Neutron readings, regardless of the lithology (shaliness and anhydrite or gypsum excluded). Second, lithology is often evident from the relative positions of the two curves. Third, gas-bearing zones stand out.

Fig. 9–2 illustrates the first two points. The beds at 14,553, 14,564, 14,650 ft and similar zones are clearly anhydrites with zero porosity. The levels at 14,504 and 14,680 ft are dolomites (somewhat radioactive, as

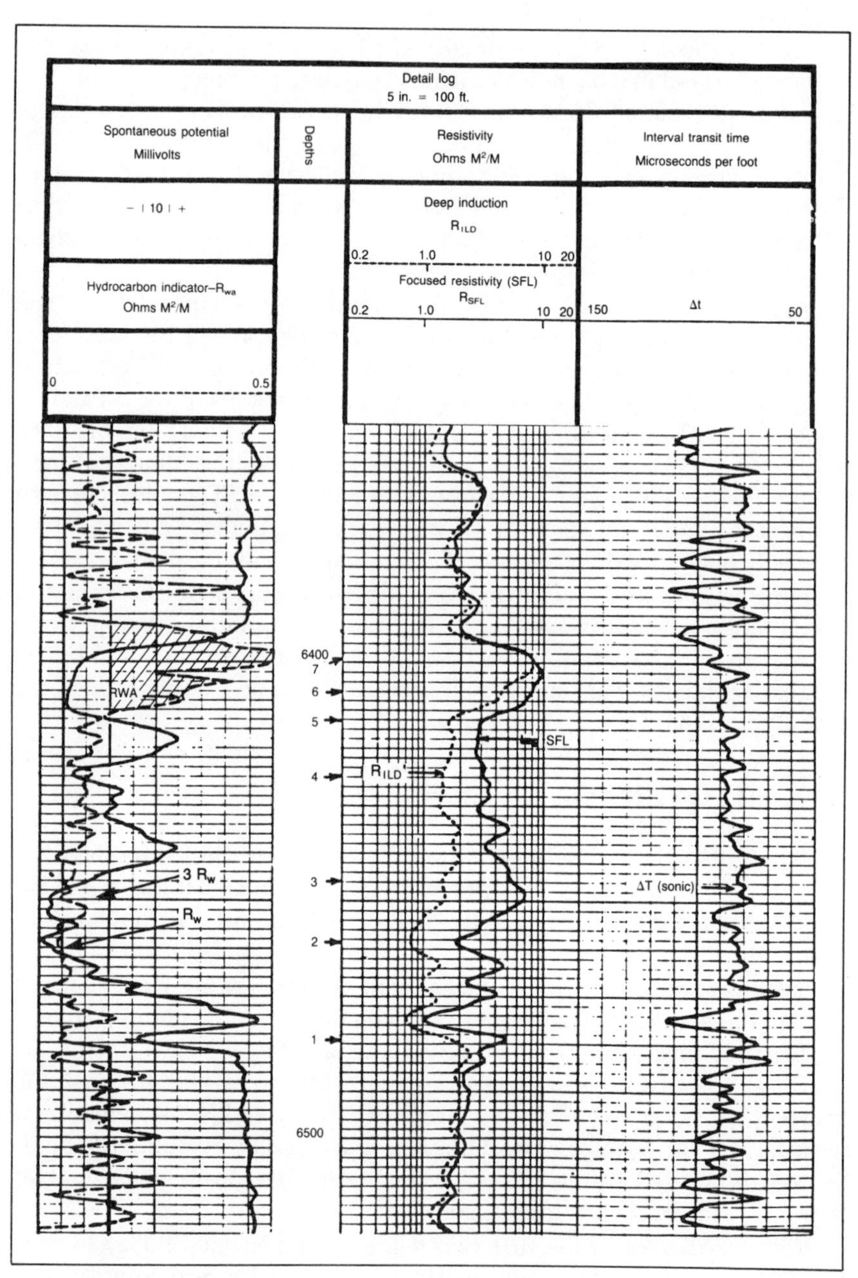

Fig. 9–1 A hydrocarbon zone indicated by the R_{wa} curve (courtesy Schlumberger)

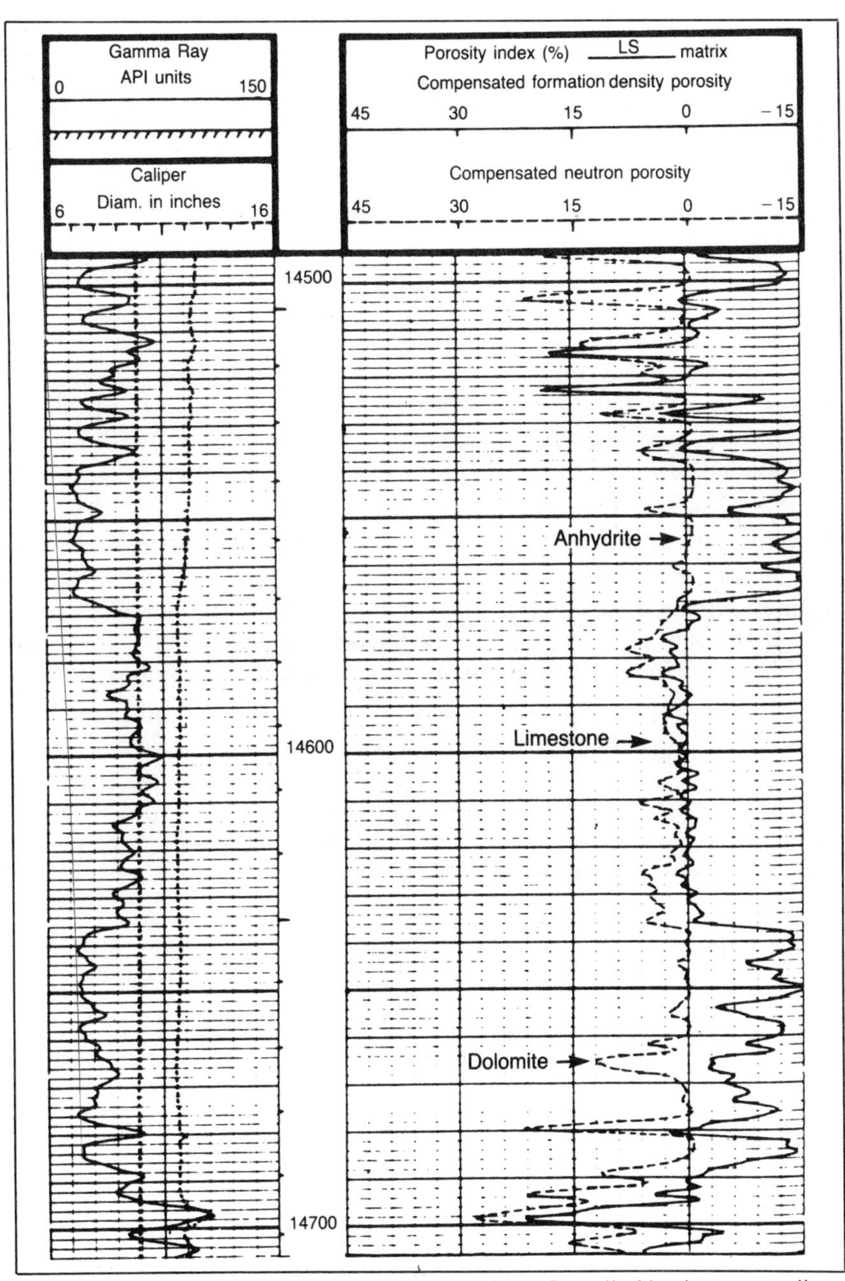

Fig. 9–2 Lithology changes indicated by a Density-Neutron porosity overlay (courtesy Schlumberger)

indicated by the GR) of porosity close to 10%. The region from 14,570–14,636 ft is primarily limestone, slightly dolomitized, with an average porosity of about 3%. These interpretations come initially from the crossplot chart, Fig. 5–21, but with a little experience the porosity overlay can be read directly.

The presence of gas, as indicated by crossover of the curves with the Neutron reading lower porosity than the Density, is strikingly shown in Fig. 9–3. All zones where the 2-ft depth lines are interrupted are gas bearing. They vary from 30 ft to 2 ft in thickness. Thin shale stringers separating the permeable zones, such as at 3,579 ft, stand out very well also because the Neutron reverts to reading much higher porosity than the Density, as it normally does in shale. As described in Chapter 5, the presense of shale or dolomite tends to suppress gas indication; the presence of sand may give false gas indication (when ϕ_D and ϕ_N are recorded on limestone matrix).

The other two porosity overlays, the Sonic-Neutron and the Density-Sonic, are not commonly used. Porosity evaluation is more dependent on the lithology choice — very much so for Density-Sonic. Gas indication is less distinct, primarily because gas effect on the sonic curve is unpredictable. In fact, both Sonic and Density logs tend to give abnormally high porosities in gas-filled reservoirs, so this combination is not at all useful for gas detection.

Sporadic attempts have been made to overlay water-filled porosity, derived from resistivity by inserting R_w and R_t values in Eq. 9.1 ($S_w = 1$) with total porosity obtained from Sonic or Density-Neutron. Separation between the curves is directly read as hydrocarbon-filled porosity. The concept is intriguing especially since it can be extended to show movable hydrocarbon porosity if an R_{xo} curve is run. However, the technique has never caught on.

RESISTIVITY OVERLAYS

The resistivity overlay is a recording of a calculated R_o log along with the deep Induction (or Laterolog) R_t log. The two curves should track wherever formations are water bearing and should depart only in hydrocarbon-bearing zones.

The R_o log is calculated using porosity from the porosity log that is run, the logging engineer's estimate of R_w (perhaps from an R_{wa} curve), and the assumption that $S_w = 1$ everywhere. From Eq. 9.1

$$R_o = (c/\phi)^2 \cdot R_w \qquad (9.5)$$

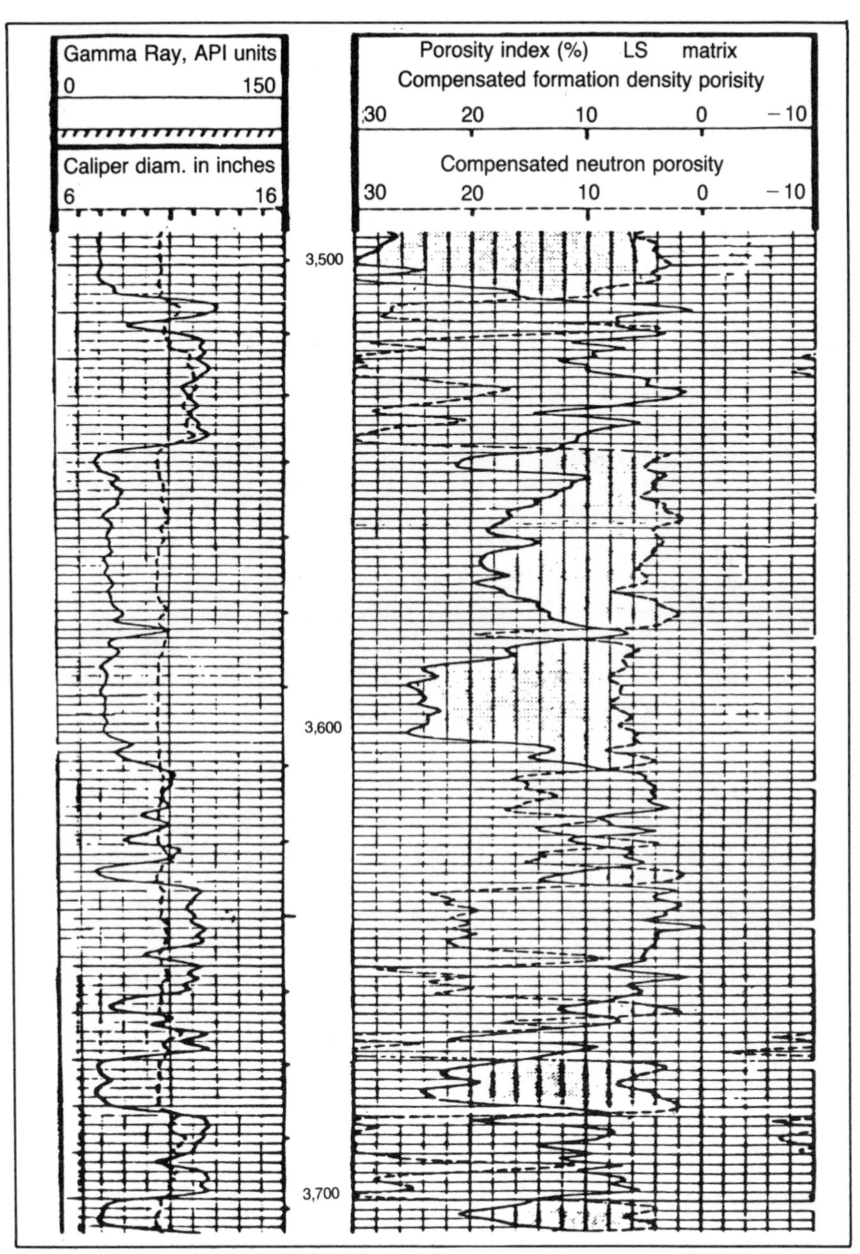

Fig. 9–3 Gas-bearing zones indicated by Neutron-Density crossover (courtesy Schlumberger)

Thus, R_o is generated continuously by obtaining ϕ from the porosity log and combining it with R_w as indicated.

A comparison of calculated R_o in a given zone with measured R_t immediately gives water saturation

$$S_w = \sqrt{R_o/R_t} \tag{9.6}$$

$$\begin{aligned} \text{If } R_t &= R_o & \text{then } S_w &= 100\% \\ &= 2\,R_o & &= 70\% \\ &= 3\,R_o & &= 58\% \\ &= 4\,R_o & &= 50\% \end{aligned}$$

Commercial production is possible if $R_t > 3R_o$. On the logarithmic scale used for resistivity, a given R_t/R_o ratio represents a constant separation between the curves regardless of the absolute resistivities. Consequently, any zone where the curve separation is greater than the distance between the 1.0 and 3.0 ohm-m lines should be carefully analyzed for possible production. Fig. 9–4 is an example of the R_o-R_t overlay when R_o is calculated using Density porosity. It shows sand A is hydrocarbon-bearing. The R_t/R_o ratio averages about 9/1.4 or 6.4, so $S_w \approx 0.4$. The lower sand B is water bearing; R_o matches R_t.

One form of resistivity overlay is the F-overlay. The F-curve is the formation factor F calculated from the porosity logs

$$F = (c/\phi)^2 \tag{9.7}$$

By definition, $F = R_o/R_w$. Consequently, when the calculated F is recorded on the logarithmic resistivity scale and is shifted to match the measured R_t curve in water-bearing zones (which essentially multiplies F by R_w), the curve becomes an R_o log. Separation between R_o and R_t shows hydrocarbon-bearing zones as described above.

As with R_{wa} curves, resistivity overlays recorded with the Density log as the source of porosity have no automatic compensation for shaliness. Water saturations in shaly sands tend to be pessimistic. Values will be more correct with Density-Neutron or Sonic porosities.

THE SP OVERLAY

The SP overlay, also called the R_{xo}/R_t or SP quick-look, is a comparison of a computed SP curve with the recorded SP log such that departures directly indicate the presence of *movable* hydrocarbons.[4] It is best used in conjunction with R_{wa} or resistivity overlays, which respond to total hydrocarbons.

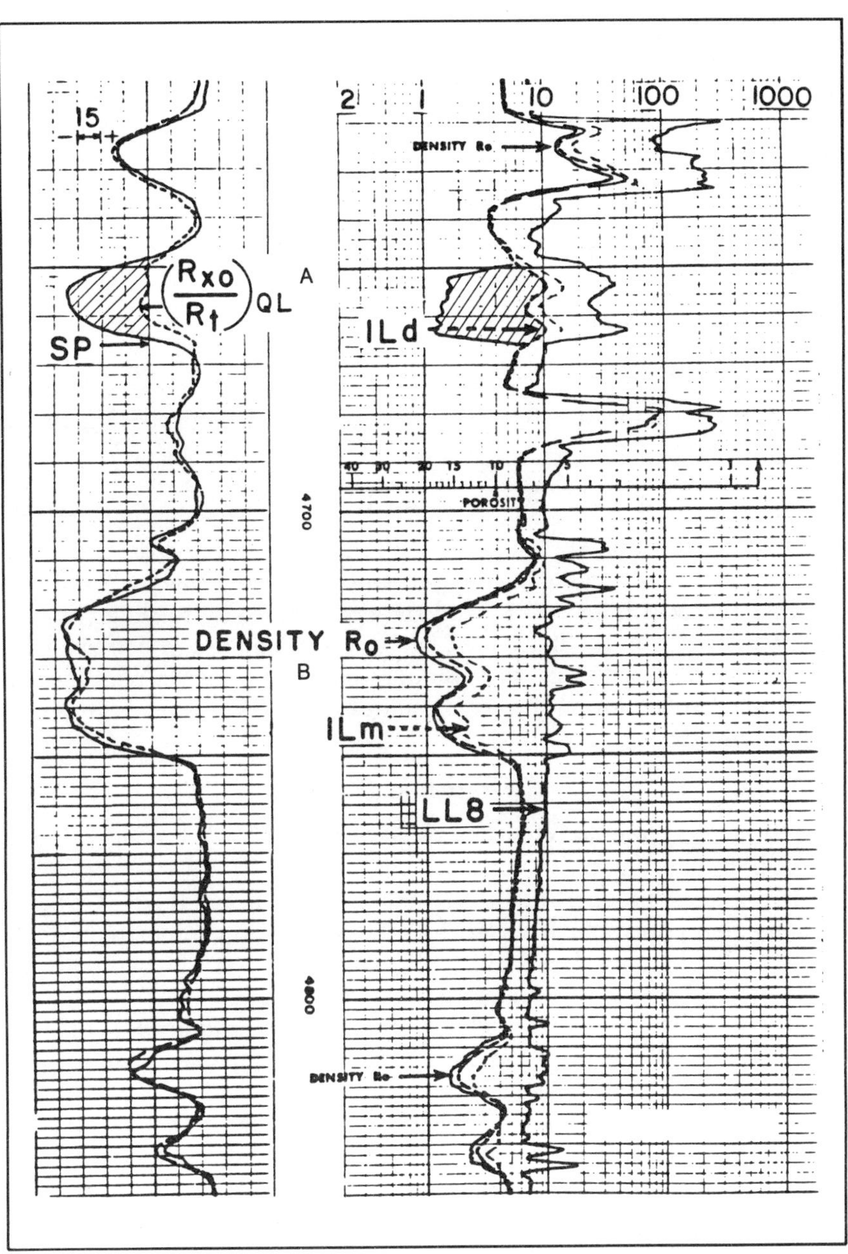

Fig. 9–4 Hydrocarbon presence indicated by resistivity overlay and movability by SP overlay (courtesy Schlumberger and SPWLA)

Movable hydrocarbons are indicated by a difference between undisturbed-zone water saturation, S_w, and flushed-zone saturation, S_{xo}. If $S_w/S_{xo} = 1$, the zone will produce water, regardless of apparent water saturation. If $S_w/S_{xo} < 1$, the zone contains movable hydrocarbons. The comparison is accomplished as follows

$$S_w = c\sqrt{R_w/R_t}/\phi \tag{9.8}$$

$$S_{xo} = c\sqrt{R_{mf}/R_{xo}}/\phi \tag{9.9}$$

Dividing both sides and rearranging

$$S_w/S_{xo} = [(R_{xo}/R_t)/(R_{mf}/R_w)]^{1/2} \tag{9.10}$$

The logarithm of the ratio R_{mf}/R_w is routinely presented in Track 1 as the SP since

$$SP \approx -K\log(R_{mf}/R_w) \tag{9.11}$$

The object is to compare log R_{xo}/R_t with the SP. Typically, the Induction tool is run and the ratio R_{xo}/R_t is obtained from the shallow resistivity reading (16″N, LL8, or SFL) and the deep resistivity reading (IL_d). A reasonable value of R_{xo}/R_t can be derived, provided the invasion diameter is between 20 and 100 in. even though the shallow curves read beyond the flushed zone. An on-site, continuous computation is made to give a quick-look SP

$$SP_q = -K\log(R_{xo}/R_t) \tag{9.12}$$

The SP_q curve, compatibly scaled, is recorded in Track 1 along with the normal SP.

In a water zone where $S_w/S_{xo} = 1$, the ratios R_{xo}/R_t and R_{mf}/R_w are equal and the curves converge. In a zone with movable hydrocarbons, $S_w/S_{xo} < 1$. Thus, R_{xo}/R_t becomes less than R_{mf}/R_w and the curves diverge, with the quick-look SP_q becoming smaller than the recorded SP.

Fig. 9–4 includes the SP quick-look curve. The separation between that curve and the SP in Sand A indicates the hydrocarbons can move.

The SP overlay has several advantages relative to R_{wa} and resistivity quick-looks. It indicates movable hydrocarbons, not just the presence of hydrocarbons. It is independent of porosity and lithology; therefore, it is more suited to variable-lithology formations than R_{wa} or resistivity overlays. Shaliness is automatically compensated for since it affects both SP and SP_q in much the same way. Finally, variations in R_w are automatically taken into account.

In spite of these advantages, the SP quick-look has not found widespread acceptance. Some drawbacks are that the method is limited to fresh muds where $R_{mf} > R_w$, that the computation is sensitive to abnormal invasion diameters (< 20 in. or > 100 in.), and that numerical values of S_w/S_{xo} cannot readily be estimated.

THE CYBERLOOK LOG

The Cyberlook log is a Schlumberger quick-look shaly formation analysis.[5] It is based on the Dual-Water model described in Chapter 7. The output consists of a set of computed curves, the most important of which are shale-corrected water saturation and porosity. The program is suitable for both sand and carbonate analysis.

The basic measurements utilized are Dual Induction-SFL (or an equivalent set), Density, Neutron, SP and GR. Optional measurements are Sonic, which is desirable in rough hole situations, and R_{xo}, which allows movable oil computation.

Processing is carried out after the logging runs are completed. First, a Pass One set of curves is optically recorded. The purpose of this is to let the logging engineer select parameters needed for the final Pass Two interpretation run. For Pass One he needs only to specify the following constants: matrix to be used for computing Neutron and Density porosities (sand or lime), mud weight, bit size, pore fluid density, bottom-hole temperature, SP drift, and depth offsets for merging.

In Pass One the basic measurements are merged, if they are obtained on different runs, and environmental corrections are applied. The Gamma Ray is corrected for hole size and mud weight effect, the deep Induction is corrected for invasion, the Neutron is adjusted for temperature and pressure effects, and the Density is corrected for borehole size variations.

Fig. 9–5 shows portions of ISF-Sonic and CNL-FDC logs recorded over an interval of Miocene sands and shales in a South Louisiana well. Fig. 9–6a is the Pass One output for these logs. In Track 3 are the corrected Neutron and Density porosity curves and an apparent total porosity curve, ϕ_{ta}, computed from the Neutron-Density crossplot. Track 1 contains the SP and corrected GR logs and an apparent grain density, ρ_{ga}, curve computed from ϕ_{ta} and bulk density. In Track 2 is the corrected deep Induction log and an apparent fluid resistivity log, R_{fa}, computed from corrected R_t and ϕ_{ta}. The scales are quite different for the R_t and R_{fa} curves; it just happens that the two curves overlap in this case.

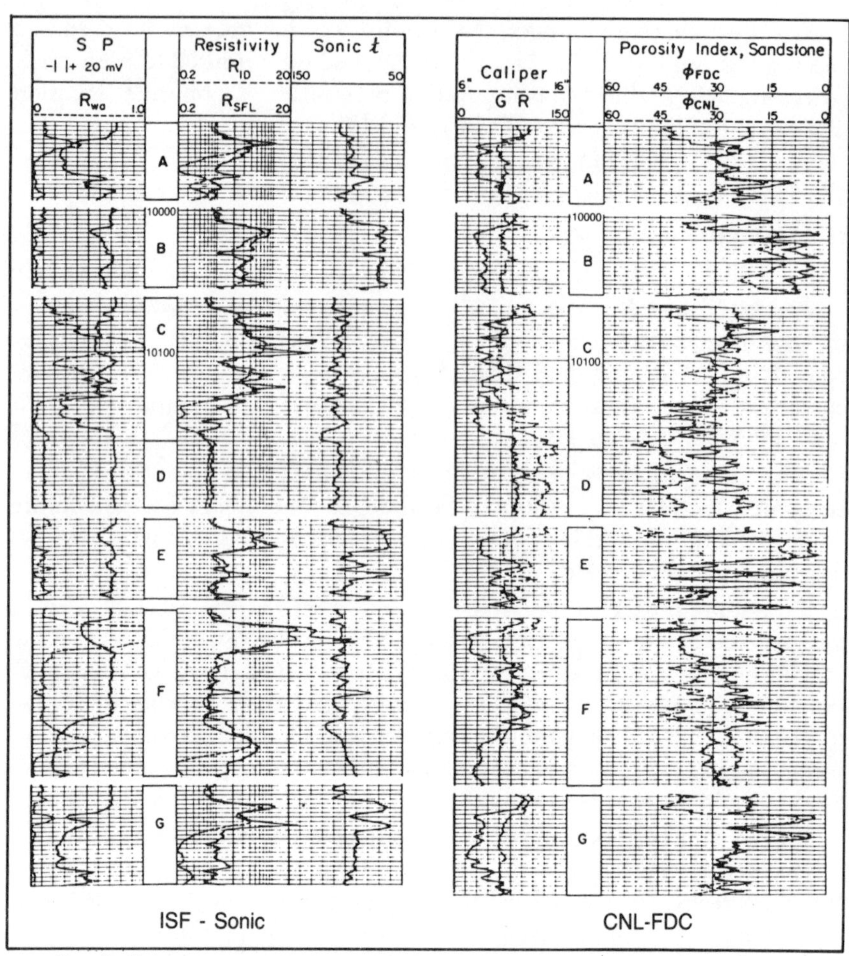

Fig. 9–5 Logs recorded in a portion of a South Louisiana Miocene sequence (courtesy Schlumberger and SPWLA)

The Pass One log is used for zoning the log into intervals, typically those in which matrix type and R_w can be considered constant, and for selecting parameters applicable to each interval. These are (1) the values needed for computation of the shale index in Pass Two, principally the clean-formation and shale levels of the Gamma Ray and SP logs; (2) the parameters needed for porosity computation, namely the Neutron matrix (sand or lime), the matrix and fluid densities, and the maximum porosity allowed (to discriminate against hole washout effects); and (3) the parameters needed for water

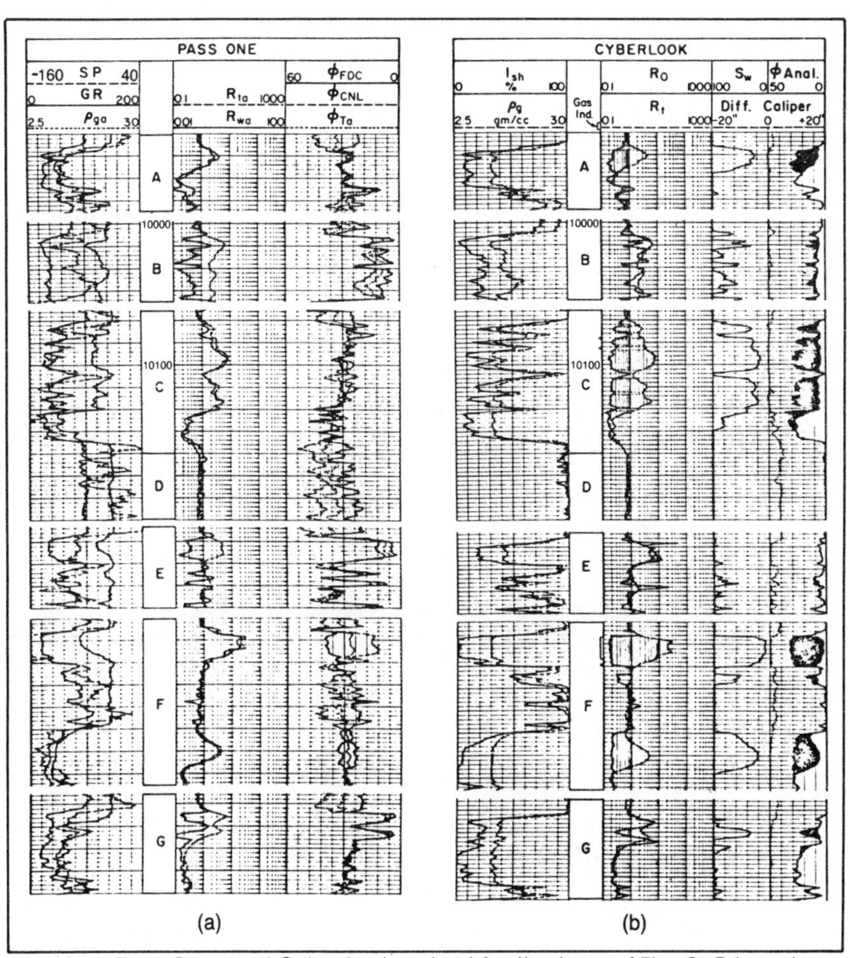

Fig. 9–6 Pass One and Cyberlook output for the logs of Fig. 9–5 (courtesy Schlumberger and SPWLA)

saturation computation, namely, free-water and bound-water resistivities. The former is the R_{fa} reading in a clean, water-bearing zone and the latter is the corresponding value in shales.

The parameter choices for the example zone are as follows:

GR_{sh} = 95 APIU, GR_{cl} = 20 APIU (intervals D and F)
SP_{sh} = −15 mv, SP_{cl} = −120 mv (intervals D and F)
Neutron matrix = sand
Matrix density = 2.65 g/cc; fluid density = 1.0 g/cc (fresh mud)

Maximum porosity = 36%

R_{wf} = 0.018 ohm-m (clean wet sections of F and G)

R_{wb} = 0.09 ohm-m (interval D)

In this case, parameter picking is fairly straightforward. However, in situations where there are no clean formations, where R_w is varying rapidly, or where matrix is not constant, selection is very judgmental. In such instances the experience of the logging engineer in the region of interest is an important factor.

With the selected parameters entered into the computer, the Pass Two run is made and the final interpretation log is produced. Fig. 9–6b is the Cyberlook for the example case. In Track 1 is the shale index, essentially the lowest of the values computed from the GR and SP (and other, if used) shale indicators. In the same track is the computed grain density. This curve reads correctly in clean formations, but the indicated grain densities in shaly formations are abnormally high because the Neutron porosity includes the lattice-bound hydrogen in clay.

In Track 2 is a four-decade logarithmic resistivity overlay. The corrected deep Induction resistivity, R_t, is reproduced as the dashed curve. Overlain on it, as a solid curve, is the wet resistivity, R_o, computed from total porosity, bound-water fraction, and free- and bound-water resistivities. This curve should overlay the R_t curve in all water-bearing zones but should show lower resistivity values in hydrocarbon-bearing zones. The latter are clearly shown in zones A, C, and F as separations between the R_o and R_t curves. If the hydrocarbon is gas, it is so indicated by a bar in the depth column, as appears opposite the upper sand of Zone F.

Water saturation is recorded in the left portion of Track 3. This is calculated using a simplified version of the Dual-Water equations presented in Chapter 7. When an R_{xo} log is run, the flushed zone water saturation, S_{xo}, can also be recorded alongside S_w; the difference between the curves indicates movable oil.

In the middle of Track 3 is a differential caliper curve that represents the difference between actual hole size and bit size. The zero is in the center of the track.

On the right-hand side of Track 3 is the porosity information. The outer envelope of the curve toward the depth track represents effective porosity; the shaded section represents that portion of bulk volume which contains hydrocarbons; and the white section represents that portion containing water.

The quick-look appeal of the computed log is quite evident, with the hydrocarbon-bearing zones standing out in black. In this case all of these intervals are indicated as good pay zones on the R_{wa} curve of the basic ISF-Sonic log. However, in shalier, lower-porosity formations such as the South Texas Wilcox, the quantitative information in the computed log is needed for completion decision.

Questions have been raised about determining bound-water fraction and equations that are used for water saturation. Nevertheless, the answers agree fairly well with those from more sophisticated programs.[6]

THE PROLOG ANALYSIS

Prolog is the Dresser-Atlas computerized wellsite log analysis system.[7] It consists of two separate packages: Wellsite SAND for clastic sediment analysis and Wellsite CRA for carbonate sediment sequences.

The input to Prolog and the final output are similar to those for Cyberlook, and the same operations of data merging and environmental log correction are applied. However, in the Prolog system a first pass or preinterpretation optical log is not made. Instead, the engineer selects a zone to be interpreted and proceeds through a dialogue with the computer in which he specifies the necessary clean and shale values for shale index determination, the matrix values for porosity determination, and the R_w value for water saturation calculation. The computer then calculates the desired output, including shale fraction, effective porosity, and water saturation, and displays the results on the video monitor. If the engineer questions the computed values, he readjusts input parameters and repeats the calculation. Only when he is satisfied with the result is the computed data committed to tape to become part of the final output. This procedure is carried out for each interval into which the well has been zoned.

To help the engineer zone the well and choose correct values of R_w and matrix constants, the Prolog system allows the generation of various types of crossplots at the wellsite, including bulk density vs Neutron porosity and porosity vs resistivity. For water saturation calculation, Wellsite SAND uses the Simandoux shaly-sand equation and Wellsite CRA uses the Fertl shaly formation relation.

Fig. 9–7 shows the Wellsite SAND output. It is a four-track format. In Track 1 are SP and shale fraction. Also shown is integrated borehole volume (VBH) as pips, each one representing 10 cu ft.

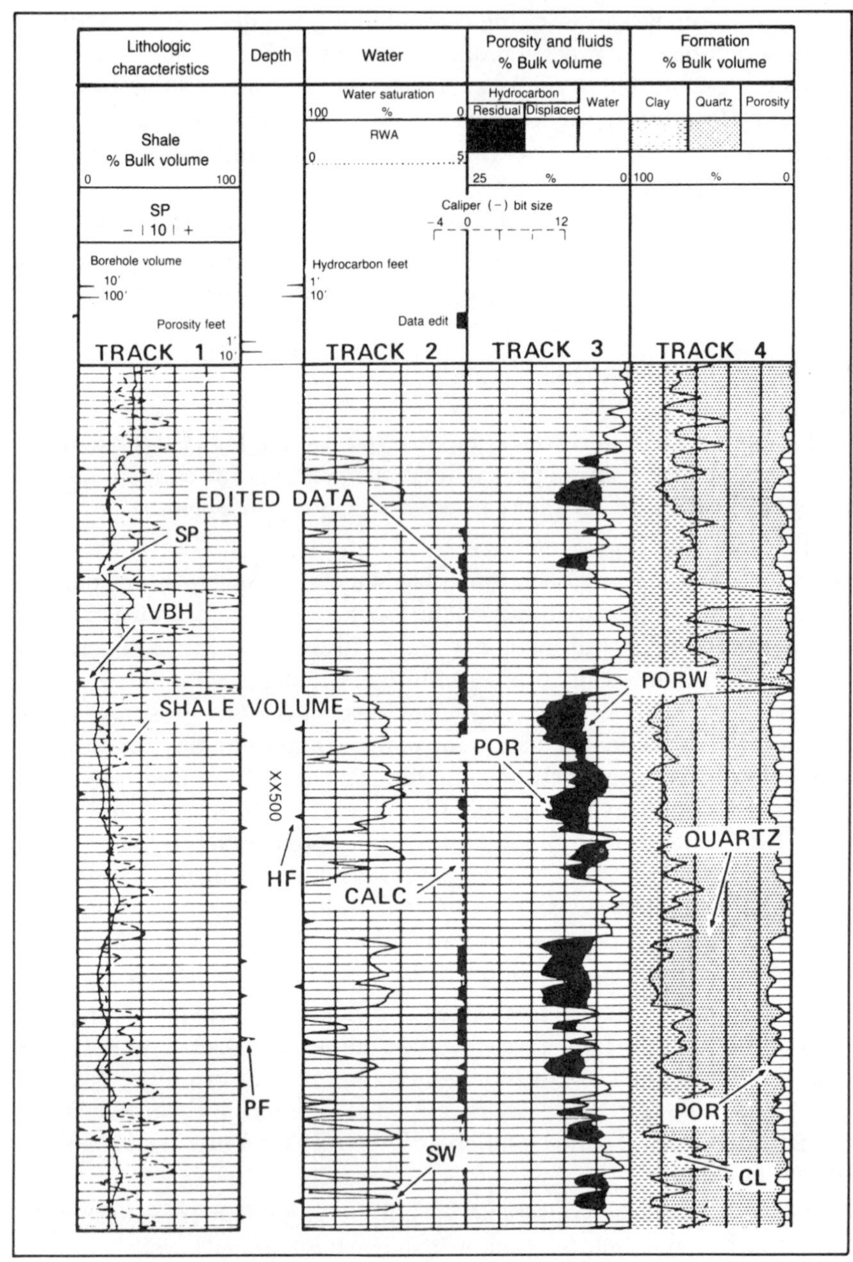

Fig. 9–7 Example of Wellsite SAND (courtesy Dresser, © SPE-AIME)

In the depth track integrated porosity feet (PF) is shown as pips on the left, and integrated hydrocarbon feet (HF) is shown as pips on the right, each representing 1 ft.

Track 2 displays water saturation (SW). The black bars on the right-hand side indicate sections where log data have been corrected for borehole enlargement.

Track 3 shows effective porosity (POR) and its breakdown into water (PORW) and hydrocarbon (blackened). Between Tracks 2 and 3 is the differential caliper (CALC).

Track 4 is the bulk volume analysis, showing the porosity, sand, and clay components.

The Wellsite CRA output format is very similar. A GR curve is included in Track 1. If an R_{xo} curve is run, Track 3 also shows the portion of pore space filled with mud filtrate (PORF). The difference between the PORF and PORW curves represents the bulk volume fraction of oil moved by filtrate invasion. Track 4 shows a bulk volume separated into porosity, shale, and other matrix components, the latter representing all clean rock, of whatever type, present.

OFFICE-COMPUTED LOGS

Wellsite-computed logs such as Cyberlook and Prolog are excellent for the purpose intended but are not meant to give final answers. Where interpretation is difficult, the best accuracy is desired, and the cost is justified, more complete log processing should be carried out at log interpretation centers. The major oil and service companies have such centers in strategic locations around the world. There, the processing is supervised by experienced analysts, interpretation algorithms are more sophisticated, and much more time is allotted to generating histograms, crossplots and intermediate playbacks for zoning and parameter selection.

For reference, office-based open-hole interpretation programs that are used by major oil or service companies and that have been described in the literature are listed below. In each case the water saturation equation used is indicated.

Shell: Lamisi (for laminated or dispersed sand-shale; Waxman-Smits relation)[8]

Schlumberger: Saraband (for shaly sands: modified Simandoux relation)[9,10]

Coriband (for shaly carbonates; Simandoux relation)[11,12]

Volan (replacing the previous two for shaly sands or carbonates; Dual-Water equation)[13]
Global (an adaptive minimum-error program for shaly sands or carbonates; choice of Waxman-Smits or Dual-Water)[14]

Dresser-Atlas: Epilog Sand (for shaly sands; Simandoux relation)[15]
Epilog CRA (for shaly carbonates; Fertl relation)[15]
Digital Shaly Sand Analysis (Waxman-Smits)[16]

Reference should be made to the indicated papers for the details of these programs; they are quite complex. However, the final computed logs are similar in presentation to those of Cyberlook or Prolog.

SUMMARY

The following quick-look curves provide wellsite direct indication of hydrocarbon presence.

1. Apparent Water Resistivity, R_{wa}
   - Requires deep resistivity and porosity input
   - Indicates presence of hydrocarbons
   - Works best in medium-soft rock, constant lithology
   - Compensation for shaliness using ϕ_s or $\phi_{N\text{-}D}$, not ϕ_D
   - Direct reading of R_w
   - Commercial production possible if $R_{wa} > 3\ R_w$

2. Porosity Overlay
   - Requires two porosity curves; Density-Neutron is best
   - Indicates presence of gas by crossover
   - Shale and dolomite suppress crossover, may obscure gas
   - False indication of gas if matrix is sandstone but recording is limestone
   - Lithology variations evident in tight carbonates

3. Resistivity Overlay, R_o vs R_t
   - Requires porosity and R_w input
   - Indicates presence of hydrocarbons
   - Works best in medium-soft rock, constant lithology

- Compensation for shaliness using $\phi_{N\text{-}D}$, not ϕ_D
- Commercial production possible if $R_t > 3\ R_o$

4. SP Overlay
    - Requires only deep and shallow resistivity input
    - Indicates presence of *movable* hydrocarbons
    - Independent of lithology and porosity
    - Inherent compensation for shaliness
    - Best used with R_{wa} or resistivity overlay
    - Restricted to fresh mud ($R_{mf} > R_w$)

5. Cyberlook or Prolog
    - Requires deep Resisitivity, Density, Neutron, and GR (or SP)
    - Provides shale-corrected water saturation and porosity
    - Indicates gas if present
    - Shows moved hydrocarbon if an R_{xo} log is available
    - Excellent for picking zones to test or sidewall core

REFERENCES

[1]Schlumberger, "Data Processing Services Catalog," (June 1981).

[2]Dresser Atlas, "Computer-Processed Open and Cased Hole Log Interpretation Services," (November 1981).

[3]M.P. Tixier, F.M. Eaton, D.R. Tanguy, and W.P. Biggs, "Automatic Log Computation at Wellsite: Formation Analysis Logs," *SPE 987*, Houston (October 1964).

[4]J.L. Dumanoir, J.D. Hall, and J.M. Jones, "R_{xo}/R_t Methods for Wellsite Interpretation," *SPWLA Logging Symposium Transactions* (May 1972).

[5]D.L. Best, J.S. Gardner, and J.L. Dumanoir, "A Computer-Processed Wellsite Log Computation," *SPWLA Logging Symposium Transactions* (June 1978).

[6]J.R.J. Studlick and W.A. Gilchrist, "A Sensitivity Study of Schlumberger's Cyberlook Computation and Its Comparison to Other Evaluation Methods," *SPWLA Logging Symposium Transactions* (June 1981).

[7]E. Frost and W.H. Fertl, "Interactive Digital Wellsite Formation Evaluation — The Prolog System," *SPE 8919* (Dallas: September 1980).

[8]R.A. Haley, "A Synergetic Log and Core Analysis Program Using a Laminated Shale-Dispersed Clay Sandstone Model," *SPWLA Logging Symposium Transactions* (June 1979).

[9]A. Poupon, C. Clavier, J. Dumanoir, R. Gaymard, and A. Misk, "Log Analysis of Sand-Shale Sequences — A Systematic Approach," *J. Pet. Tech.* (July 1980), pp. 867–881.

[10]J.R. Ratliff, W.H. Throop, F.G. Williams, and J.D. Hall, "Applications of the Saraband Sand-Shale Techniques in North America," *SPWLA Logging Symposium Transactions* (May 1971).

[11]A. Poupon, W.R. Hoyle, A.W. Schmidt, "Log Analysis in Formations with Complex Lithologies," *J. Pet. Tech.* (August 1971), pp. 995–1005.

[12]A.W. Schmidt, A.G. Land, J.D. Yunker, and E.C. Kilgore, "Applications of the Coriband Technique to Complex Lithologies," *SPWLA Logging Symposium Transactions* (May 1971).

[13]G.R. Coates, R.P. Schulze, and W.H. Throop, "VOLAN — An Advanced Computational Log Analysis," *SPWLA Logging Symposium Transactions* (July 1982).

[14]C. Mayer and A. Sibbit, "Global, A New Approach to Computer Processed Log Interpretation," *SPE 9341*, (Dallas: September 1980).

[15]Dresser Atlas, Ibid.

[16]N. Ruhovets and W.H. Fertl, "Digital Shaly Sand Analysis Based on Waxman-Smits Model and Log-Derived Clay Typing," *The Log Analyst* (May–June 1982).

Chapter 10

RECOMMENDED LOGGING SUITES

The question of what logs to run in a given situation is often raised. Many factors influence the choice, including type of formations, prior knowledge of the reservoir, hole size and deviation, cost of rig time, and availability of equipment.

To provide a general framework for decisionmaking, recommended logging suites are listed in Tables 10–1 and 10–2. Table 10–1 is for conditions of fresh mud ($R_{mf} > 2R_w$) and medium-to-soft rock ($R_t < 200$ ohms). These are conditions appropriate for Induction logs. Table 10–2 is for conditions of saltier mud ($R_{mf} < 2R_w$) or hard rock ($R_t > 200$ ohms), for borderline cases where the bit size is large (> 10 in.), or where invasion is deep. These are situations appropriate for Laterologs.

In each category three possible logging suites are listed: one for infill wells, one for development wells, and one for exploration wells. For infill wells it is assumed the reservoir has been completely delineated. The main requirement for a new well is to ascertain the exact depth and thickness of the hydrocarbon-producing zone. In this case a minimum logging suite is sufficient, provided lithology is not too variable.

Most wells fall in the development well category. Sufficient logs must be run to distinguish gas from oil, to handle lithology variations, and to cope with shaliness. Basically, this means running the Density-Neutron combination for porosity-lithology determination and gas indication.

For exploration wells, particularly rank wildcats, all information about subsurface structure, lithology, porosity, and hydrocarbon saturation is desired. It is important to correlate seismic sections with Sonic-Density synthetic seismograms for selecting offset locations and to obtain as much pore pressure information as possible to optimize the offset drilling. All of this requires a full suite of applicable logs.

The number of logging runs required in each case is shown in Tables 10–1 and 10–2. For each run the order in which the logging tools are listed represents the order in which they are combined in the logging array, from bottom to top. Resistivity tools are always on bottom with porosity tools above if they are run in combination. Logging arrays vary from 30–80 ft in length. This means the first reading for a given curve (indicated as FR on the bottom of the log) may vary anywhere from about 3–70 ft off bottom. If it is important to obtain all curves in a target formation close to bottom, the well must be drilled about 80 ft beyond the target or the logging tools must be run separately.

TABLE 10-1 RECOMMENDED LOGGING SUITES FOR MEDIUM-TO-SOFT ROCK, FRESH MUD

1. Infill Wells

   One run : Induction/SFL-Sonic

   - 6 curves: SP, IL_d, SFL, t, R_{wa}, Tension
   - Fast (5,000 ft/hr); no pads to stick
   - Adequate in clean formations when lithology known
   - Inadequate in uncompacted, shaly, or variable lithology formations
   - Unreliable gas indication
   - Insufficient for computer processing

2. Development Wells

   One run : Dual Induction/SFL-Density-Neutron-GR

   - 10 curves: SP, GR, IL_d, IL_m, SFL, ϕ_D, ϕ_N, CAL, R_{wa}, Tension
   - Slow (1,800 ft/hr); pads may cause sticking
   - Handles uncompacted, shaly, or variable lithology formations
   - Excellent gas indication
   - Inadequate in very rough hole
   - Sufficient for computer processing at wellsite or office

   Additional runs: Multiple Formation Tester, Dipmeter, or Sample Taker as required

3. Exploration Wells

   Run 1 : Dual Induction/SFL-Sonic

   Run 2 : Litho-Density-Neutron-Microlog-Electromagnetic Propagation-Spectral Gr

   - Run 1: 5,000 ft/hr

     6 curves: SP, IL_d, IL_m, SFL, t, R_{wa}

     Run 2: 1,800 ft/hr

     12 curves: CAL, GR, U, Th, K, P_e, ϕ_D, ϕ_N, ϕ_{EP}, MINV, MNOR, Tension
   - Handles variable lithology, shaliness, or rough hole
   - Excellent gas indication
   - Movable oil determination
   - Depth calibration of seismic
   - Complete data for full range of computer processing

   Run 3 : Dipmeter

   Run 4 : Multiple Formation Tester

   Run 5 : Sidewall Sample Taker

TABLE 10-2 RECOMMENDED LOGGING SUITES FOR HARD ROCK OR SALT MUD

1. Infill Wells
   Run 1 : Dual Laterolog-R_{xo}
   Run 2 : Sonic-Gamma Ray
   - Run 1: 5,000 ft/hr
     5 curves: SP, LL_d, LL_s, MSFL, CAL
   - Run 2: 1,800 ft/hr
     2 curves: GR, t
   - Adequate when lithology well known
   - Inadequate in variable lithology
   - Movable oil determination
   - No gas indication
   - Insufficient for computer processing

2. Development Wells
   Run 1 : Dual Laterolog-R_{xo}
   Run 2 : Litho-Density-Neutron-Spectral GR
   - Run 1: 5,000 ft/hr
     5 curves: SP, LL_d, LL_s, MSFL, CAL
   - Run 2: 1,800 ft/hr
     9 curves: CAL, GR, U, Th, K, P_e, ϕ_D, ϕ_N, Tension
   - Good lithology determination
   - Movable oil determination
   - Excellent gas indication
   - Inadequate in very rough hole
   - Allows log interpretation at wellsite or office

   Additional runs: Multiple Formation Tester, Dipmeter, or Sample Taker as required

3. Exploration Wells
   Run 1 : Dual Laterolog-R_{xo}
   Run 2 : Dual Induction SFL-Sonic
   Run 3 : Litho-Density-Neutron-Spectral GR
   - Run 1: 5,000 ft/hr
     5 curves: SP, LL_d, LL_s, MSFL, CAL
   - Run 2: 5,000 ft/hr
     5 curves: SP, IL_d, IL_s, SFL, t
   - Run 3: 1,800 ft/hr
     9 curves: CAL, GR, U, Th, K, P_e, ϕ_D, ϕ_N, Tension
   - Handles low porosities, variable lithology, rough hole
   - Movable oil and secondary porosity calculable
   - Excellent gas indication
   - Depth calibration of seismic
   - Allows log computation at wellsite and office

   Run 4 : Dipmeter
   Run 5 : Multiple Formation Tester
   Run 6 : Sidewall Sample Taker

FRESH MUD, MEDIUM-TO-SOFT ROCK LOGGING SUITES

These suites are typically run in most of the U.S. and Canada except for the Permian Basin and in parts of South America, the North Sea, Nigeria and the Far East.

Infill Wells

The Induction-Sonic combination has long been a popular minimum logging suite and is adequate when formations are fairly clean and are sufficiently compacted. It can be run fast and provides the very useful R_{wa} quick-look hydrocarbon indication. Fig 10–1 is an example, with SP and computed R_{wa} in Track 1, the resistivity curves in Track 2, and the travel time and tension curves in Track 3. The R_{wa} curve clearly shows one hydrocarbon-bearing zone, about 6 ft thick, at level F. The R_w value, from zones D and G is 0.02. Consequently, S_w at level F, by application of Eq. 9.4, is $\sqrt{0.02/0.18}$ or 0.33. It is not obvious from the logs whether the hydrocarbon is gas or oil, but as an infill well, this would be known from nearby producers. The other high readings on the R_{wa} curve, at zones A, B, C, E, and H are spurious anomalies caused by lignite streaks that show on the Sonic log as excursions to long travel times.

Development Wells

For development wells the Sonic tool is replaced by the Density-Neutron combination. All logs can still be obtained in one run, but logging speed is about three times slower than with Sonic. The advantages of Density-Neutron are gas indication, independence of porosity determination to lithology variations, and ability to apply shale corrections.

Fig. 10–2 is an example of the indicated logging suite. The curves SP, GR, R_{wa} and $\Delta\rho$ are in Track 1; the resistivity curves are in Track 2; and the porosity curves are in Track 3.

The R_{wa} curve shows a single hydrocarbon-bearing interval 10 ft thick at zone C. The R_w is about 0.05 ohm-m (zone D) so water saturation at level C averages $\sqrt{0.05/0.4}$ or 0.35. The hydrocarbon is oil, as indicated by the lack of crossover of the Density-Neutron curves. The high porosity of zone C, 24%, coupled with the fact the sand appears to be very clean—since GR and SP are low and D and N are overlaying—indicates a high permeability for that zone. The high R_{wa} values at zones A, B, and E should be disregarded because they again are caused by lignite streaks.

Exploration Wells

For exploration wells two basic logging runs are required. The Sonic log is needed in addition to the Density-Neutron for seismic depth calibration and as a porosity backup in case of rough hole. If large holes (> 14 in.) and

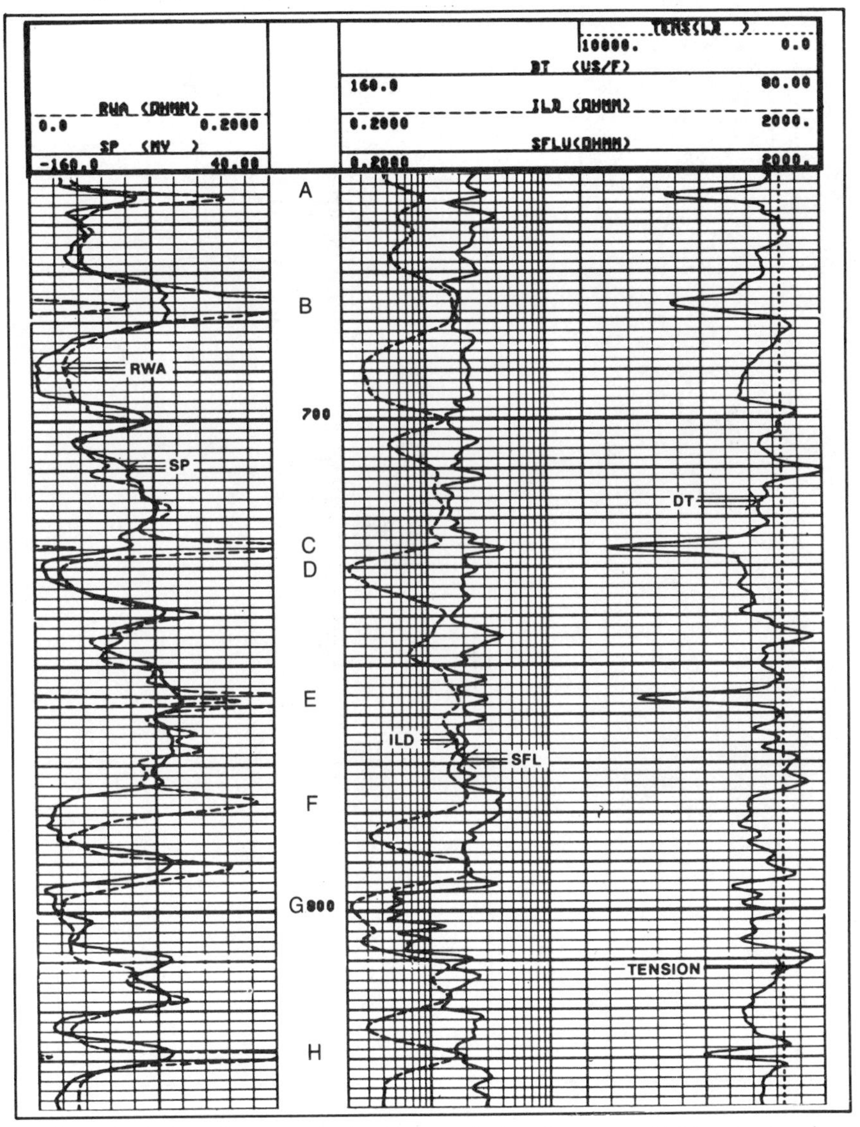

Fig. 10–1 Simultaneous Induction-Sonic log (courtesy Schlumberger)

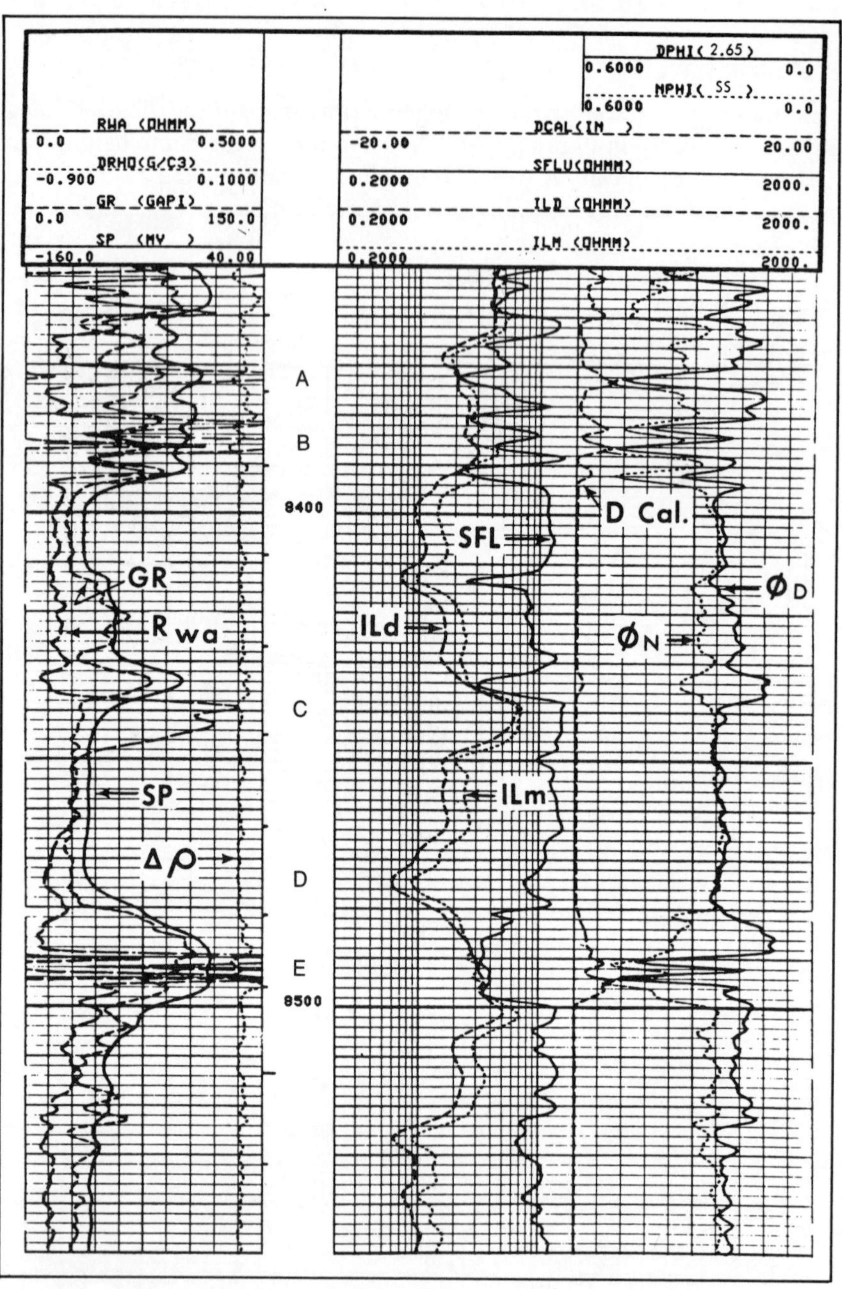

Fig. 10–2 Simultaneous Dual Induction-SFL-Density-Neutron-GR log suite in Gulf Coast sand-shale series (courtesy Schlumberger)

shale alteration are expected, the Long-spaced Sonic should be run in place of the BHC. The Litho-Density log is desirable for lithology identification, particularly in gas-bearing formations. The spectral GR aids in identifying clays and in distinguishing abnormally radioactive sands or dolomites from shaly formations. The EPT-Microlog combination allows estimation of movable oil and pinpointing of permeable zones. If mud cakes greater than ⅜ in. in thickness are expected, the Proximity-Microlog combination should be substituted for the EPT-Microlog. This would require a third basic logging run, which may be desirable even with the EPT-Microlog because of the large number of curves to be recorded when it is run with Litho-Density, Neutron and Spectral GR.

Auxiliary runs with the Dipmeter, Multiple Formation Tester, and Sidewall Sampler are indispensable in important exploration wells. Dipmeter logs provide structural and stratigraphic information.[1,2] Pressure data from the RFT can define reservoir continuity as soon as other wells in the reservoir are drilled. Sidewall cores provide lithology and permeability information.

HARD-ROCK, SALT-MUD LOGGING SUITES

Hard-rock, salt-mud logging suites are typically run in the Permian Basin of the U.S. and in areas of the North Sea, North Africa, and the Middle East where low porosities or low water saturations result in R_t values well above 200 ohm-m. Generally, these conditions occur in carbonate reservoirs. In the Eastern Hemisphere boreholes tend to be large and R_{mf}/R_w ratios low, which also favors use of Laterologs.[3]

Infill Wells

For infill wells the Sonic log along with the Laterolog-R_{xo} combination is sufficient if formations are clean and lithology is well known. Two runs are required since the two tools are not yet combinable. However, if formations are not clean or lithology is quite variable from well to well, the development well suite may be preferable. A case has been documented where spectral GR information obtained on new infill wells led to a 55% increase in net pay of the reservoir on reanalyzing the older logs.[4]

Infill wells may often be drilled preparatory to waterflooding. In this case knowledge of inter-well communication is important so pressure measurements with the Multiple Formation Tester are desirable.

Development Wells

Two basic logging runs are required, the first with the Dual Laterolog-R_{XO} tool and the second with porosity tools. For the latter the Litho-Density-

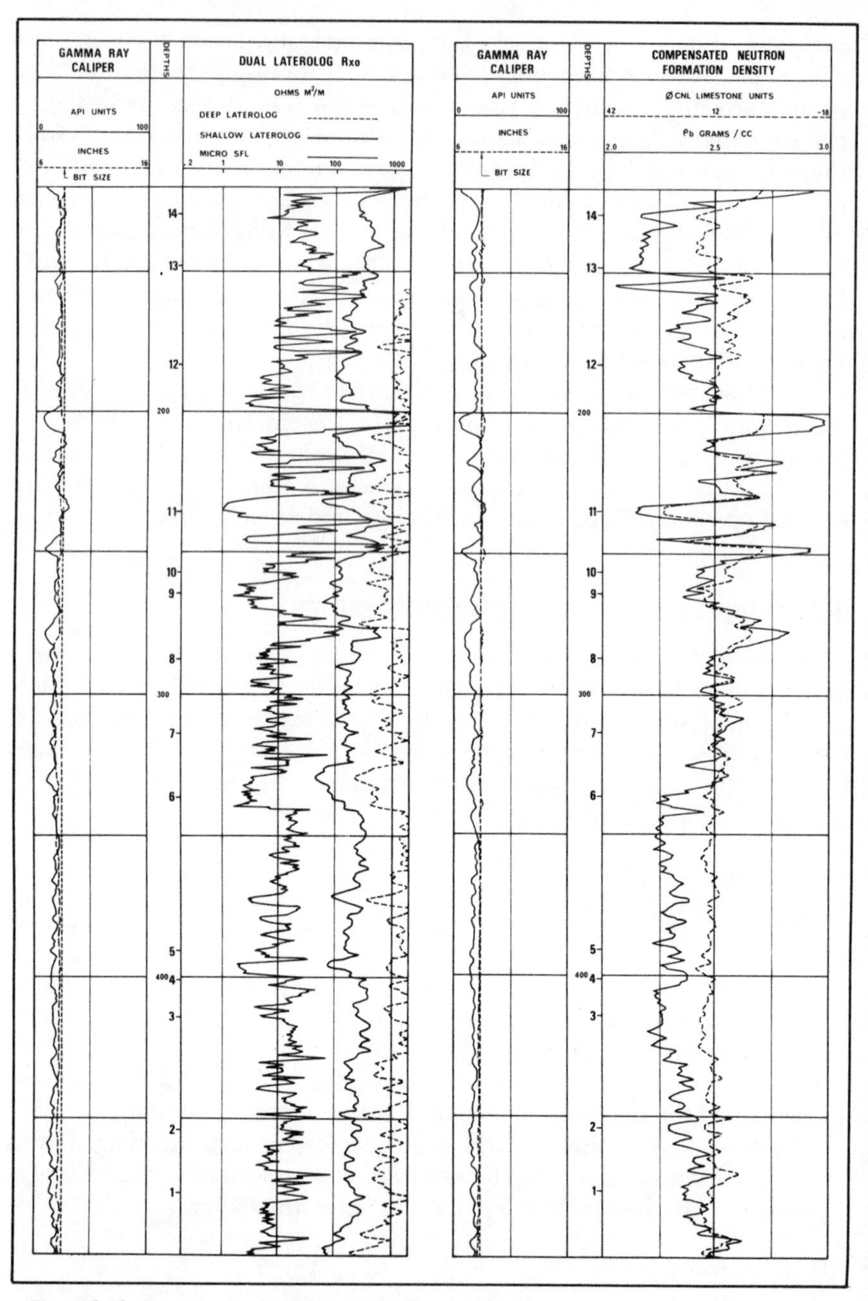

Fig. 10–3 Dual Laterolog-R_{xo}-Density-Neutron-GR log suite in Middle East carbonate series (courtesy Schlumberger)

Neutron-Spectral GR combination is recommended in place of the conventional Density-Neutron-GR. The Litho-Density log is needed for lithology identification, particularly in gas-bearing carbonates. The spectral GR aids in distinguishing radioactive dolomites from shales.

Fig. 10–3 is an example log suite run in a Middle East carbonate series. The GR log shows the whole section to be extremely clean; a few exceptionally low readings such as those at 205 and 250 ft (relative depths) correlate with density readings of 2.9–2.95 g/cc, indicating those intervals to be anhydrites.

The R_t values are 1,000–2,000 ohm-m, in which range the Induction log would be extremely inaccurate. The ratios R_{LLd}/R_{LLs} and R_{LLd}/R_{xo} average about 8 and 100, respectively, indicating $R_t \simeq 1.3\ R_{LLd}$ and that invasion diameter is approximately 30 in. (Fig. 4–18). The R_w for the formation is known to be 0.013 ohm-m and R_{mf} to be 0.045, giving $R_{mf}/R_w = 3.5$. Quick-look water saturation between levels 7 and 8, where $R_{xo} = 10$ and $R_t = 1.3 \times 1{,}000 = 1{,}300$ ohm-m, is given by Eq. 4.7 as

$$S_w = \left(\frac{10/1{,}300}{3.5} \right)^{5/8} = 0.02$$

Thus, the resistivity logs indicate that the whole section, with the exception of the anhydrite streaks, has extremely high hydrocarbon saturation.

The Density and Neutron logs are scaled such that the curves would overlay in liquid-filled limestone. Over much of the section there is a large Neutron-Density crossover indicating the hydrocarbon is gas. This being the case, porosity at level 3, for example, is given by Eq. 5.13 as

$$\phi = \sqrt{(0.15^2 + 0.28^2)/2} = 0.22$$

The R_t for level 3 is $1{,}000 \times 1.3 = 1{,}300$ ohm-m. Applying Archie's equation 2.8 with $c = 1.0$ gives

$$S_w = \frac{1.0}{0.22} \sqrt{\frac{0.013}{1{,}300}} = 0.014$$

This verifies the quick-look value. Both values are abnormally low, probably because cementation and saturation exponents of about 2.5 are more appropriate than 2.0 for these formations; they would give $S_w = 0.045$, a more realistic figure.

Between levels 7 and 8 the Density and Neutron curves overlay. This interval could be interpreted as liquid-filled limestone. However, since there is gas above and below the zone and no intervening shale barriers exist, the interval is probably gas-bearing dolomite. The P_e curve of the Litho-Density log would resolve the ambiguity as in Fig. 5–23 for a similar case.

Exploration Wells

In exploration wells where both hydrocarbon-bearing zones of very high resistivity and water-bearing zones of quite low resistivity are encountered, it is desirable to run both Laterolog and Induction tools. The former will read more correctly in the hydrocarbon-bearing zones; the latter, more correctly in the water-bearing zones (where R_w must be determined). This is illustrated in Figs. 10–4 and 10–5. In the hydrocarbon zone A of Fig. 10–4, the IL_d reads a factor of two lower than the LL_d. In the water-bearing zone B of Fig. 10–5, the LL_d reads too high by a similar factor. An extra logging run for the DIL-SFL is not required since it can be run in combination with the Sonic. Once R_w values are established, there is no longer the need to run the Induction. In any event a total of three runs is required for the basic logs.

In the case of the porosity tools, the forthcoming Dual Porosity Compensated Neutron, described in Chapter 5, should be run along with the Litho-Density and Spectral GR, as soon as is available. It will undoubtedly provide better porosity values in tight dolomitic carbonates. The Sonic log is important for secondary porosity information as well as seismic tie-in.

Additional logging runs with the Dipmeter, Multiple Formation Tester, and Sample Taker are required for the same reasons as for soft rock exploration wells. Where detection of fractures is important, the Fracture Identification Log obtained with the Dipmeter tool should also be run.[5]

SPECIAL SITUATIONS

Local borehole or formation conditions often dictate variations in usual logging suites. Two cases are worthy of special mention: that of oil-base mud and of heavy-mineral-bearing formations.

Oil-Base Mud

Oil-base or inverted-oil-emulsion muds are used in some areas to improve drilling efficiency and to maintain good hole conditions, particularly where swelling shales and high temperatures pose drilling problems. Such muds are electrically nonconductive and, therefore, preclude running any type of resistivity curve except the Induction log. Therefore, even with high formation resistivities the Induction log is the only choice. Fortunately, borehole and invasion corrections are at a minimum with nonconductive mud.

Density, Neutron, and Sonic logs all perform quite well in oil-base mud so the choice between them can be made on the basis of the guidelines previously outlined. However, the Electromagnetic Propagation Log is not applicable for reasons explained in chapter 5.

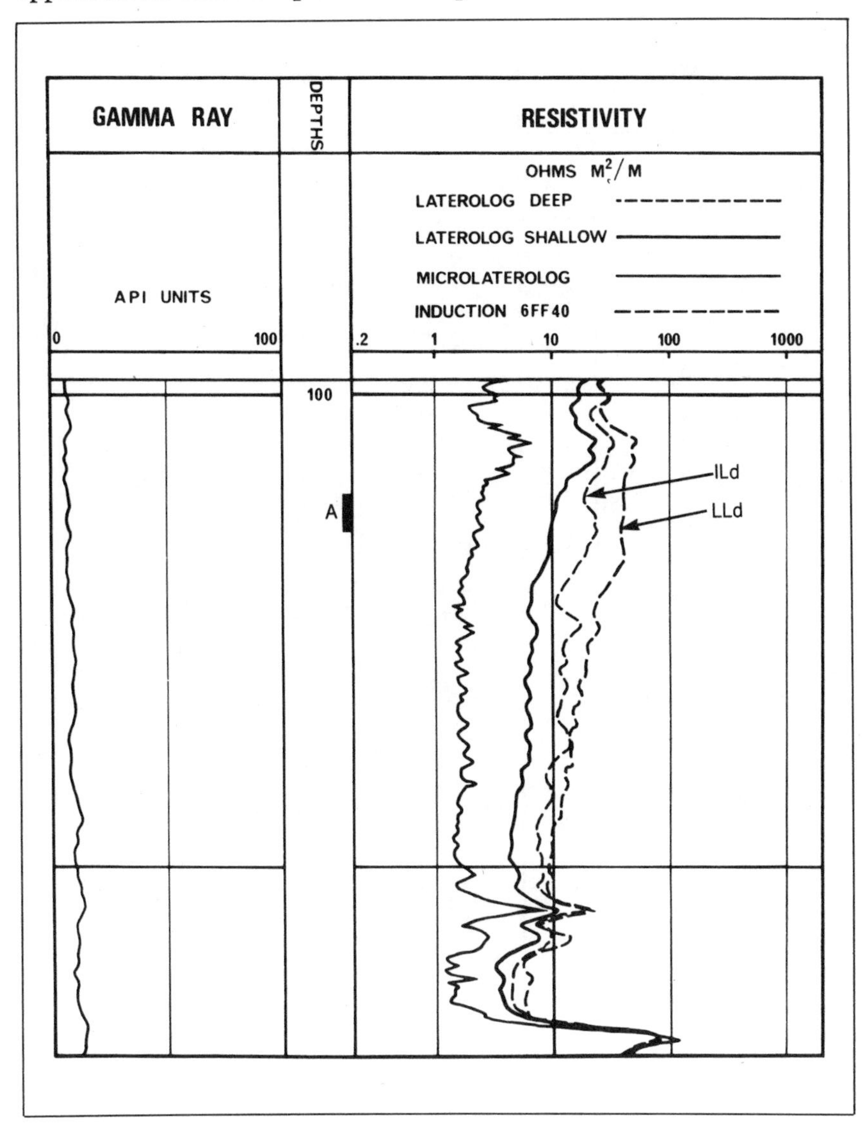

Fig. 10–4 Laterolog showing correct resistivity in hydrocarbon-bearing zone (courtesy Schlumberger)

Interpretation-wise, the standard Archie equation 2.8 can be applied to the uninvaded formation. However, a significantly different approach to estimating the movable hydrocarbon is required because mud filtrate invasion is that of oil rather than water.[6]

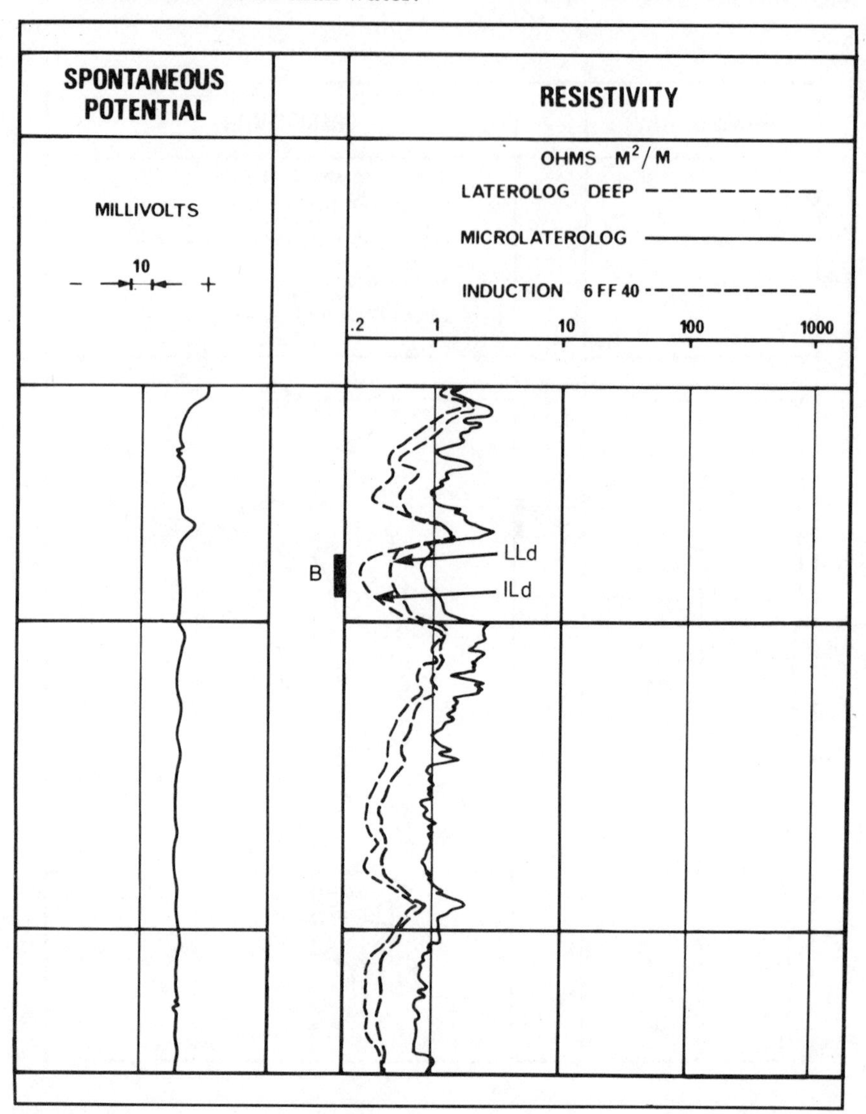

Fig. 10–5 Induction log showing correct resistivity in a water-bearing zone (courtesy Schlumberger)

Heavy Minerals

Heavy minerals, particularly pyrite (FeS_2) and siderite (Fe_2CO_3), have been found in important oil-producing reservoirs in Alaska, the North Sea, Canada, and the U.S. These minerals are electrically conductive and, when present in sufficient quantity, significantly affect resistivity logs. The higher the frequency of the log measurement, the more the log is affected. Consequently, Laterologs, which are recorded at frequencies below 1 kHz, are preferred to Induction logs, which are recorded at 20 kHz.

In addition, Density logs read abnormally high densities in pyritic formations, which leads to porosities that are too low. In some areas such as Prudhoe Bay, this forces fallback to the Sonic log as the most reliable porosity indicator.[7]

REFERENCES

[1]Schlumberger, *Dipmeter Interpretation*, volume 1 (1981).

[2]Dresser-Atlas, *"Relating Diplogs to Practical Geology,"* reprinted 1979.

[3]P. Souhaite, A. Misk, and A. Poupon, "R_t Determination in the Eastern Hemisphere," *SPWLA Logging Symposium Transactions* (June 1975).

[4]J.E. Hall, "The Importance of a Complete Set of Logs in an Old Reservoir," *SPE 10297*, presented in San Antonio (October 1981).

[5]J. Beck, A. Schultz, and D. Fitzgerald, "Reservoir Evaluation of Fractured Cretaceous Carbonates in South Texas," *SPWLA Logging Symposium Transactions*, (June 1977).

[6]D.P. Edwards, P.J. Lacour-Gayet, and J. Suau, "Log Evaluation in Wells Drilled with Inverted Oil Emulsion Mud," *SPE 10206*, San Antonio (October 1981).

[7]C. Clavier, A. Heim, and C. Scala, "Effect of Pyrite on Resistivity and Other Logging Measurements," *SPWLA Logging Symposium Transactions* (June 1976).

INDEX

A

B

C

D

E

F

G

H

I

K

L

M

N

O

P

R

S

T

U

V

W